FORECASTING AND MANAGEMENT OF TECHNOLOGY

FORECASTING AND MANAGEMENT OF TECHNOLOGY

Second Edition

ALAN THOMAS ROPER
SCOTT W. CUNNINGHAM
ALAN L. PORTER
THOMAS W. MASON
FREDERICK A. ROSSINI
JERRY BANKS

WILEY

JOHN WILEY & SONS, INC.

This book is printed on acid-free paper. ♾

Published by John Wiley & Sons, Inc., Hoboken, New Jersey

Published simultaneously in Canada

Library of Congress Cataloging-in-Publication Data:

Forecasting and management of technology / Alan Thomas Roper ... [et al.]. – 2nd ed.
p. cm.
Includes index.
ISBN 978-0-470-44090-2 (hardback); 978-0-470-95161-3 (ebk); 978-0-470-95178-1 (ebk); 978-1-118-04798-9 (ebk); 978-1-118-04816-0 (ebk); 978-1-118-04818-4 (ebk); 978-1-118-04821-4 (ebk)
1. Technological forecasting. I. Roper, A. T. (Alan Thomas), 1936-
T174.F67 2011
601′.12–dc22

2011012199

Printed in the United States of America
10 9 8 7 6 5 4 3 2 1

Tom R.: To my wife, Karen Roper, without whom I would still be interested only in airplanes

Scott: To my wife Lucie, my son Loïc, and my mother Joan

Alan: To my wife, Claudia

Tom M.: To my wife, Sandy Mason, for sharing my life with me, and to all the innovators who have inspired me

Fred: To my wife, Ann Mahoney

Jerry: To Danny, Zack, Emma, Grace, and Zoey

CONTENTS

ACKNOWLEDGMENTS

The authors gratefully acknowledge the contributions of Dr. Jill Slinger, at the Delft University of Technology, for her contributions to the discussion of simulation and forecasting in Chapter 6. The authors also appreciate the extensive contributions of Ying Guo to chapter 13, a case study on technology forecasting for solar cells.

1 INTRODUCTION

Chapter Summary: This chapter gives a preview of this book, its motivation, audiences, major themes, and differences from the first edition. It is intended to give the reader an overview and a framework within which to place the chapters that follow.

1.1 ABOUT THIS BOOK

The intent of this book is to make better managers for the twenty-first century. Almost any organization succeeds or fails because of the decisions of its managers. Evidence shows that the most critical role of managers is to anticipate and drive changes in both their organization and perhaps the world with which it interacts. For example, Henry Ford's decision to produce a motorcar for the masses dramatically altered life in the twentieth century. Technology is a primary cause of change. If managers are not successful in anticipating and rapidly adapting, the constantly changing environment will render their carefully designed structures unproductive. This is well known by the current managers of Ford, who have had to struggle for the company's survival.

The first edition of this book emphasized that technology is the key to productivity and change is a fact of life. Thus, technology managers must be able to forecast and assess technological change to obtain competitive advantage. Managers now embrace this view, and add global thinking and continuous, at times radical, technical change as essential survival skills. Much has been written about the potential and the difficulty of forecasting and managing technology as well as about the importance of knowledge as the basis of national, corporate, and individual prosperity. This book focuses on practical tools to produce information for making effective decisions about the future and on the actions needed to mold that information.

The intended audiences for this book range from upper-level undergraduate and graduate students to experienced managers: present and future decision makers who want more rigorous techniques to guide decisions rather than relying solely on intuition and conventional wisdom. The tools and discussions presented here should be accessible to those who have studied business or social science, as well as to those with science and engineering backgrounds. While some books treat technology as one factor in management, this book provides the tools to make future technology a major component of strategy development for both executive and operational decisions.

The tools presented here are consistent with those in the first edition; however, their context has been updated to reflect the complexities faced by today's global managers. Changes in this edition reflect progress in thinking about the management of technology. The book's most important enduring feature is its presentation of usable tools to aid in the assessment of technology and technological change. Most examples given in the first edition have been updated and new examples added. Software used in various calculations now emphasizes generally available packages rather than proprietary approaches presented in a toolkit. While the broad range of the impacts of technology is still addressed, this edition puts greater emphasis on the contexts within which managers make and implement decisions.

This book follows the approach most likely to be used by those who must develop forecasts of technology and act on their implications. The introductory chapters, Chapters 1, 2, and 3, address what technology forecasting is, its methodological foundations, the most frequently used methods, and structuring and organizing the forecasting project. In many cases, these three chapters introduce concepts that are more fully explored in the book as a whole.

Ensuing chapters provide additional methodological depth. Chapters 4 and 5 explore information gathering and some tools that can be used to analyze and generate results. Chapter 6 describes tools for analyzing information using techniques of simulation or modeling. Chapters 7 through 11 focus on the results of the forecast and show how they can be applied to reach conclusions about market potential, economic value, impacts of the technology, risk, and cost-benefit trade-offs. The book concludes with techniques and reflections for effectively implementing forecasting results. The appendix provides a case study demonstrating a complete life cycle of technology forecasting.

1.2 TECHNOLOGY AND SOCIETY

While a lot of attention is paid to how technology affects society, those who manage technology and the businesses that use it must recognize that the interaction goes both ways. Society affects the paths of technological change, and many innovators have failed because they have tried to sell solutions to problems that their customers were not ready to solve. The time lapse between discoveries of new knowledge and commercialization can be short or long, depending upon the willingness of society to embrace the change. For instance, it took decades for television to really catch on, but social networking sites like Facebook have exploded, playing prominent roles in millions of lives within a matter of months. And while information technology could dramatically increase the efficiency and effectiveness of health care, medical professionals and institutions have been slow to adopt it. Yet the applications of new drugs, medical devices, and diagnostic equipment grow very rapidly as soon as they are available. Understanding the social and cultural dimensions of technological change is complex and uncertain, but technology managers must deal with the systems of forces that will affect the results of their efforts.

1.2.1 Social Change

Some forces that affect a new technology will be the traditional resource directions of the market, but others will arise from social change and related political forces. For instance, rapid rises in energy prices in the 1970s led to dramatic energy-saving innovations in the 1980s. The run-up in oil prices in the first decade of the twenty-first century had a similar effect, as evidenced by sales of hybrid cars and the development of other alternatives. In some cases, social changes open new opportunities. For instance, women's increased labor force participation created demands for time-saving appliances like the microwave oven. Today young people who once met in town squares or restaurants to socialize use the latest technologies to stay in touch via online and texting communities. Some social forces acting on technologies can be puzzling. Europeans reject genetic engineering of their food supplies but embrace irradiation for the sterilization of milk, while Americans have tended to do the opposite.

Social changes can produce political actions, such as government spending, subsidies, taxes, and regulations. The debate over stem cell research in the United States illustrates how values can put restrictions on government funding of research and slow change. The desire for energy independence has led to huge U.S. subsidies for corn-based ethanol plants and research spending on fuels from cellulose, trash, and algae. U.S. nuclear power generation was at first promoted and then stopped by major expensive regulatory requirements, but it may be promoted again in the twenty-first century. Of course, new technologies have their own impacts on markets, social movements, and government policies. The lines of causation go both ways, and easy generalizations are elusive. Islamic fundamentalist groups like the Taliban and al Qaida seemingly reject modern ways of life but have made very effective use of satellite phones, modern weapons systems, and the Internet. However, the complexities of these relationships do not reduce their importance to managers who must make decisions about technologies. This book will not provide the answers, but it will provide a framework for asking many of the right questions and organizing the resulting information.

1.2.2 Technological Change

The significant reduction in employment in the U.S. manufacturing sector illustrates the dramatic implications of the global impacts of technology. Conventional wisdom is that manufacturing is going offshore and that it is playing a declining role in the American economy. However, data from the Census of Manufacturers and the U.S. Bureau of Labor Statistics show that this is *not* what is happening. True, there has been outsourcing to both foreign and domestic operations by both firms classified as manufacturers and others in different classifications that formerly were integrated into manufacturing companies. However, output of manufacturing as measured by contributions to the gross domestic product increased by over 20% in the decade ending in 2007. While manufacturing employment fell by 24% over the decade, worker productivity went up over 58%. Increasing automation has meant that fewer workers are needed in manufacturing processes (Krugman and Lawrence 2008). Even outsourcing has been enabled by information technology. Firms that were once dominant but failed to anticipate and change with the opportunities offered by new technologies have discovered their vulnerability. Clearly, advances in technology have totally transformed the way we make and distribute goods.

The issue of vulnerability is even clearer if one considers the threat posed by radical innovations. Typically, successful organizations engage in continuous innovation that improves technologies by a few percent every year. If a radically different approach emerges, existing producers often find reasons to consider it irrelevant. Examples abound. In the mid-twentieth century, for instance, producers of electronics products like stereos and TVs used vacuum tube technology that they consistently and incrementally improved. They viewed the first Sony transistor radios as cheap novelties. However, it was not long before those producers struggled and failed while Sony grew to global dominance. American automobile producers initially ignored technology changes in production. Other examples could be drawn from the computer industry, where the personal computer completely transformed a relatively young industry, or the steel industry, where new methods destroyed the dominant leadership of an old industry. Retailing is not manufacturing, but Sam Walton's Walmart showed how technology could completely change that business. And finally, throughout the twentieth century, North American agriculture experienced similar declines in employment while production capabilities rose.

As Clayton Christensen described in *The Innovator's Dilemma: Why New Technologies Cause Great Firms to Fail* (Christensen 1997), the very strengths that make an organization successful can become obstacles to success in a new technology paradigm. Sometimes the problem is arrogance. Sometimes it is the inability to quickly and smoothly adopt technical skills. Christensen and Overdorf (2000) described the demise of Digital Equipment Corporation (DEC), arguably the leading computer company in the 1980s but absorbed by Hewlett Packard in the 1990s. The personal computers (PCs) that ended DEC's dominance could easily have been designed by their talented engineers and scientists, and DEC had a great brand name and a lot of cash to shift into the new business. However, their internal operating procedures were designed to spend two or three years perfecting each new generation of mini-computer to be sold at a high profit margin to engineering organizations. By contrast, the PC business was focused on the assembly of outsourced modular components with rapid design for low-margin, high-volume sales to the masses of customers who went to retailers to buy them. It was not the basic technology of computers that defeated DEC; it was the whole range of technologies throughout the value chain from components to sales and service.

These examples illustrate what Schumpeter called *creative destruction* (Schumpeter 1937). The new products and huge increases in productivity resulting from technology advances should not be surprising. Robert Solow, Nobel laureate in economics, showed decades ago that much of the improvement in American living standards in the mid-twentieth century was due to technological progress. His work applied to other developed countries as well (Solow 1957). Increasingly, the job of effective managers has become to continuously find ways to make their currently profitable businesses obsolete and to position their firms to be dominant in the next wave of technology. For many, this is a process of continuous improvements. On the other hand, 3M Corporation has set a goal of 30% of all sales from products that are no more than four years old (Kanter, Kao, et al. 1997, p. 55; von Hippel, Thomke, et al. 1999).

Opportunities to innovate with new technologies will abound, but only those who can adapt to the unforeseen changes will be really successful. In his book *Mastering the Dynamics of Innovation* (1996), James Utterback concluded, "Innovation is not just the job of corporate technologists, but of all major functional areas of the firm." In the case of radical innovation, he went on to say that "the responsibility of management is nothing less than corporate regeneration" (p. 230).

The next decade will present both opportunities and pitfalls for those who want to exploit new possibilities for technology. The recent period of rapid rise in energy prices seemed to make the development of biomass fuels, solar power, electric vehicles, fuel cells, and methods of increasing efficiency and conservation inevitable. Indeed, billions have been invested in ethanol and biodiesel production and in research on alternative energy sources. However, the credit crisis and the subsequent global recession have made the economic viability of new approaches much less clear. Similar fluctuations in expectations have affected many other technologies. At the same time, the growing global consensus about the impacts of human activities on global warming, as well as persistent concerns about terrorism and political insecurity, continue to motivate exploration and investment in more sustainable ways to live. Moreover, discoveries related to the human genome and nanotechnology seem likely to generate more commercial opportunities, and evolving computer and communication technologies keep opening new pathways for products and services. Expanding numbers of aging retirees and longer life expectancies in the developed world are creating greater demand for medical and day-to-day care. At the same time, Paul Polak (2008) and others are showing that the billions of barely subsisting people can use new, dramatically low-cost technology to improve their living standards and thus create markets for novel products and services.

The methods, tools, and perspectives of this book are useful to nearly every manager. In a world of rapid change and global competition, being the best at any management function will be a short-lived advantage. Effective twenty-first-century managers must constantly have a vision of the future that guides their actions today; we call this managing "from" the future. Only by managing from the future will they encourage the new ideas, develop the flexible processes, and invest in the collection and management of knowledge that will allow them not only to adapt and survive, but to be part of the changes that create that future. This book provides frameworks of thinking and practical tools to more systematically anticipate the road to a successful future.

1.3 MANAGEMENT AND THE FUTURE

A standard college textbook defines management as

> A set of activities (including planning and decision making, organizing, leading and controlling) directed at an organization's resources (human, financial, physical and information) with the aim of achieving organizational goals in an efficient and effective manner. (Griffin 1999, p. 7)

This definition is a static way of looking at what managers do. Scholars of organizational economics point out that managers must assign rights to make decisions, decide on rewards for making and implementing good decisions, and implement ways to evaluate the performance of both people and business units (Brickley, Smith et al. 2004, p. 5). However, the context in which managers act has become both more intense and more complex. Virtually every business and many not-for-profit organizations depend upon technology strategy for survival. While some industries are called *high tech* even a supposedly low-tech business, like retailing, is dominated by companies like Walmart, whose strengths were built upon highly sophisticated systems for logistics and inventory management. Thus, all managers need to realize that technologies are pervasive in all of their activities.

Several decades ago, Peter Drucker (1985) talked about strategic planning, which he described in the following way:

> It is the continuous process of making present entrepreneurial (risk-taking) decisions systematically and with the greatest knowledge of their futurity; organizing systematically the efforts needed to carry out these decisions; and measuring the results of these decisions against the expectations through organized systematic feedback. (p. 125)

Drucker stressed knowing what the business was, what it would be, and what it should be. In subsequent years, major resources were poured into elaborate planning efforts, especially by large organizations. However, lessons from these planning exercises were seldom disseminated to day-to-day decision makers. Richard Florida and Martin Kenney (1990), among others, pointed out that in the 1980s, corporate America's bureaucratic approaches were ineffective in making their companies globally competitive. In fact, many bright innovators left large employers to launch new technologies in start-up ventures.

Today markets and technologies change so rapidly that even large companies look for entrepreneurial approaches that are simple to grasp and easy to change when external changes demand. Apple, amazon.com, and Google are cited as representatives of the new wave of management thinking. Wikipedia discusses, for instance, the novel lattice organization and the approach to problem solving used by W. L. Gore and Associates, a manufacturing firm (Harder and Townsend 2000). These firms and others are the laboratories for the new manager in a time of falling entry barriers, growing buyer power, and very efficient competitors fostered by the Internet. New management schemes reward employees for initiative, creativity, and *passion*, another word for *engagement*. A major concern of the new manager is how to get employees to become fully engaged in the firm's enterprises.

Guy Kawasaki (2004, p. 5) was particularly critical of large corporate planning activities in *The Art of the Start*. He pointed out that the typical corporate mission statement, the starting point for planning, often was a collection of meaningless generalities. He advocated that organizations search instead for a mantra—a few words that capture the essence of what the firm is trying to do in a way that will keep people focused and passionate. Whether the organization is an established firm or a start-up, its resources need to be applied with the flexibility to change with markets and technology. That flexibility is best used in a framework that provides for creative responses within a context of a strategic vision of the ultimate future goal.

The fact that corporations no longer place high value on complex long-range plans does not mean that they or start-up ventures can function well without a systematic view of the future to guide day-to-day decisions, as well as major investments, alliances, and other strategic decisions. Since technology is integral to almost all management activities, technology planning is not separate from overall planning. Planning begins with a vision of the future toward which the organization is moving. It also provides intermediate goals as milestones to assure decision makers that they are on the right track. Reaching those goals requires a strategy. For example, one might try to be the first to implement new technology so as to grow rapidly and assure a dominant market position. Or one could take the strategic position of letting rivals rush ahead to establish a market and reveal their vulnerabilities before moving in with superior products and services. These alternative strategies have very different implications for organizational tactics. The people who are attracted and assigned to various functions, and the resources that are deployed for them, will be subject to the strategy and tactics that are pursued.

1.3.1 Management and Innovation Processes

Presentation of the frameworks mentioned above requires some explanation of the processes used by managers to produce innovation. These processes range from strategic management to the specifics of scheduling resources and reviewing project performance. While complete coverage obviously is beyond the scope of this book, some discussion of management is needed to show how to implement the suggested approaches. Jay Conger (1998) has pointed out that leaders must be champions of innovation. Thus, the innovation process must begin at the top. Executives are responsible for developing and implementing strategies that lead to continuing success. The first requirement for this is vision, both of what the future holds and the role that the organization will play in that future. In addition to providing vision, leaders must align the organization's resources and mobilize as well as motivate people to meet the challenges of ambitious goals for the future they envisage. To provide direction for the use of resources, it is essential that leaders grasp the potential of incremental change and the threat that radical innovations, approaches, or products will change the market. Once the vision is embraced, it has to be implemented by tangible changes in processes and products.

Two decades ago, Richard Florida and Martin Kenney (1990) suggested that the belief that breakthroughs alone are sufficient to keep firms—and even nations—on a competitive footing is an illusion. They emphasized that better integration of shop floor activities with R&D, and empowering production workers to innovate, are critical. Much progress has been made on these fronts, as the previous discussion of manufacturing productivity noted. Designing for quality and efficiency, and applying tools such as those under the six sigma banner, have greatly improved American competitiveness. As new technologies rapidly become commodities, their production will still shift to the lower-cost regions of the world. However, products are increasingly becoming highly configurable to individual tastes and are changing so rapidly that innovation and production have become much more closely linked. Yet, the notion that high-technology breakthroughs alone bring great prosperity remains an illusion in most situations.

Timmons and Spinelli (2008) pointed out that Ralph Waldo Emerson's poem about the world beating a path to the door of the better mousetrap's creator is just not true. It generally takes years for discoveries to become innovations, and trying to force new technology to become a market success is extremely difficult. A manifestation of the difficulties of moving from breakthrough to commercial success is the small fraction of university research output that has actually produced new products or processes. There is a "valley of death" that seems to trap many great ideas and even patented inventions between breakthrough and production. There are technical and business reasons for this. Getting from the lab to a prototype can involve challenging creativity and effort that often does not excite world-class researchers, even when they have economic incentives. Moving from prototype to production versions generally implies another series of hurdles. Moreover, evidence shows that transforming new knowledge into a successful product or process must reflect the input of the ultimate customers. All of this requires a great deal of patience, persistence, time, and money.

1.3.2 The Role of Technology Forecasting

Forecasting has been done since people first started making long-term investments. The earliest farmers cleared land and planted because they expected to harvest. Certainly the pyramids and other ancient structures are testaments to the builders' belief that the world as the builders knew it would continue for a long, long time. The most fundamental forecast is that things will happen in the future in pretty much the same way they have happened in the past. While this is referred to as *naive forecasting*, it is still prevalent and powerful. How many companies or communities have delayed taking necessary actions because they believed that their businesses' way of life would not really change?

Extrapolation of the past to the future is an intuitive approach, and while it may be dangerous, it is often correct. Economics and other fields have built complex models of extrapolation, sometimes with hundreds of causal relationships. However, even these elaborate applications of sophisticated statistics estimate those relationships from historical data. Nonetheless, the dangers inherent in extrapolation are real. In 2008, the world experienced a financial crisis unlike any since the 1930s. Institutions and regulations had been established to prevent such a thing, and although there were regional meltdowns and Japan experienced a lingering financial malaise, the systems of the United States and other economic powerhouses seemed more than up to the task. Information technology and confidence enabled many innovations, and the whole world seemed to be booming as a result.

By the late fall of 2008, spring forecasts for a mild U.S. recession and continuing boom conditions in China and other emerging economies suddenly seemed wildly optimistic. With all of the powerful computer and information technology and the sophistication of world financial professionals, how could the outlook have shifted so dramatically? The quick answers include the volatility of human emotions and the fundamental requirement for trust to make the systems work. However, a brief examination of the triggers for the near collapse of the financial system is instructive for a discussion of forecasting. Investment banks had put together complex financial instruments based upon home mortgages so that institutions like insurance companies, pension funds, and banks could earn high rates of return. These instruments seemed secure because of the stature of the organizations that originated them. Therefore, institutions did a lot of borrowing to acquire them. Based upon data compiled since World War II, these schemes should have been safe. However, the problem with even the most mathematically elegant predictions is that things change. In this case, pressure for low and even zero down payment mortgages to encourage home ownership, and unwillingness to burden financial markets with rigorous regulation, spawned a boom in risky mortgages. Inevitably, the boom in home values peaked and, for the first time in most memories (and data sets), house values declined, destroying the security of the instruments and leading to solvency problems for financial institutions all over the world.

While the story of the 2008 financial disasters illustrates the dangers of extrapolation in a changing world, the lessons for technology forecasting are clear. In a world of rapid and dramatic change, it is hard to forecast. And it is even harder to forecast the progress of really new technology, because there are no past data upon which to draw and no real understanding of the impacts of the technology itself. Nevertheless, forecasts will be needed if planning is to be done and investments in innovations are to be made.

This book addresses the dilemma of adaptive management under rapid change, as described above. Forecasting technology certainly should use extrapolation, but there is much more that must be brought to bear to produce reasonable views of the future. For instance, qualitative assessments of the technology and analogies to other technologies with similarities will be needed, as well as structured approaches to gathering information on the technology itself and on supporting and competing technologies. Much of this book describes how the problems of forecasting technology can be formulated, how creative approaches can be designed, and how information can be explored, evaluated, and focused for useful decision making. At each step, the discussion will be framed by the strategic context in which the forecast will be useful. Understanding that context is crucial both for effective results and for scoping the investigation to meet time and resource constraints for good management decisions.

1.3.3 The Importance of Technology Forecasting

An important part of technology forecasting is to assess the impacts of implementing a new technology on both the firm and its external environment. Ignoring possible negative effects can have disastrous consequences for the firm as well as for people who are neither involved in decisions about the technology nor are likely to benefit from it. Societies around the world have reacted to such disasters by holding businesses accountable, even to the point of bankrupting them. Governments also have established rigorous regulations that can stifle change. It was probably fears of unintended consequences that stopped the growth of U.S. nuclear electricity generation. More and more regulatory requirements for design and operation apparently made the option uneconomic in the minds of electric utility executives. Requirements for acceptance and agency approval of such things as environmental impact statements also have been used to protect natural resources, important species, and public health. Unfortunately, the phrase *impact assessment* is often associated in business with notions of bureaucratic compliance. The principles and tools described here are motivated by a very different purpose, although they may be complementary to regulatory requirements.

Joseph Coates (1976), one of the pioneers in holistically viewing the effects of technology, advocated

> *the systematic study of the effects on society, that may occur when a technology is introduced, extended, or modified, with emphasis on the impacts that are unintended, indirect, or delayed. (p. 373)*

A "systematic" study involves an approach that is as orderly and repeatable as possible, one in which information sources and methods are defined.

Exhibit 1.1 Consequences of Social Networking Sites

First: People have a new way to connect with others all over the world.

Second: People are physically at home but are virtually engaged in cyberspace.

Third: The power of the Internet enables people to find others whose interests are aligned more closely with theirs.

Fourth: People find it easier and perhaps more enjoyable to deal with others on the Internet than to try to reinforce relationships with those around them.

Fifth: Internet relationships become more and more intimate with time.

Sixth: Increased divorce rates result when marriages are unable to adjust to spouses who meet their emotional needs on the Internet.

Although the study focuses on the "impacts" of the technology on society, a comprehensive study also will address the reverse effects of social forces on technological development. "Unintended, indirect, or delayed" effects extend beyond the direct costs and benefits traditionally considered in technical and economic analyses. Exhibit 1.1 captures the idea of indirect, or higher-order, impacts that are unanticipated.

There are major global issues associated with development of technologies, some of which have been discussed for decades. For example:

- Global climate change, while still debated, is motivating policy changes that affect technology in many parts of the world.
- Energy sources and uses, as well as their economic and geopolitical implications, continue to dominate thinking and decision making by nations, firms, and individuals.
- Pollution continues to plague cities, especially in the countries with emerging urban economies that continue to attract millions of people to unhealthy environments.
- Finding adequate water resources is a problem for both the rich and the poor.
- Technology has extended life and created health care cost burdens for even the richest societies.
- Falling birth rates have caused shortages of young people in countries like Japan, while the poorest of the poor continue to have too many children born and too many die.

The continuing public debate over global warming shows that even determining past causes and effects is hard. So, what hope is there to determine future ones? *Certainly there is none of providing certainty*.

There is little hope of predicting the precise effects of a change in technology, even less of predicting the magnitude or timing of those effects, and still less of foreseeing the manner in which the effects will interact among themselves and with other forces. Instead of trying to achieve certitude, it is more helpful to seek to reduce the uncertainty and to know more about the interrelationships of the systems involved. The technology planner can profit from identifying possible impact vectors. Knowing what is possible, and assessing what is relatively likely, can lead to better plans.

The alternative to forecasting is to cover one's eyes and jump into the future unguided. It is far better to "look before you leap," even if future vision is considerably less than 20/20. Technology managers need to understand likely patterns of acceptance and resistance to a changing technology, and how opportunities and challenges may arise, and include their implications in planning.

While the problems are both complex and potentially devastating, they also can offer opportunities for managers in their technology planning. This demands awareness of the methods used to forecast and analyze technologies and their impacts; these issues are discussed in the chapters to follow.

1.3.4 The Role of Social Forecasting

Technology managers quickly learn that social and political forces can dramatically affect patterns of technological change. Therefore, looking ahead must include social as well as technology and economic forecasts. This may seem a somewhat arbitrary distinction since both technology and the economy are elements of the social context; however, the division is convenient because forecasting sociopolitical factors involves different concepts and problems than projecting either technological or economic ones. Social forecasts often deal with deference values, such as respect and power, rather than with welfare values, such as income, wealth, or well-being. Ascher (1978) identified five issues that make these factors more volatile and therefore more difficult to forecast:

1. The factors often can be easily altered through human volition since material resources frequently are only marginally important.
2. There is seldom a consensus on a preferred direction of sociopolitical change.
3. Social attitudes are far less cumulative than material growth patterns.
4. Single discrete events are often the central focus.
5. A single factor is apt to be meaningless without reference to the entire sociopolitical context.

These issues make it difficult to assess the validity of sociopolitical forecasts. Since some approaches to social forecasting can be very expensive, this question of validity can limit their attractiveness to managers. However, making no social forecasting effort can be equivalent to assuming no changes in the status quo, and there is a lot of evidence that this can be even more expensive. For example, the predicted and validated aging of populations in developed countries is bound to have significant economic and political impacts that managers would be foolish to ignore. Therefore, a prudent manager should look for cost-effective methods of social forecasting and interpret results in light of their limitations.

Sociopolitical forecasts are likely to rely heavily on qualitative approaches. Exceptions include social indicator and demographic projections, regression analyses, and certain simulations. Indeed, even simulation models that produce quantitative output rely on quantifying qualitative input about the interaction of important variables. Ascher (1978) suggested that two sociopolitical forecasting techniques had special promise: scenarios and social indicator projections. To these should be added expert opinion and, in some contexts, simulation models. While these techniques appear to be the most promising, no method is without problems or limitations.

1.4 CONCLUSIONS

While it may be interesting, educational, and fun to explore the future of technologies for the sake of the knowledge itself, applying the methods presented in this book requires resources. The use of those resources must be justified by the value the forecast produces for the organization. That value will come from better decisions, even though those decisions occur in an environment laden with risks and uncertainties. The test of the validity of forecasting only really comes with the passage of time. Yet, decisions need to be made in the present, and delaying them can also generate significant, even disastrous, costs.

Experience with the methods described in this book shows that better decisions can result when careful consideration of the future and its uncertainties is included in making and implementing decisions. Therefore, those who make decisions about the future of technologies must balance the desire for more and better information about the future with their limited resources and inevitable time constraints. If this book succeeds in producing better strategies and tactics, ones that have benefited from an informed look into the future that expands knowledge and reduces uncertainty, then its outcomes will have been worthwhile.

REFERENCES

Ascher, W. (1978). *Forecasting: An Appraisal for Policy Makers and Planners.* Baltimore, Johns Hopkins University Press.

Brickley, J., C. Smith, et al. (2004). *Managerial Economics and Organizational Architecture.* New York, Irwin McGraw-Hill.

Christensen, C. (1997). *The Innovator's Dilemma: Why New Technologies Cause Great Firms to Fail*. Boston, Harvard Business School Press.

Christensen, C. and M. Overdorf. (2000). "Meeting the Challenge of Disruptive Change." *Harvard Business Review* **78**(2): 66–76.

Coates, J. F. (1976). "Technology Assessment: A Tool Kit." *Chemtech* **6**: 372–383.

Conger, J. A. (1998). "Leadership." In *The Technology Management Handbook*, ed. R. C. Dorf. Boca Raton, FL, CRC Press: 7–97.

Drucker, P. (1985). *Innovation and Entrepreneurship*. New York, HarperCollins.

Florida, R. and M. Kenney. (1990). *The Breakthrough Illusion: Corporate America's Failure to Move from Innovation to Mass Production*. New York, Basic Books.

Griffin, R. W. (1999). *Management.* Boston, Houghton Mifflin.

Harder, J. and D. R. Townsend. (2000). "W. L. Gore & Associates," Case Study, *Harvard Business Review*, Product # UV3176-PDF-ENG, Retrieved 3 April 2011 from http://hbr.org/product/w-l-gore-associates/an/UV3176-PDF-ENG.

Kanter, R. M., J. Kao, et al., eds. (1997). *Innovation: Breakthrough Ideas at 3M, DuPont, GE, Pfizer and Rubbermaid.* New York, HarperCollins.

Kawasaki, G. (2004). *The Art of the Start*. New York, Penguin.

Krugman, P. R. and R. Z. Lawrence. (April 2008). "Trade, Jobs and Wages." *Scientific American* **270**(4): 44–49.

Polak, P. (2008). *Out of Poverty*. San Francisco, Berrett-Koehler.

Schumpeter, J. (1937). *Economic Development*. Boston, Harvard University Press.

Solow, R. M. (1957). "Technical Change and the Aggregate Production Function." *Review of Economics and Statistics* **39**: 312–320.

Timmons, J. and S. Spinelli. (2008). *New Venture Creation: Entrepreneurship for the 21st Century.* New York, McGraw-Hill.

Utterback, J. M. (1996). *Mastering the Dynamics of Innovation*. Boston, Harvard University Press.

von Hippel, E., S. Thomke, et al. (1999). "Creating Breakthroughs at 3M." *Harvard Business Review* **77**(5): 47–57.

2

TECHNOLOGY FORECASTING

Chapter Summary: Chapter 1 sketched the roles that planning and forecasting play in improving decision making in organizations dealing with significant change. This chapter deals with technology forecasting in more depth. Models of technological growth and diffusion are introduced, and the methodological foundations, including the technology delivery system, are established. A range of related concepts in technology forecasting and impact assessment are introduced. The chapter ends with an overview of forecasting methods and some guidance in selecting among them.

2.1 WHAT IS TECHNOLOGY FORECASTING?

In this text, the definition of technology forecasting is broader than intuition might suggest. *Technology* is defined as systematized knowledge applied to alter, control, or order elements of the physical or social environments. This includes not only the hardware systems usually equated with technology, but systems of analysis, regulation, and management as well.

People have been adapting to fairly rapid technological changes for a long time. Managers in the latter stages of their careers have seen computer and information technology dramatically change the way they work and live. Their grandfathers may have been born in the era of horse-drawn transportation and steam locomotion and yet lived to see astronauts walk on the moon. However, despite the evidence of technology change and its impacts, organizations and individuals have not learned very much about how to anticipate and plan for it.

Technology forecasting is focused on changes in technology, such as its functional capacity, timing, or significance. It is distinct from forecasts in which technology plays a role but is not the central issue, such as population projections Ascher (1978). Forecasting of any kind is difficult or weather forecasting would be a lot more accurate. Meteorologists at least have data from years of observing weather patterns to help them. Technology forecasters deal with new concepts, with little historical evidence to draw upon. Like weather forecasting, the context of technology forecasting is very complex.

Technology forecasting activities masquerade under many names. This book adopts a broad definition that incorporates competitive technical intelligence, foresight, impact assessment, risk assessment, and technology road mapping. These approaches all adopt a systematic view for analyzing sociotechnical systems and draw upon a common set of methods. All are intended to aid in sound decision making. They differ in their intended audience, problem conceptualization, and mode of providing guidance.

- *Competitive technical intelligence* (CTI) emphasizes corporate or private sector applications. Analysis of open-source or "gray" literature frequently is central, and there is often a focus on numerical analyses and trends. CTI may emphasize downstream technologies that have reached the marketplace.
- *Foresight* often is adapted to public sector and governmental concerns. It emphasizes achieving desirable futures through policy implementation rather than accepting the future as a given. Foresight activities may emphasize upstream or fundamental aspects of new technology.
- *Impact assessments* are a class of studies that evaluate the environmental and social effects of a technology. Environmental impact assessment (EIA) began in response to the requirements of the U.S. National Environmental Policy Act of 1969 (NEPA). Since then, it has spread throughout the world and its concerns have grown beyond the physical environment. Social impact assessment (SIA) emphasizes impacts on people, cultures, and institutions. It has been applied to concerns as wide-ranging as the effects of modern technology on indigenous peoples and the effects of texting on highway safety. Technology assessments (TAs) are broad-spectrum attempts to foresee all impacts of a new technology.
- *Risk assessment* addresses the probability of bad results ensuing from a technological decision. While it probably has been applied most often to technologies related to food and drugs and to the financial prospects for new ventures, it has been used to evaluate many other public health and safety concerns as well. These issues are complex and often involve subjective judgments as much as objective measures of probability.
- *Road mapping* emphasizes techniques for coordinating complex technologies distributed across multiple stakeholders or components of an organization. It frequently relies heavily on visual approaches. Clear alignment and consensus about priorities are often quick benefits of such studies.

Whatever the name, there are misunderstandings about sound technology forecasting, often fostered by past inadequate forecasting efforts.

Technological forecasting is not deterministic, that is, it does not seek to project a single certain future. Rather, good forecasts project a range of possible futures, of which some may be more likely than others. A good technology forecast may be quantitative, qualitative, or, frequently, a mixture of both. Since forecasts are done to help decision makers choose from a range of desirable futures, or to avoid the least desirable ones, it definitely has a normative component. The claim that technology forecasting has paid too little attention to the social aspects of new technology contains an element of truth. Thus, this book gives extensive emphasis to the social context of a technology and ways to include social concerns in the forecasting process.

To forecast technology, one must understand what is known about how technologies develop and mature. The growth of technologies is strongly affected by changes in the social/political context in which they are embedded and by the growth of supporting and competing technologies. Not only is this context dynamic, it affects different technologies in different ways. Thus, there is no single growth pattern that describes the development and diffusion of all technologies. There are general concepts of how technologies develop, however, and these useful guides are described in the following section

2.1.1 Models of Technology Growth and Diffusion

The attributes of technology most often forecast are:

1. Growth in functional capability
2. Rate of replacement of an old technology by a newer one
3. Market penetration
4. Diffusion
5. Likelihood and timing of technological breakthroughs

Regardless of the attribute to be forecast, it is important to understand both the technology and the process of conception, emergence, and diffusion that characterizes its growth. Measures of functional capacity may differ for technologies that appear similar. For example, maximum speed is one legitimate measure of fighter aircraft performance because of its mission. However, speed alone is not a legitimate measure for transport aircraft because it captures only part of the aircraft's functional capacity: to rapidly deliver a payload. Often the forecaster must understand not only the technology in question, but also earlier ones that fulfilled the same need. Such understanding is required to develop trends that are defined by successive technological approaches. A firm understanding of basic principles also is required to identify competitive technologies, as well as technologies that are necessary to support the subject technology or that may be supported by it.

Technologies generally follow a growth pattern that is S-shaped, as shown Figure 2.1. When the technology is *emerging,* growth is slow as innovators

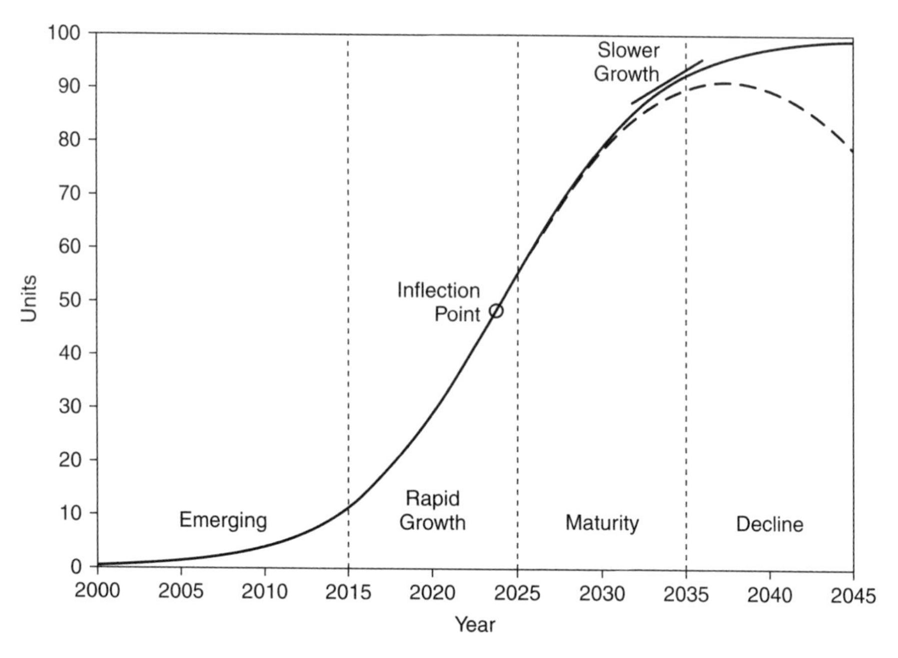

Figure 2.1. Growth Phases and the S-Shaped Curve

develop prototypes and try to determine the configuration of the product based upon the technology's functionality. Once the product is established, there is a period of *rapid growth*, followed by an inflection point and slower growth as the product enters a period of *maturity*. Eventually, the technology becomes obsolete and its use *declines*. This growth model is discussed further in Chapter 6.

Each stage involves a different type of management. The emerging stage is dominated by R&D in conjunction with collection and integration of market information. The rapid growth period is a time of slow product change, but accelerated output, as the organization tries to dominate the industry. During the mature phase, management decisions usually are about evolutionary improvements in features, quality, and costs. In declining stages, there usually is consolidation and downsizing of operations. These patterns of growth can take a long time. For example, technologies such as fuel cells and alcohol fuels from cellulose materials have been "emerging" for decades. On the other hand, innovations like Facebook and Twitter emerged and grew very rapidly, to make enormous impacts on life around the globe.

2.1.2 Technology Forecasting in Context

The speed of adoption, growth, and decline of a technology is often affected by technical challenges that take a long time to resolve. Moreover, the dynamics of the economic and social/political contexts fundamentally affect its development.

Therefore, technology forecasts require background forecasts in economic, social/political, and other areas as well. A forecaster lacking resources or expertise may be forced to rely on existing contextual forecasts. This can cause problems. Wheeler and Shelley (1987), for example, investigated forecasts of demand for innovative high-technology products and found them to be uniformly optimistic by 50% or more. They attribute this to a lack of forecaster expertise in consumer behavior, overenthusiasm for high technology, and poor judgment. Moreover, existing background forecasts may embody core assumptions that are not explicitly cited or are no longer valid. Ascher (1978) refers to the latter problem as "assumption drag." This difficulty is especially acute in social/political arenas that often exhibit considerable volatility. In such cases, the manager needs some approximate estimate of the magnitude of errors that inaccuracies in the contextual analyses might cause.

In critical background forecasts, recent projections are apt to be more accurate than earlier ones *regardless of their relative levels of methodological sophistication*. Several rough background forecasts can be made at a cost comparable to that of a single sophisticated forecast with the expectation that they will be worthwhile. Finally, the interdependence of background forecasts suggests the need for several disciplinary specialties within the forecasting team.

Ascher (1978) noted that the selection of a broad forecasting method and of more specific techniques is much more than "a technical choice or a matter of convenience" (p. 196). He also suggested that one should consider the following factors in preparing a technology forecast:

1. Dependence on basic scientific breakthroughs
2. Physical limits to the rate of development
3. Maturity of the science and applications of the technology
4. Sensitivity of the pace of innovation to high-level policy decisions
5. Relevance of R&D funding
6. Extent of substitutability by other products or by parallel innovations
7. Relevance of diffusion
8. Opportunities to borrow advances from related technologies

The discussion above emphasizes the importance of understanding that technology is one element in a larger social, physical, and institutional system. Technology does not operate in isolation; choices made across a system affect its timing and delivery. Acceptance of a new technology will be strongly affected by the positive and negative impacts perceived as the results of its implementation. For example, there was strong political support for corn-based ethanol in the United States both as an alternative energy source and as a boost to farm incomes, but rising food prices brought a public backlash.

Clearly, there are many factors to consider when forecasting the future of a new technology. Much of this book describes how the problems of forecasting can be formulated, how creative approaches can be designed, and how information can be explored, evaluated, and focused to aid good decision making. At each step, the discussion will be framed by the strategic context in which the forecast will be used. Understanding that context is crucial to producing effective results and to scoping the forecast to fit time and resource constraints.

2.1.3 What Makes a Forecast Good?

In the end, the goodness of a forecast is measured by whether or not it leads to the right decisions. Exhibit 2.1 relates a story told by Peter Drucker (1985, pp. 46–47) that illustrates an unsuccessful use of forecasting to implement a technology. The new lock in the exhibit was better technology. The company had forecast that increasing affluence would lead to the desire for superior functionality, that is, a lock that actually required a key. This was a bad forecast by almost any criteria. It was clearly wrong. To manage from the future, it is important to assess accurately what that future might be. And that is very difficult.

Consider the forecasts in Table 2.1 that were produced two decades ago by Abrams and Berstein (1989). None of them came to pass, even a decade after the latest implementation forecast. In some instances, the technology did not work or better technologies were discovered. Lasik surgery, for example, was probably superior to implanted rings. In other cases, the technology does exist but has not become a commercial success, as in the case of smart houses, which few consumers value highly enough to pay for.

The fact that forecasting is difficult and not always very accurate does not mean that it lacks value. The process of doing and following up forecasts can lead to good decisions, even if the forecasts themselves are incorrect. Sometimes a bad forecast is recognized for what it is and provides valuable information about the future that can be incorporated in current decisions. For example, Vanston (1985) cites a 1978 survey by the Technical Association of the Pulp and Paper Industry

Exhibit 2.1 The Lock

A cheap lock exported to British India was a firm's best-seller. As Indian personal income rose in the 1920s, lock sales declined. This was interpreted as unhappiness with the quality of the lock, so a new, superior model was designed to sell at the same price. The new lock was a disaster. It seems that Indian peasants considered locks to be magic; no thief would dare open a padlocked door. Therefore, the key was unimportant and was often misplaced or lost. So, to gain entry to their homes, peasants needed a lock that was easily opened. The new lock was not, nor was it strong enough to discourage a thief at the homes of the more well-to-do. The new lock could not be sold, and the firm went out of business four years later.

Source: Based on Drucker (1985).

TABLE 2.1 What's Ahead for the 1990s?

Trend	How Likely?	When?	How Much?
X-ray-less mammogram (breast exams using light rays)	100%	1995	$35/exam
Intracorneal rings (corrects nearsightedness)	75%	1994	$2000
Poison-ivy vaccine	50%	1994	Unknown
Walking TV(follows the viewer from room to room)	50%	1995	$5000
Smart houses (computer controls all Electronic components)	90%	1997	$7000–$10,000

Source: Abrams and Bernstein (1989).

in which manufacturers were asked to estimate the paper volume that they would produce in 1985. When the estimates were totaled, it was found that predictions would require that every tree then growing in the United States and Canada be harvested every two years. Even though the estimate clearly was inaccurate, decision makers realized that forestry programs would have to be expanded and other sources of cellulose would have to be found. The lesson here is that the *process* of *making and reviewing* the implications of forecasts is useful.

Of course, technology forecasts are sometimes accurate! George Wise (1976) reviewed the accuracy of a range of technology forecasts after separating them by technology domain. Some were too vague to be assessed; of the remainder, he concluded that 38 to 51% (by domain) were correct.

While the accuracy of a forecast can only be judged retrospectively, there are guidelines to help produce a good result. First, good forecasts usually draw upon a range of perspectives and methods. Linstone (1989) provides explicit guidelines for considering a range of different perspectives, both technical and social. Second, applying more than one method increases the probability of an accurate forecast. Mixing and matching techniques allows the forecaster to balance their strengths and weaknesses. Cook and Campbell (1979) offer a concrete way to identify and balance the methodological strengths and weaknesses of different forecasting methods.

2.1.4 Common Errors in Forecasting Technology

There can be no exhaustive list of ways to fail at technology forecasting. People are adept at finding new ways and repeating old ones. However, experience shows that there are some sources of error that regularly occur: contextual oversights, bias, and faulty core assumptions.

Contextual errors arise because the forecaster does not consider changes in the social, technical, and/or economic contexts in which the technology is embedded. Changes in these areas affect the assumption of continuity between past and future that lies at the heart of empirical forecasting. Contextual changes can produce discontinuities in behavior. For example, in the recent past, federal government deregulation policies produced major changes in financial instruments and the technology for developing and trading them. However, the world economic meltdown of 2008 and 2009 likely will reverse this relaxation of regulation and produce more change. Sometimes the development of a competitive or supporting technology also can produce discontinuity. Martino (1983, p. 230) notes that digital computers would not have been possible "without the transistor or something that shared its properties of low cost, high reliability, and low power consumption."

Sometimes failure to fully appreciate the context leads to underestimation of the time it takes to implement a technology. New technologies are seldom immediately embraced. Rogers (1983) pointed out that they go through a diffusion process that involves the following stages:

- *Knowledge Stage*: Although the Internet allows almost instantaneous distribution of knowledge, it still takes time to get people's attention and inform them of a new technology.
- *Persuasion Stage:* Information overload increases the time people require to realize that they want what the technology offers.
- *Decision Stage*: New approaches bring uncertainty, particularly when they affect other aspects of business and life. Individuals require time to weigh benefits and costs, and organizations may take even longer to decide to buy.
- *Confirmation Stage*: The period for adopters to engage in repeat buying and/or recommend the innovation to others can be critical to a new technology.

Forecasts of the spread of a new technology often underestimate the time required for these stages. This is a contextual error, as it results from factors other than the technology itself.

Geoffrey Moore (1999) raised another contextual issue about the growth of technologies that are disruptive. While some buyers quickly see the potential of new technologies, these early adopters are only a small part of the market. Success with them seldom leads to sustainable growth. Ways must be found to get more pragmatic customers to embrace the technology. Demonstrating the advantages does help, but it can take considerable time and should be factored into forecasts.

Errors also arise from *bias*. Bias can be conscious, but more frequently it is unconscious. Intentional bias often results from having personal, political, ideological, or corporate "turf" to protect. While conscious bias is most easily recognized, it is not always easy to correct. Unconscious bias is subtler, as it is unrecognized by the forecaster and may be very difficult for decision makers to detect. Occasionally, bias even may result from overcompensation for biases that forecasters recognize in themselves (Martino 1983). Unconscious bias may be manifest as under- or overemphasizing recent trends at odds with historical behavior; unwillingness to be the bearer of bad news; belief in or reaction to the

TABLE 2.2 Megamistakes

1. *Fascination with the exotic*: Technology forecasters exhibit a bias toward the optimistic and a disregard for the realities of the marketplace.
2. *Enmeshed in the Zeitgeist*: Everyone sees the same technologies as hot (devaluing expert consensus) and emphasizes the same pressing societal needs.
3. *Price-performance failures*: Many technologies deliver lower benefits at greater cost than anticipated.
4. *Shifting social trends*: Changing demographic trends and social values are not well considered and may change users' desires and market opportunities.
5. *Ultimate uses unforeseen*: Rarely do forecasters fully anticipate applications.

Source: Based on Schaars (1989).

"technological imperative"; and worldview. The best safeguards against bias are forecast team and method diversity.

Assumptions fill gaps where no data or theory exists. Thus, they are especially critical in forecasting. The forecaster will do well to internalize the commonsense observation that "It's not what you don't know that hurts, it's what you know is true that isn't." *Core assumptions* are ones that derive from the forecaster's basic outlook. Problems deriving from them are particularly troublesome. Core assumptions, like unconscious biases, often are not recognized by the forecaster. However, they are so central to the forecast that they strongly influence the result.

Ascher (1978) noted that core assumptions are major determinants of forecast accuracy. If the core assumptions of a forecast are correct, the choice of method is either obvious or secondary; if they are not, the result cannot be corrected by method selection. If the forecaster begins with a preconceived notion, like the relation between lock quality and sales in Exhibit 2.1, the data and methodology usually can be made to bear it out. As with many human endeavors, if you are not careful, the end is determined by where you begin.

Schnaars (1989) reviewed forecasts, largely from the 1960s or later. Table 2.2 summarizes some of his conclusions about specific sources of error.

2.2 METHODOLOGICAL FOUNDATIONS

It should be clear how technology forecasting can inform management decisions about a crucial aspect of a business. However, before developing forecasting tools, there is a need to provide more foundation for their validity. While *methodology* sometimes is used as a synonym for *method*, it is really about why methods are legitimate. Understanding the scientific rationale for the approaches used in forecasting can add credibility and perspective.

Two systems that are critical for the formulation and design of technology forecasts are introduced in this section. The first is the technology delivery system (TDS), a simple diagrammatic technique that is used throughout the book to help frame the technology within its broader environment. The second, the inquiring systems approach to knowledge generation, is useful for selecting the appropriate questions and methods to use in a forecast.

2.2.1 The Technology Delivery System

Technological activity and the larger societal context of which it is a part interact through a complex and only partially understood system of relationships. Decades ago, Wenk and Kuehn (1977) recognized the need for a model to address the most important relationships in the process of sociotechnical change. The framework they suggested (Figure 2.2) still captures the dimensions of the problems of projecting and managing technological change. However, it is important to stress that while this simple diagram organizes the broad context, it greatly aggregates the many complex relationships that operate in the global marketplace of the twenty-first century.

TDS is a simple "boxes and arrows" model in which direct information flows are shown by solid lines and feedback loops by dashed ones. Institutions directly involved in developing the technology—the *technological enterprise*—are displayed along the horizontal axis. Those in the larger society that influence and are influenced by the technology are arrayed across the top. Each TDS is especially developed for the technological innovation being considered.

There are four elements to the TDS:

1. Inputs to the system, such as capital, natural resources, tools, knowledge from basic and applied research, and human values
2. Public and private institutions that play roles in the operation of the TDS or that modify and control its output
3. Processes by which institutions interact through information linkages, markets, and political, legal, and social means
4. Outcomes, including both direct (intended) and indirect (unintended) effects on the social and physical environments

The TDS is a microdescription of sociotechnical change. It is dynamic and changes with time to reflect the ongoing process of technological development. While there will be few changes to the main elements of a well-constructed TDS with time, their relative importance and the relationships among them may significantly change.

Studies in system engineering and policy analysis (Sage and Armstrong 2000; Walker 2000) have produced modifications to the original TDS representation. There are three basic elements in this modified system:

1. *System*: Key productive elements of the TDS that may produce a single innovation or a range of current or potential innovations
2. *Arena*: Social, political, and decision-making entities in which decisions about the technologies within the system are made
3. *External environment*: Influences and events that may affect the system, which are not in either the system or the arena

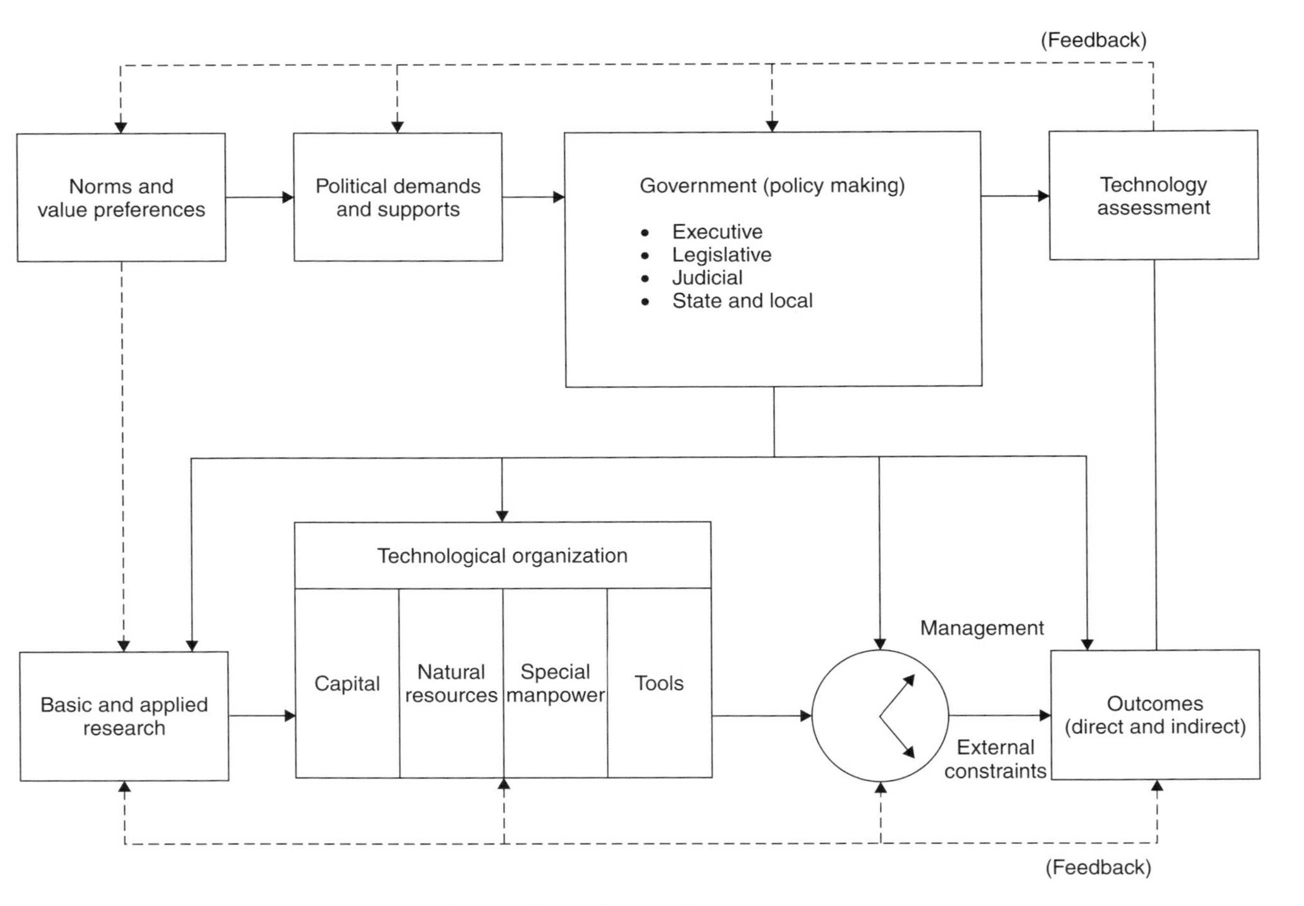

Figure 2.2. Wenk and Kuehn's TDS

When setting the boundaries between the arena and the system it is often helpful to consider for whom the forecast is being made. As the examples below show, depending on the perspective, TDS representations can differ for the same technology innovation.

There also are three basic flows in the system model:

1. *Forces*: Shocks to the system from the external environment that often are uncertain or even unknowable
2. *Outcomes*: Direct and indirect impacts of new technologies flowing from the system to the arena that produces policies in response
3. *Policies*: Consequences of decision making flowing from the arena to the system

A generic TDS comprised of three systems and three flows is shown in Figure 2.3.

The arena in this figure is mostly composed of national, state, and local governmental institutions. Stakeholders and interest groups that are most capable of affecting governmental decision making are shown as well. The system includes industrial participants with subsystems for management and production. The available labor, capital, and necessary capital goods are identified as key elements in the delivery of the system. The external environment is mostly characterized by shocks to surrounding fields of science and technology.

The situation portrayed in Figure 2.4 is one in which government policy is presumed to be the primary determinant of new technological activity. This case

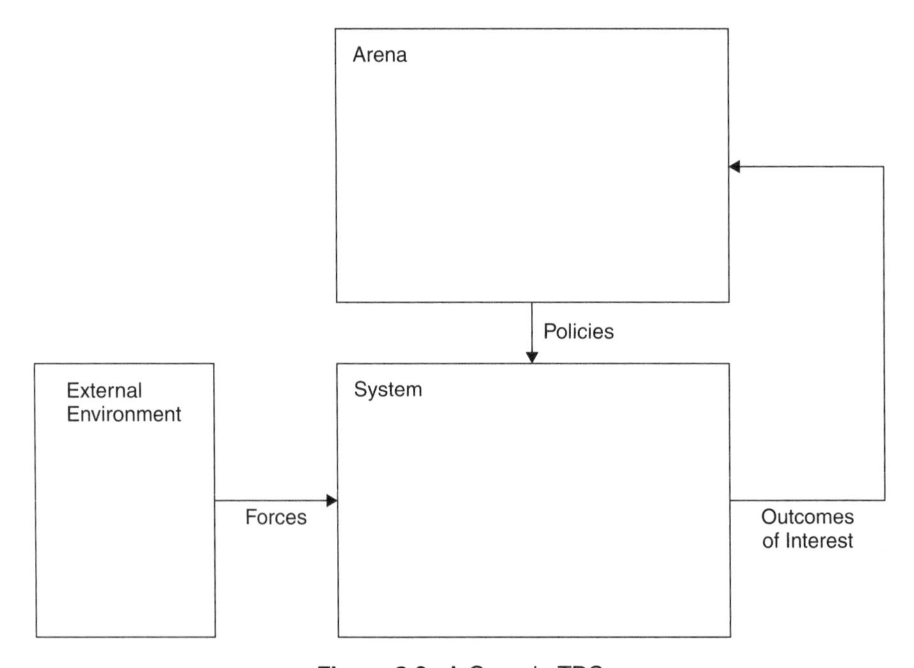

Figure 2.3. A Generic TDS

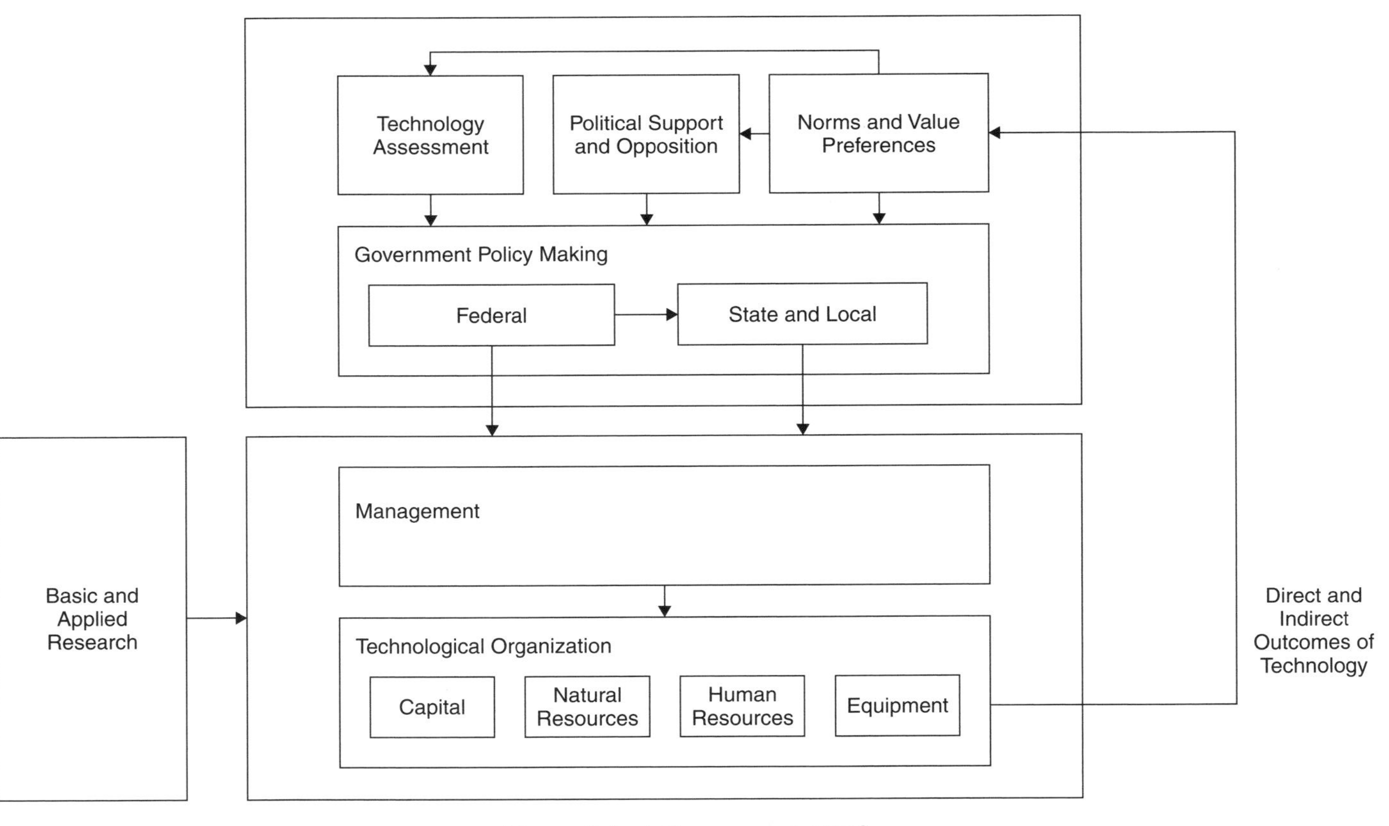

Figure 2.4. A Governmental TDS

illustrates one benefit of the TDS approach; the forecaster can advance different hypotheses about the delivery of new technology as part of early problem structuring. These hypotheses may be advanced and revised as the forecast progresses.

The external environment in this TDS is consumer-driven. Consumer tastes, preferences, and demographics can exert huge influences on the development of new technologies. Certainly consumer preferences are neither in the arena nor the system and therefore are a key external force. Unfortunately, these influences often can be puzzling. Nokia is a company that tries to reduce this uncertainty by hiring "corporate anthropologists" to discover exactly how their technologies are being used (Palmer 2008). By understanding their customers, Nokia can design new products to better meet their needs.

Neither the governmental nor the industrial TDS is necessarily more representative of the wide range of technologies one may be called upon to forecast. What is key is that the consideration and customization of the TDS can be modified to match the technology and governance structures in play.

The TDS provides several critical inputs to the forecaster, including:

- Framing the questions of *what* might be done with the technology to generate innovations and for *whom* (what customers in what sectors)
- Arraying the essential *enterprise* components necessary to take new R&D advances to market
- Mapping key contextual institutions and individuals that can affect the development or be affected by it
- Spotlighting leverage points

The TDS offers an important framework for gathering and organizing information and drawing conclusions about the implications that can be used for decisions. It also helps the forecaster organize and communicate the critical problem-structuring phase of a forecast. The TDS recurs as a unifying theme throughout this book.

2.2.2 Inquiring Systems

Understanding the scientific rationale for the approaches used in forecasting can add both credibility and perspective for the forecaster. Table 2.3 summarizes the inquiring systems used to produce the knowledge on which forecasts are based. This classification, given by Mitroff and Turoff (1973), is based on the general approach first articulated by Churchman (1971).

In the *a priori* system, the inquirer builds logical structures or models that relate to the real world. These are intended to represent the major features of some part of the world. The model may be expressed in many ways: sophisticated computerized simulations, simple boxes and arrows diagrams, theories, equations, or physical models. It is not necessary to model the entire real system or even every feature of it: models are simplifications. It is important, however, that the model incorporate the features and dynamics that are important to the inquiry. Knowledge is generated by the process of model building and by observing the

TABLE 2.3 Five Underlying Approaches for Knowledge Generation

Approach	Description	Best Suited for Problems That Are:
A priori	Formal models from which one deduces insights about the world, with little need for raw data	Possible to define well conceptually
Empirical	Beginning with data gathering, one inductively builds empirical models to explain what is happening	Possible to define well with data
Synthetic	Combines the a priori and empirical approaches so that theories are based on data, and data gathering is structured by preexisting theories and models	More complex but amenable to multiple forms of analysis
Dialectic	Opposing interpretations of a set of data are confronted in an active debate, with the goal of seeking a creative resolution	Ill-structured, and when conflict is present
Global	A holistic broadening of inquiry by questioning approaches and assumptions	Nonstructured, requiring reflective reasoning

Source: Based on Mitroff and Turoff (1973).

behavior of the model under various conditions. The assumptions upon which the model rests, and the internal consistency of the model manipulations, are critical. However, assumptions may come from any source so long as they are credible to the modeler. An obvious weakness of the a priori approach is that it does not necessarily require data. Thus, models may be weakly founded and open to criticism on empirical grounds. Still, in areas for which data are lacking, modeling can provide a means of approximation that is quite useful. Models also can be the bases for hypotheses to which data can be applied to support or reject. Models for technology forecasting are discussed in later chapters.

In its pure form, the *empirical* system consists entirely of data that relate to some aspect of the world. Its strength stems from the fact that these data are concrete and closely linked to very specific features of the world. Moreover, the empirical system is very close to being the system being studied. Its major weakness is that data alone do not provide principles or rules for structure or selection. Unless there is some underlying notion of structure, data selection may be arbitrary and a wasted effort. Issues about the categories of data that should be measured, relationships among data, and the conditions of measurement cannot be settled either by the system or by the data alone. For example, a lot of data are compiled and presented in debates about global warming and its causes, but data alone are insufficient to produce agreement about policy prescriptions. Yet the incredible richness of empirical approaches makes them enormously powerful in inquiry. For example, examination of data often reveals inconsistencies unexplained by existing models. This can motivate the use of other inquiry approaches to resolve the questions that have been raised. The opinion of experts often is valued because of their command of the data.

The *synthetic* inquiring system combines the a priori and empirical systems to overcome some of the limitations of each. This system provides an interplay in which frameworks, concepts, relationships, and variables are determined a priori and are measured empirically. Frameworks guide the measurement of data, which are then analyzed to modify the framework. This iteration between theory and concept, on the one hand, and measurement and observation, on the other, allows systematic inquiry. Much of the inquiry concerning the natural and physical sciences uses this synthetic approach. In forecasting, synthetic inquiry is useful to develop assumptions for model building and to integrate forecasts that are made by different techniques or that require complex parameter selection. Synthetic inquiry also is useful in monitoring approaches because it helps to structure and filter data to avoid information overload.

The *dialectic* inquiring system poses a view of the world that is then countered by a diametrically opposed view. The goal is to synthesize these views in a way that resolves their intrinsic conflict and moves understanding to a new plane. This system stresses the roles of conflict, controversy, compromise, and consensus in developing knowledge. Its use is exemplified by the legal system and the party systems of government. In forecasting, the dialectic inquiring system plays a role in some forms of expert opinion forecasting (see the discussion of Delphi methods in Section 5.1.2). Panels also provide opportunities for opposing positions to be presented and defended. The resulting forecast may use information drawn from both conflicting views.

The *global* system of inquiry sweeps information from a wide variety of diverse sources into the system. It establishes a wide perspective without investigating a lot of details. The coherence of the pattern developed in the inquiry and the overall robustness of the knowledge base help to ensure that all major issues are included. This builds user confidence and forecast viability, and it can make conflicting information or analyses evident. However, lack of depth is a weakness. In forecasting, the global inquiring system is used by monitoring systems (Chapter 4) and often is employed in scenario construction (Chapter 7).

Forecasting social change and the adoption of new technologies tends to lack the robust data environment and sophisticated models that support scientific inquiry. In fact, the nature of the available data and the likelihood that important parameters will change over time make the use of complex models suspect. In the end, the forecaster has to use judgment and intuition, as well as available knowledge and data. The techniques presented in this book can help to organize the information that does exist to reduce uncertainty and bias and to enhance understanding of the issues. This can lead to sounder decisions.

The inquiring systems in Table 2.3 are neither exclusive nor exhaustive. Moreover, any real inquiry will combine or approximate these systems (and possibly other systems as well). The discussion in this section shows that there are different ways of approaching knowledge generation; each has strengths and weaknesses that can complement each other. Forecasting can benefit from applying a variety of inquiring systems, adapted and used as the situation requires. In general, the more methods that can be applied to a forecast, the greater the confidence that can be placed in its results.

2.3 TECHNOLOGY FORECASTING METHODS

Forecasting methods can be classified as either *extrapolative* or *normative*—that is, by whether they extend present trends or look backward from a desired future to determine the developments needed to achieve it. Although this is a useful distinction, many methods can be considered either normative or extrapolative, depending on how they are applied. Further, the classifications can be confused with the perspective of the planning or decision-making activity that the forecast is intended to inform.

Porter and Rossini (1987) suggest that the hundreds of forecasting techniques fit into five families:

1. Monitoring
2. Expert opinion
3. Trend extrapolation
4. Modeling
5. Scenarios

This system has some limitations. Monitoring is not a forecasting method per se but rather a systematic method used to accumulate and analyze data from which forecasts are made. Further, it is unclear where to place forecasts made by analogy or those that employ lead-lag indicators. It can be helpful to categorize methods by whether they are direct, correlative, or structural. Table 2.4 indicates

TABLE 2.4 Categorizing Technology Forecasting Methods

Category	Definition	Methods
Direct	Direct forecast of parameters that measure an aspect of the technology	Expert opinion, Delphi, surveys, nominal group technique, naive time series analysis, trend extrapolation, growth curves, substitution curves, life cycle analyses
Correlative	Correlates parameters that measure the technology with parameters of other technologies or other background forecast parameters	Scenarios, lead-lag indicators, cross-impact analyses, technology progress functions, analogy
Structural	Explicit consideration of cause-and-effect relationships that effect growth	Causal models, regression analysis, simulation models, gaming, relevance trees, mission flow diagrams, morphology

how some of the more common techniques might be categorized in the latter framework.

Direct methods forecast parameters that measure functional capacity or some other relevant characteristic of the technology. These methods do not explicitly consider correlations with technological, economic, social, and political contexts, nor do they consider structural relationships within those contexts. Thus, they involve major assumptions about the nature and permanence of context and structure. Expert opinion methods, however, do allow subjective consideration of contextual change through the implicit mental model that each expert has internalized about the nature and likelihood of change.

Correlative methods relate a technology's development to the growth or change of one or more elements in its context or in contexts thought to be analogous. Lead-lag correlation techniques, for example, seek to identify a technology for which growth precedes that of the technology to be forecast. Martino (1983, pp. 100–103) presents such an analysis for combat aircraft (lead) and transport aircraft (lagged) speeds. Likewise, forecasting by analogy asserts that development of the technology will follow the pattern established by an earlier technology even though no scientific or technological tie of the kind assumed for lead-lag correlations exists. Scenarios often are used to forecast major portions of the context and the technology, although specific statements of the structural relationships are implicit. Cross-impact methods begin with a matrix that arrays some set of factors against another to examine their interactions. For instance, one might explore how gains in one energy technology would affect prospects for another. Cross-impact analysis is explicit about the impacts of elements of the technology and context but is not explicit about the cause-and-effect structure that produces them. All correlative methods make formal or informal assumptions about the relationship between the forecast technology and elements of its context. They also involve the implicit assumption that the relationship does not change (i.e., they are structurally static).

Structural methods formally consider the interaction between technology and context. To varying degrees, they must be explicit about the structural relationships between the technology and the elements of its context. Some methods (e.g., relevance trees and mission flow diagrams) merely portray the paths that connect the various elements to each other and to the technology. These methods are most often used in normative forecasts. Simulation models, however, must quantify the relationships among elements. Regression analyses seek to structure those relationships.

Regardless of their sophistication or complexity, all structural models simplify reality to make problems tractable. Therefore, they are valid only if they retain the relationships critical to accurately predict the growth of the technology. For a simulation model, not only the structure but also the mathematical formulations it embodies must be valid. It is important to realize that a model may satisfy these conditions for a technology at a given time but not at other times or for other technologies. Thus, when a structural method is chosen, the forecaster makes the core assumption that the structure it embodies is appropriate and that it will remain so over the time horizon of the forecast. While the most common models assume that change can be explained by factors internal to the system that produces the technology or by economic factors, others consider social, political, and other factors, as well as policy interventions. *Morphology* is a technique used to probe the structure of a problem to help generate ideas for innovation and/or discovery (Shurig 1984). Morphology has been used to investigate a range of diverse problems from possible jet engine types (Zwicky 1962, 1969) to Kondratieff's long wave business cycle (c.f. Volland 1987). Morphological analysis is discussed in Section 4.3.5.

2.3.1 Overview of the Most Frequently Used Forecasting Methods

The five most frequently used forecasting methods are monitoring, expert opinion, trend analysis, modeling, and scenario construction. Exhibits 2.2 through 2.6 briefly describe the underlying assumptions, strengths, weaknesses, and uses of each. The conditions appropriate to the use of each also are discussed, as well as ways to integrate them. More detailed discussions and specific examples of each are given in subsequent chapters.

Strictly speaking, *monitoring* is not a forecasting method. However, it is by far the most basic and most widely used of the five methods. Since it is routinely used to gather information, it is fundamental to almost all forecasts. While the primary sources of information are still technical and trade literature, the Internet has dramatically expanded information access (see Chapter 5). In fact, the presence of countless websites on almost any topic has made information overload likely. Thus, qualifying sources and filtering information are increasingly important parts of monitoring activities. Despite the dangers of misleading, deceptive, or false information, the World Wide Web has been a tremendous boon to monitoring.

Exhibit 2.2 Monitoring

Description: Scanning the environment for information about the subject of a forecast. It is a method for gathering and organizing information. The sources of information are identified; then information is gathered, filtered, and structured to use in forecasting.

Assumptions: There is information useful for a forecast, and it can be obtained.

Strengths: It can provide a lot of useful information from a wide range of sources.

Weaknesses: Information overload can result without selectivity, filtering, and structure.

Uses: To maintain current awareness of an area and the information with which to forecast in order to provide information useful for structuring a forecast and for the forecast itself.

Exhibit 2.3 Expert Opinion

Description: The opinions of experts in a particular area are obtained and analyzed.

Assumptions: Some individuals know a lot more about some topics than others; thus, their forecasts will be substantially better. If multiple experts are used, group knowledge will be superior to that of an individual expert.

Strengths: Expert forecasts can tap high-quality models internalized by experts who cannot or will not make them explicit.

Weaknesses: It is difficult to identify experts. Their forecasts are often wrong. Questions posed are often ambiguous or unclear, and design of the process often is weak. If interaction among experts is allowed, the forecast may be affected by extraneous social and psychological factors.

Uses: To forecast when identifiable experts in an area exist and where data are lacking and modeling is difficult or impossible.

Exhibit 2.4 Trend Analysis

Description: Mathematical and statistical techniques used to extend time series data into the future. Techniques vary in sophistication from simple curve fitting to Box-Jenkins techniques.

Assumptions: Past conditions and trends will continue in the future more or less unchanged.

Strengths: It offers substantial data-based forecasts of quantifiable parameters and is especially accurate over short time frames.

Weaknesses: It often requires a significant amount of good data to be effective, works only for quantifiable parameters, and is vulnerable to cataclysms and discontinuities. Projections can be very misleading for long time frames. Trend analysis techniques do not explicitly address causal mechanisms.

Uses: To project quantifiable parameters and to analyze adoption and substitution of technologies.

Expert opinion techniques assume that experts can forecast developments in their fields better than outsiders. However, individual experts often have produced amazingly poor forecasts. For instance, Lord Rutherford, the leading nuclear physicist of the mid-1930s, forecast no serious future for nuclear energy in his

Exhibit 2.5 Modeling

Description: A simplified representation of the structure and dynamics of part of the real world. Models range from flow diagrams, simple equations, and physical models to computer simulations.

Assumptions: The basic structure and important aspects of parts of the world can be captured by simplified representations.

Strengths: Models can exhibit future behavior of complex systems simply by separating important system aspects from unessential detail. Some models offer frameworks for incorporating human judgment. The model-building process can provide excellent insight into complex system behavior.

Weaknesses: Sophisticated techniques may obscure faulty assumptions and give spurious credibility to poor forecasts. Models usually favor quantifiable over nonquantifiable parameters, thereby neglecting potentially important factors. Models that are not heavily data based may be misleading.

Uses: To reduce complex systems to manageable representations. The dynamics of a model can be used to forecast some aspects of the behavior of the system.

lifetime. Nevertheless, experts often can be good sources on the evolution of an existing technology, although they are less good at foreseeing its future when a radical innovation is emerging. This uncertainty has led forecasters to consult a wide range of experts rather than relying solely on one individual. Further discussion of expert opinion appears in Chapter 5.

Trend analysis requires reliable time series data about well-defined parameters. When these data do not exist, as is often the case, trend analysis must be ruled out. However, with adequate data, there are powerful statistical techniques that allow useful projections to be made. These techniques range from simple bivariant regression to more sophisticated methods such as the Box-Jenkins technique. In technology forecasting, data often cover limited time periods and/or are expressed in terms of somewhat arbitrarily defined parameters. Thus, sophisticated techniques may prove to be overkill. "Eyeball fitting" or straightforward regression are often the most useful techniques. However, when there are data, techniques like Fisher-Pry and Gompertz methods can generate the S-shaped projections that are often applicable to the growth cycle of technologies. Chapters 6 and 8 address specific uses of trend analysis.

Typical forecasting *models* are either computer based (such as simulations) or judgment based. In either case, the quality of the assumptions that underlie the model is critical to its success as a forecasting tool. Therefore, it is important to recognize the modeling assumptions. Quantitative parameters typically are used in

Exhibit 2.6 Scenarios

Description: Snapshots of some aspect of the future and/or paths leading from the present to the future. A set of scenarios can encompass the plausible range of possibilities for some aspect of the future.

Assumptions: The richness of future possibilities can be incorporated in a set of imaginative descriptions. Usable forecasts can be constructed from a very narrow database or structural base.

Strengths: They can present rich, complex portraits of possible futures and incorporate a wide range of quantitative and qualitative information produced by other forecasting techniques. They are an effective way of communicating forecasts to a wide variety of users.

Weaknesses: They may be more fantasy than forecast unless a firm basis in reality is maintained.

Uses: To integrate quantitative and qualitative information when both are critical, to integrate forecasts from various sources and techniques into a coherent picture, and to provide a forecast when data are too weak to allow the use of other techniques. They are most useful in forecasting and in communicating complex, highly uncertain situations to nontechnical audiences.

computer-based modeling. Thus, qualitative subtleties that may have substantial effects elude the modeler. Judgment-based models rely on the forecaster's ability to make good assumptions and to make sound judgments about how they affect the forecast. When there is no available theoretical framework within which to develop a model, or when it is difficult to make sound assumptions, it is best not to use these techniques. In many instances, the major benefits of modeling come from the insights gained in the process of building the model rather than from using it when it has been completed.

Scenario construction can be used whether or not good time series data, experts, and useful models exist. Scenarios are stories about the future and/or sets of credible paths leading from the present to the future. They are very good ways to communicate the results of other forecasting techniques and can also contribute to the analysis. For example, construction of the complete story of a future state or of events leading to it often will reveal holes in an analysis. Scenarios can be used to integrate quantitative data with qualitative information and values. They may employ literary artifice and imaginative descriptions or even multimedia techniques to effectively deliver forecasts to diverse audiences, and often they are the best way to communicate with decision makers who are not familiar with forecasting techniques. They often can be used when no other technique is viable.

2.3.2 Method Selection

Now that the most commonly used forecasting techniques have been described, a rationale for selecting among them can be outlined. In making selections, *forecasters should use as many different approaches as practical within resource limitations*. Monitoring can provide the information base for the forecast and is the usual starting point for judgments about constructing forecasts. If experts and time series data are available, expert opinion and trend analysis can be effective. If models exist that incorporate the main features of a forecast topic, and if the forecaster is confident about the quality of the assumptions that drive them, then modeling is viable. Scenarios can be used to integrate results and communicate them in a nontechnical way. They also can be used to forecast when no other techniques can be applied.

Forecasters often have wished for a straightforward procedure for selecting methods on the basis of an algorithm that leads from a statement of the problem to an array of appropriate methods. Unfortunately, no such unambiguous mapping is possible. Creating a meaningful forecast is a design challenge that the forecaster must approach with sound judgment and a clear vision of the final role for which the forecast is intended—*supporting good decisions*.

If the forecast is to fulfill its role effectively, *the means chosen to communicate the forecast results are as important as the means chosen to conduct the forecast*. The two should be complementary. Since the value of the forecast depends on how effectively it is communicated to decision makers, the means chosen to communicate results must be tailored to their characteristics and needs. If they wish to evaluate the bases of the forecast or the implications of the forecast for their organization, then substantial and continued interaction may be necessary. Once again, scenarios can be powerful and persuasive vehicles to deliver forecast results, especially to those unfamiliar with forecasting techniques.

2.4 CONCLUSION

The forecasting environment, like that of other activities, has changed. While long-term forecasts that give large, established organizations a view of the future remain important, rapid change and the increasing importance of small and medium-sized high-growth organizations probably has reduced the use of long-range projections. The continued role of the financial market in driving the short-term focus of major company executives also has inhibited long-range planning. However, forecasts based on the TDS approach continue to be valuable for start-up companies and for small and medium-sized technology-based companies that are vulnerable to all of the forces included in that model. Moreover, the changing names in Fortune's top 500 companies shows that even large firms with global reach and enormous financial power can blunder if they do not foresee and act on changes in their technologies and markets.

Expectations for forecasts should be realistic. Forecasting the future of new technologies is particularly difficult given the inherent lack of data and the complexity of the forces that will be relevant over time. Still, experience with the methods discussed in this book has shown that uncertainty can be reduced, broad indications of trends can be deduced, and areas of opportunity and threat can be identified through forecasting. Even imperfect results can be used by a technology manager to build better decisions. While technology forecasting provides some tools, these must be supplemented by economic, market, and social forecasting methods. All depend on appropriate data and suitable assumptions. The key models, data issues, and techniques introduced in this chapter will be detailed in later chapters.

Forecasting is the foundation of planning. To plan and allocate resources, organizations must have a well-defined view of the possible future states of technology and society. The breadth of the forecasting task and its uncertainty dictate that a collection of tools will be needed to produce such a view. This chapter concluded with specific sets of methods that support a wide range of investigation, are based upon systems of inquiry that have scientific validity, and can provide practical information for decisions. Forecasting to develop knowledge of the many dimensions of the TDS is a substantial undertaking that typically is done in phases. Subsequent chapters provide a more in-depth discussion of the tools introduced here and show how the resulting foresight can improve analysis, planning, and decision making for the future.

REFERENCES

Abrams, M. and H. Bernstein (1989). *Future Stuff*. New York, Viking.

Ascher, W. (1978). *Forecasting: An Appraisal for Policy Makers and Planners*. Baltimore, Johns Hopkins University Press.

Churchman, C. W. (1971). *The Design of Inquiring Systems*. New York, Basic Books.

Cook, T. D. and D. T. Campbell. (1979). *Quasi-Experimentation: Design and Analysis Issues for Field Settings*. Boston, Houghton Mifflin.

Drucker, P. (1985). *Innovation and Entrepreneurship*. New York, HarperCollins.

Linstone, H. A. (1989). "Multiple Perspective: Concept, Applications and User Guidelines." *Systems Practice* **2**(3): 307–331.

Martino, J. P. (1983). *Technological Forecasting for Decision Making.* New York, McGraw-Hill.

Mitroff, I. and M. Turoff. (1973). "The Whys Behind the Hows." *IEEE Spectrum* **10**(3): 62–71.

Moore, G. (1999). *Crossing the Chasm: Marketing and Selling Disruptive Products to Mainstream Customers.* New York, HarperCollins.

Palmer, J. (2008). Interview: The Cellphone Anthropologist. *New Scientist*. 11 June 2008, issue 2660, pp. 46–47.

Porter, A. L. and F. A. Rossini. (1987). "Technological Forecasting." In *Encyclopedia of Systems and Control*, ed. M. Singh. Oxford, Pergammon: 4823–4828.

Rogers, E. (1983). *Diffusion of Innovations*. New York, Free Press.

Sage, A. P. and J. E. Armstrong. (2000). *Introduction to Systems Engineering*. Hoboken, NJ, Wiley-Interscience.

Schnaars, S. P. (1989). *Megamistakes: Forecasting and the Myth of Rapid Technological Change*. New York, Free Press.

Shurig, R. (1984). "Morphology: A Tool for Exploring New Technology." *Long Range Planning* **17**(3): 129–140.

Vanston, J. H. (1985). *Technology Forecasting: An Aid to Effective Technology Management*. Austin, Texas, Technology Futures, Inc.

Volland, C. S. (1987). "A Comprehensive Theory of Long Wave Cycles." *Technological Forecasting and Social Change* **32**: 123–145.

Walker, W. E. (2000). "Policy Analysis: A Systematic Approach to Supporting Policy-making in the Public Sector." *Journal of Multi-Criteria Decision Analysis* **9**: 11–27.

Wenk, E., Jr., and Kuehn, T. J. (1977). "Interinstitutional Networks in Technological Delivery Systems." In *Science and Technology Policy*, ed. J. Haberer. Lexington, MA, Lexington Books: 153–175.

Wheeler, D. R. and C. J. Shelley. (1987). "Toward More Realistic Forecasts for High Technology Products." *Journal of Business and Industrial Marketing* **2**(3): 36–44.

Wise, G. (1976). "The Accuracy of Technological Forecasts: 1890–1940." *Futures Research Quarterly* **8**: 411–419.

Zwicky, F. (1962). *Morphology of Propulsive Power*. Pasadena, CA, Society for Morphological Research.

Zwicky, F. (1969). *Discovery, Invention, Research Through the Morphological Approach*. Toronto, Macmillan.

3 MANAGING THE FORECASTING PROJECT

Chapter Summary: This chapter discusses how to conduct and manage a technology forecasting project. It discusses the need for technology forecasting, approaches to planning a project, methods for organizing the project in teams, and scheduling the project. The chapter concludes with a preview of the chapters to follow.

Managing a forecasting project demands most of the same management qualities required of other projects: sound goals, objectives, and constraints; careful scheduling and cost accounting; and good communication and people skills. However, there are differences as well. These arise from the uncertainty of forecasting and because the people needed for the task may exhibit a wide variety of personal characteristics. For example, some people deal better with the uncertainties of extending existing knowledge than others. Moreover, a forecast often requires individuals from a variety of disciplines (e.g., science, engineering, economics, and social sciences), all with different disciplinary approaches, vocabularies, and perspectives. Clearly, this complicates communication and cooperation. For these reasons and others, forecasting projects may require different organizational and communication structures than other projects.

3.1 INFORMATION NEEDS OF THE FORECASTING PROJECT

Good forecasts are ones that lead to good decisions. Providing the information to make good management decisions implies stretching present knowledge into the future—forecasting—and hence, it also involves uncertainty. However, forecasts alone will not produce enough information for sound decisions. They provide raw material about possible futures to combine with factors such as business objectives and values to fashion wise decisions.

Burgelman, Christensen, and Wheelwright (2009, pp. 4–9) assert that general managers may not need to have in-depth technical knowledge, but they must learn enough to frame strategic questions about technologies and their businesses. They cite Michael Porter (Porter 1985, pp. 1–33), who pointed out that process and product technologies should enable pursuit of the four generic strategies: cost leadership, differentiation, focus segment cost leadership, and focus segment differentiation. He addresses the "value chain"–the entire production cycle from raw materials, component parts, product to retailing of the product and perhaps even the provision of services related to the product. New technologies obviously will impact product market strategies, but they need to be implemented with an understanding of the entire value chain associated with the businesses. This requires not only current information but also projections of technology life cycles and the ability to forecast technology futures using techniques such as those presented later in this book.

While product manufacturers must be critically aware of unfolding trends in technology for their business sector, it would be wrong to think that only managers in that sector need tools for forecasting and management. Over two decades ago, Eric von Hippel showed that sources of innovation are not where intuition might guide us to look for them (von Hippel 1988). In an extensive study of innovations in a variety of late-twentieth-century industries, he and his colleagues found that innovations were functionally related and that their origins were determined by who got the most benefit from them. Table 3.1 shows some of their results.

Note that innovations in the traditional tractor and related industries generally originated with manufacturers, while the less mature semiconductor and scientific instrument industries saw more innovations originated by users. The differences shown in the table are due to the stage of the industry that was most likely to capture profits from the innovation. The lesson for twenty-first-century managers is that opportunities to create and apply new technology exist throughout the value chain.

It also is important to remember that firms benefit from applying technology and that the creator of new technology is not always, or even usually, the one who benefits most. Successfully implementing new technologies requires the right

TABLE 3.1 Innovation Origin Varies by Industry

Industry	Users (%)	Manufacturers (%)	Suppliers (%)	Other (%)
Scientific instruments	77	23	0	0
Semiconductors	67	21	0	12
Travel shovel related	6	94	0	0
Wire termination equipment	11	33	56	0

Source: Von Hippel, Thomke et al. (1999, p. 4).

combinations of marketing, production, distribution, and service. Microsoft used superior marketing rather than superior technology to dominate global sales of operating systems and software applications. Henry Ford became the dominant automobile producer by radically changing the way cars were produced, not by inventing cars, and Michael Dell accepted the technology developments of others and changed the way personal computers and related products were distributed.

Some business writers emphasize the importance of understanding the internal capabilities and vision of the organization so as to continually improve its performance in the marketplace. For example, Collins and Porras (1997, pp. 10, 185–200) noted that their six-year study of successful, visionary companies showed that they "focus on primarily beating themselves" rather than the competition. Wheatley and Wilemon (1999) investigated the "fuzzy front end" of decisions about go/no-go decisions on new products. While their discussion includes consideration of external factors, their recommendations emphasize internal management attributes and pay only some attention to projecting to the future. Brown and Eisenhardt (1998) emphasized that the environment is not controllable or even very predictable. Yet, their sixth strategic rule for competing on the edge is to reach into the future.

3.1.1 The Technology Manager's Needs

The technology manager must organize and manage the search for information upon which to base sound decisions. Using the technology delivery system (TDS) to structure the search will help ensure that the relevant variables and their likely interactions are considered. The first step is to decide what decisions the forecast is to inform. For example, the manager may wish to know how technological advances will affect the profitability of existing plants, equipment, or products; what new technologies offer opportunities or challenges; what technologies can be brought to market sooner by increasing R&D resources; or what the competitive or regulatory environments of the future may hold. The decision that is to be made shapes the needs for forecast information and the methods that will be used.

Today organizations are under increasing pressure to account for the social and environmental consequences of their activities. Sound responses to these concerns are every technology manager's ethical and moral responsibility. On a different level, the viability of any technology may strongly depend on society's responses and those of its regulatory agencies. But what are the likely positions of society and regulators during the product's life? What new concerns about quality of life and environmental and health effects are likely to emerge? What will be the impact of the technology on the ecosystem? The answers to such questions require some of the most vexing and important information any manager is likely to need. Predictions of such social and political factors are among the most uncertain forecasting tasks, and the cost of being wrong can be high (ask anyone from the nuclear power industry).

Managers also must decide if the decision will be best served by extrapolative or normative forecasting perspectives. The former asks what the future may bring if trends continue; the latter asks what actions, advances, or breakthroughs may be needed to shape the future. Some decisions will require both perspectives. Early in the project the manager also must decide what information is needed, the degree of specificity required, and the amount of uncertainty that can be tolerated.

Since forecasting requires time and resources, the manager must determine the scope and depth of the forecast based upon judgment about the value that a picture of the future will provide. Will the potential benefits of having the information justify the costs of obtaining it? Timing is critical. *When* will the information be needed? Finally, the manager must determine what human and financial resources will be available. Perspective, information needs, potential benefits, timing, and resources shape the project and the choice of methods.

The need for vision and the ability to make assessments of future environments is nowhere more evident in technology management than in making decisions about intellectual property. Kevin Rivette and David Kline wrote *Rembrandts in the Attic* (2000) to encourage managers to use patents as a competitive weapon. And while patents provide some competitive protection, the authors emphasize how they also can provide competitive information. If the new technology is forecast to have decades of commercial viability, it may make sense to inform competitors how it works and count on legal protection. However, if the manager judges that an innovation will be obsolete in a few years, it may make better sense to use trade secrets and rapid market deployment. Patent planning requires knowledge beyond the patented technology. Even mining patent information for insights on technology strategies has to be done within broad long-term contexts.

3.1.2 The Forecast Manager's Needs

One of the first tasks faced by the forecaster is to determine the information needed to make the forecast. To do this, the forecast must be *bounded*—that is, significant thought must be given to which factors will be considered and which won't. Bounds are strongly influenced by the information needs of the decision the forecast is to inform and by the timing of that decision. They also are affected by the likely costs of poor decisions and the nature of the factors that will be forecast. For example, forecasting the sales potential for a new technology implies that the dynamics of the marketplace peculiar to it must be considered. Incorrect estimates can mean the difference between success and failure of the product and even of the organization. While the study must be bounded early to allow work to begin, initial bounds should be set broadly and remain flexible as long as possible. This provides the opportunity to incorporate factors whose importance is recognized after work has begun. Maintaining breadth and flexibility will require willpower; the pressures exerted by time and resource constraints are relentless, and at some point the investigation must be narrowed and conclusions drawn

Vanston (1985) suggests that five types of information are required by the technology manager that the forecaster must anticipate and meet. Because of the need to understand the broader dimensions of the TDS, the authors have added a sixth:

1. Projections of rates at which new technologies will replace older ones
2. Assistance in managing technical R&D
3. Evaluation of the present value of technology being developed
4. Identification and evaluation of new products/processes that may present opportunities or threats
5. Analysis of new technologies that may change strategies and/or operations
6. Probable responses of regulatory agencies and society to a new product, process, or operation

Unless otherwise noted, examples in the following paragraphs are taken from Vanston (1985, 2008).

To forecast the rate at which a new technology will substitute for an old one, basic characteristics of both must be understood. Further, substitution must have proceeded long enough to establish a trend. The substitution rate is as important to old technology producers as to potential producers of the new technology. All need it to allocate resources. New technology producers will be especially concerned about financing and expanding plant facilities and developing strategies to speed substitution. Potential producers need to time their market entry accurately. Old technology producers must decide how to respond to the incursion in their markets. They may wish to develop strategies to retain profitability as long as possible; plan an orderly production halt; or introduce new products or perhaps even "leap frog" to an even newer development. Substitution forecasting is an established tool of technology management, and there are many examples of its use.

Information from forecasts based on the TDS also can be used to support decisions about R&D. The high costs of R&D increasingly require that it be tied to relatively quick market success. For example, Firat, Woon, et al. (2008, pp. 12–14) found that high-technology firms that emphasize R&D tended to find technology forecasting crucial to their operations. They cite the specific case of Glaxo Smith Kline and French, which merged with Beecham and used forecast information to alter allocation of their combined R&D resources (Norling, Herring et al. 2000).

In "Practical Tips for Forecasting New Technology Adoption," Vanston (2008) addressed the information needed to evaluate a new technology. He suggests that issues such as when the technology is likely to come to market and what drivers and constraints will affect it need to be answered. This is the type of information that a well-developed forecast using the TDS can provide.

3.1.3 Information about Team Members

Most forecasting efforts will require persons with several disciplinary skills. This, of course, presents inherent complications that have already been alluded to.

That a typical forecasting effort will involve several individuals implies more. A successful effort must be managed with awareness that team members will exhibit different personality characteristics.

Individual and organizational personality types will affect how the technology forecasting project is managed. Probably the most widely used personality typing scheme is the Myers-Briggs Type Inventory (MBTI). It can be effective in distinguishing the various modes of thinking among those involved in multi-skill projects such as technology forecasting. There are many websites that discuss personality types. For example, Personality Pathways (2010) describes the MBTI and provides a simple illustrative self-test, although it emphasizes the importance of professional follow-up. The "Big Five" provide a directly comparable personality traits framework to the MBTI. However, the MBTI dimensions seem more helpful in showing how the differences are relevant in technology forecasting. Another such site is Human Metrics (2010). The characteristics of various personality types that might be exhibited in forecasting team members are given and contrasted in Table 3.2. All can contribute to a good forecast.

Myers and Briggs developed their personality typology as a celebration of diversity in human personality. A similar diversity of people and approaches can enrich technology forecasting. The multidimensional demands of a good forecast are best met by the different perspectives brought by the various personality types. Therefore, the manager should design the project to build on the attributes of each personality type, just as he or she should design it to accommodate different disciplinary approaches. For instance, while forecasting methods tend

TABLE 3.2 Myers-Briggs Types and Appropriate Technology Forecasting Approaches

Extraversion (E)		Introversion (I)
Team approach	vs.	Solo or single-partner approach
Discussion needed for processing ideas		Introspection needed for processing ideas
Experimentation, then reflection		Reflection, then experimentation
Sensing (S)		Intuition (N)
Past experience and facts are valued	vs.	Theories and patterns valued
Incremental approaches used		Imaginative approaches used
Current, concrete perspectives		Ambiguous, fuzzy perspectives on the world
Thinking (T)		Feeling (F)
Oriented toward facts	vs.	Oriented toward values
Values the discussion of new technology		Values the discussion of people
Appreciates objectivity in presentation		Appreciates subjectivity in presentation
Judging (J)		Perceiving (P)
Emphasis on planning projects and events	vs.	Emphasis on allowing projects to unfold
Settled and decided		Open to late-breaking information
Emphasis on closure on decisions		Emphasis on gathering of information

to be oriented toward the thinker, there are ways to make the project appealing to the feeling individual. Applying the normative scenario approach (Chapter 7) is one way.

3.2 PLANNING THE TECHNOLOGY FORECAST

Technology forecasting projects in some ways are similar to the processes used to design new technologies. However, only on rare occasions can forecasters adapt existing information to their organization's purpose. More often, they must adopt a problem-solving perspective, adapting methods or recombining information to serve new purposes. The forecasting process, like the engineering process, is iterative. Decisions made at one stage may affect choices made later. Like the design process, the technology forecasting project may need to broker diverging interests within an organization. The structure of new technology, and the design of the organizations that create new technologies, are often complexly interlinked (De Sanctis and Poole 1994).

This book proposes that technology forecasting typically involves a three-phase approach, which this book characterizes as *cold, warm,* and *hot.* The names given to the phases differ from author to author. Ben Martin (1995) recasts technology forecasting as *foresight*, a use of the word that encompasses the broader issues and context of the TDS model used in this book. The fact that foresight refers to technology forecasting in this broad sense is supported by Ruff (2004, p. 44), who notes that "technology foresight has evolved from an earlier narrow focus on technology forecasts to a broader definition, which takes political, economic and societal factors and their interactions into consideration." While Martin was concerned with technology foresight for national planning and policy, Ruff's focus is on the use of the methods to enhance corporate strategy and resources allocation.

The cold phase of the forecasting project can be characterized as *exploring* (Chapter 4). In this first phase, forecasters seek to identify the problem and to specify its boundaries. Agreement about the need to look ahead is reached, and a consensus about how the information will be used is developed. These results will help the forecaster develop a design for the next phase.

The *cold phase* begins by describing and understanding the technology, its key variants, and how these could translate into bona fide innovations. All possible information about the technology and its context is sifted to quickly check if information that would normally be excluded may have unrecognized significance. It is unwise not to check for *black swans* (significant, very-low-probability, very-high-impact possibilities) (Taleb 2007) lurking in an otherwise useless expanse of information. Besides acting as a quick filter, the first phase builds a broad map of the study and sketches in most parts of the map. The cold phase often may result in refocusing the forecast. Beyond that, it is a convenient point to make a go/no-go decision about further development or forecasting of a technology. Monitoring is the principal method used in this phase.

The second, or *warm, phase* deepens essential parts of the technology/context description and forecast. It identifies pivotal developmental possibilities for further study. The early parts of this phase require strategic thinking about social and economic forces, as well as technology, so that the design of the project will reflect the decision makers' needs. The warm phase involves the bulk of the forecasting about how the future states will evolve. The investigation determines the specific areas, audiences, resources, and approaches that will be most useful. In this book, the warm stage is described in terms of its principal activity, *analyzing* (Chapter 6). This phase is characterized by analytical techniques such as systems analysis, trend analysis, and simple modeling.

The "scientific method" and the "policy analysis process" offer detailed insight into what happens during this stage. Alternatives are generated, looking for both creative and logical solutions. The solutions are analyzed using a range of techniques. While scientists conduct experiments, technology forecasters often use simulation or modeling. The various alternatives are challenged by evidence and by uncertainty.

In the *hot phase,* the most important paths forward are chosen and analyzed in as great a depth as is consistent with time and resources. These paths may engage more specialized analyses particular to the pivotal issues (and largely beyond the scope of this book). Scenarios could well be constructed around principal dimensions of the technology and its context to integrate findings and help communicate key possibilities.

In a management context, the hot phase involves using what has been learned about the technology, its likely futures, and its impacts. Strategies about markets and resource allocation are developed in this phase. This book suggests that there should be a narrowing of attention to specific technologies and their implications before the forecast is implemented. This *focusing* stage is an effort to highlight the most significant issues for managers who have to make decisions. The hot or focusing phase is dealt with in Chapters 7 through 11.

The hot phase may be the end of the forecasting project, but it is not the end of forecasting. There should be ongoing monitoring to allow the midcourse corrections in planning that will be required to keep the organization moving toward its goals.. New information must be continuously gathered. The path to the future must be updated. Indeed, the vision that guides the path to the future may also need to be questioned and revised.

3.3 TEAM ORGANIZATION, MANAGEMENT, AND COMMUNICATIONS

Technology forecasting and the communication of its results occur both between organizations and within them. The management of twenty-first-century organizations increasingly involves alliances and collaboration throughout the supply chains of most industries. This section focuses on considerations for adapting a technology forecast to a specific organization. Later in the book (Chapter 11), the use of technology forecasting to coordinate activities between related organizations and stakeholders is discussed.

Individuals are confronted by technical change and may benefit from information about the future of technology. However, for some organizations this information is crucial to their prosperity, even to their survival. These organizations often are expressly designed to manage large technical projects. Organizational design involves sharing and hiding information, as well as allocation and delegation of tasks. Some organizational designs are better than others for forecasting technology futures and using the information those forecasts produce.

Mintzberg (1980) argued that the design of organizations is contingent on several factors, including the power of different parts of the organization, the organizational environment, and the coordination mechanisms it uses. Five basic organizational structures emerge from Mintzberg's analysis. Technology forecasters who are aware of these structures can adapt the design and the communication of their forecasts to enhance its impact. These five structures are briefly discussed below in terms of their need for technology forecasting as well as their ability to absorb forecast results. Recommendations for internal communication, based on Mintzberg's framework, are presented.

Simple structures are flat organizations operating under direct supervision of the top general managers (the *strategic apex*). Such organizations may seem to have little need for formal technology forecasts, since they limit the complexity of their environments by specializing in a limited number of goods or services. These organizations often are young and small, with close connections among the principals and the staff. This may limit their future orientation if they are outside the technology arena. High-technology start-ups, on the other hand, typically aim for future success as the payoff for current activities. These firms need at least an informal and agreed-upon view of their TDS to develop and execute a technology-based business plan. All simple organizations can act directly upon the information provided by a forecast. In fact, even a limited forecast can help a simple organization develop in a coordinated way. Personnel receive direct supervision from executives, enabling a rapid and centralized response to the strategic mission being pursued. Communication about the future is often best conducted informally and face-to-face with one of the company executives. Technology forecasting therefore is likely to be done by the executives with whatever outside help they can afford.

Machine bureaucracies concentrate organizational power in their technical staff and rely upon the standardization of tasks as a means of coordination. These organizations need technology forecasts since they are generally highly regulated organizations, with specialized personnel and considerable resources dedicated to action planning. Nonetheless, they may find it challenging to act upon forecasts given their emphasis on formalized, bureaucratic, and insular modes of action. In fact, such organizations struggle to provide needed cross-functional integration of technical staff with the marketing, finance, and other vital parts of the business even in regard to current operational issues. Communication about the future or about systematic changes within the technology environment often occurs as part of an organizational mandate. Thus, future planning can provide the integrative focus to improve the organization's execution. Management should encourage forecasters to join in this problem-oriented mandate. However, management may need to have forecasters take extra pains to document findings and to create results that can be audited.

Professional bureaucracies, like simple structure organizations, believe they have little need for forecasts and have little capability to assimilate forecast results. These organizations concentrate their power in the hands of front-line personnel who deliver goods and services to customers. These individuals are highly skilled and have specialized tasks and responsibilities. Such organizations often expect them to be aware of future trends in technology and to have a high level of de facto knowledge about future trends gained through networking and external contacts. In reality, this reliance can make them vulnerable to disruptive innovation. Professional bureaucracies often change their personnel and hiring profiles to obtain necessary competencies, and they are likely to try to adapt to anticipated change by changing their leadership. Communication about the future in these organizations is often best performed as part of on-the-job training or reeducation.

Divisional forms are like machine bureaucracies in the sense that they have a strong need for formal analyses of new technology, yet find it challenging to absorb such information. Such organizations concentrate power in their mid-level management and are adept at making a range of standardized products. Communication about the future often takes place as part of market valuation of existing technologies or as part of the design of new ones. Managers should include forecasters in these activities and look for opportunities to insert the results of their forecasts. Forecasters may need to take extra care to document anticipated threats as well as their technology solutions in order to demonstrate their value to the divisional managers tasked with implementing change.

Adhocracies is a term coined by Mintzberg (1980) to describe a certain "make and make do" style of organizational design and governance. Despite the relatively freewheeling style of the adhocracy, adhocracies also need technology forecasting because of their exposure to a highly dynamic and complex environment. The range of skills they employ, and their relatively open and organic structure, suggest that they may routinely conduct internal, informal technology scanning activities, alleviating the need for formal activities. Mutual adjustment of activities in light of new information is a hallmark of the organizational structure. Introducing technology forecasting within an adhocracy may involve creating opportunities for group learning or participation, or introducing forums to address emerging issues. Managers in such organizations should encourage forecasting throughout the structure rather than assigning forecasting to a particular group.

Unfortunately, due to their organizational structure, government and other bureaucracies, as well as divisional form corporations that most need technology forecasting, find it the most difficult to assimilate the kind of information that forecasters produce. Ruff (2004) provides a useful contrast between the information needs of government and industry. One major difference concerns the actual conduct and capitalization of research. There is a corresponding difference in time scales: governments typically examine time frames ranging from 5 to 50 years, while commercial organizations have a shorter time frame of 2 to 15 years. Moreover, government foresight projects are typically larger enterprises, lasting from one to three years, while commercial organization activities typically span three months to one year of project time.

3.3.1 Organizing and Managing the Technology Forecast

As noted, forecasting projects usually involve factors traditionally thought of as being within the purview of different disciplines (and organizations). For instance, market and economic factors generally are studied by economists and finance experts and technological factors by engineers and scientists. Social and political considerations introduce still more disciplinary specialties. Such projects have been characterized as *multiskill* (Porter and Rossini 1986). Management arrangements and communications between team members are critical in multiskill projects, especially if team members have no history of collaboration. In addition to disciplinary differences, forecast and implementation project teams often will be made up of individuals from different areas of the organization. These persons also come with various perspectives, methods, worldviews, and turf to protect (e.g., production, R&D, and marketing). Thus, the forecasting and implementation projects must be managed to allow these individuals to cooperate and communicate effectively. Finally, because the result will be a multiskill product, special attention must be given to management structures and communication patterns that incorporate substantive knowledge and contributions from a variety of fields.

Activities in which individuals with different disciplinary backgrounds cooperate to produce an integrated information product have been characterized in several ways. Differences in these products rest primarily on the success with which individual work has been integrated to produce a seamless result. A weakness of this approach, of course, is that it must be retrospectively applied. A more productive approach seems to be to emphasize the intellectual skills involved and their organization rather than the disciplines or the character of the results.

When a relatively large number of individuals are involved, the manager usually will find it expeditious to establish a *core or coordinating group* to direct activities. Ideally, members of this group are expert in the various areas required for the project. If multiple units of an organization are involved, however, political realities usually dictate that the group will be composed of individuals designated by the various units. Regardless of how the core group is constituted, the project manager must work hard to foster communication and cooperation among its members. He or she also must ensure that they understand the special requirements of the project.

Porter and Rossini (1984) suggest that there are two distinctly different kinds of administrative organizations: *open* and *closed*. The former is structured to tolerate, even foster, a range of knowledge and techniques. The structure of the latter discourages such diversity. Whether or not the organization is open or closed is usually determined by the nature of the problems it must solve. For example, open structures are characteristic of organizations (e.g., forecasting and implementation groups) that deal with problems that straddle broad categories of knowledge. Closed structures often are found in groups that consider a more narrowly focused problem in great depth (e.g., heat transfer in the turbine blades of a jet engine).

The manager must account for both internal and external openness in developing a method for organizing the project. Thus, the structure of forecasting and implementation project groups must be open. It is perhaps less obvious that even in an open structure, a pecking order based on factors such as the quantification or "practicality" of individual skills may exist. Thus, a group structure that appears open may actually act like a closed one, denying itself important contributions from members whose skills are implicitly discounted. In technological organizations, this informal structure often discourages input from so-called soft disciplines. The manager must ensure that this situation is avoided, for unpleasant surprises await those who use forecasts in which social or political factors are dominant but underplayed.

Rossini and Porter (1981) identified four generic management approaches used to structure technology assessment projects that share many of the same concerns as forecasting and implementation projects. A fifth has emerged in the Internet era:

1. Common group learning
2. Group model building
3. Negotiation among experts
4. Integration by leader
5. Collective intelligence

These five structures are diagrammed in Figure 3.1. Although developed from other types of projects, they provide sound guidance for managing the forecasting project.

In *common group learning* the information product is generated by a group that learns and acts as a whole. Thus, its output is the common intellectual product of the group (Kash 1977). In this approach, project tasks are divided among group members, usually according to substantive skills, personal interest, or organizational identity. Preliminary analyses are generated by each member and then critiqued and modified by the full group. Finally, each task is redone by a new group, often one that is not expert in the area. This iterative process continues until the group and the manager are satisfied with the result. While the final result is well integrated, the process is time-consuming and tends to achieve integration at the expense of technical sophistication and analytical depth. Such projects often run over their allotted schedule (Rossini and Porter 1981). Because of the costs associated with this method, in large projects only the core group generally employs the common group learning approach. The approach is

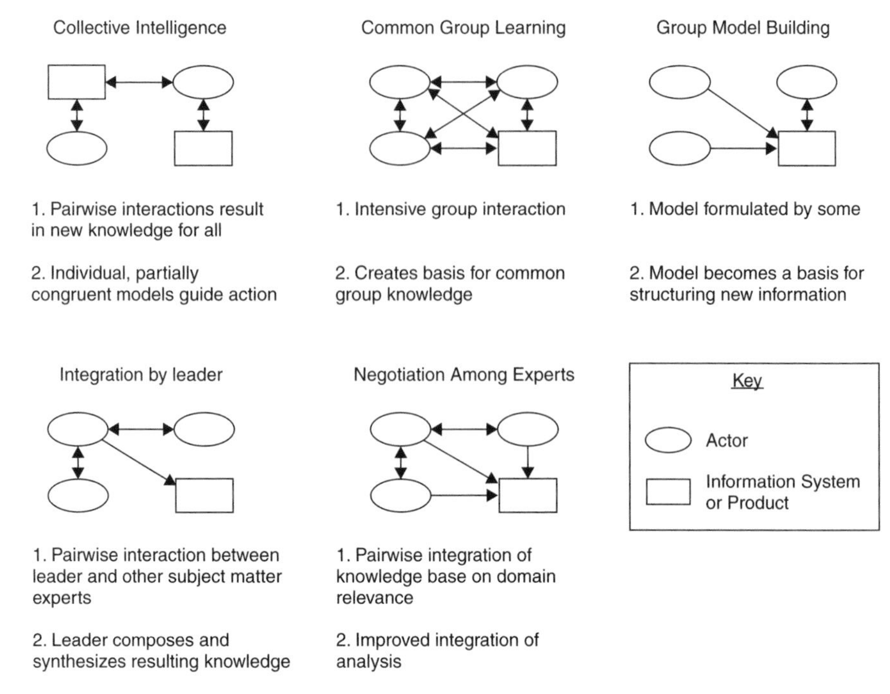

Figure 3.1. Five Approaches to Structuring a Technology Forecasting Team

commonly used in the related field of integrated resource management, where issues of consensus building are paramount (Pahl-Wostl and Hare 2004).

Group model building involves cooperation in the creation of one or more models. Rossini and Porter (1981) note that models can provide a common ground where disciplinary contributions can meet. That is, construction of a new model or operation of an existing one can provide a platform upon which to integrate the dynamics that will shape the outcome. However, models often are highly quantified representations that require significant computer resources. Although such models provide focus and a platform, they also can narrow the perspective by tending to undervalue factors that are difficult to quantify. Few forecasting or implementation projects that the manager will encounter will be amenable to highly quantified models that can be employed "off the shelf." Even more rarely will time and resources permit their construction. When such models are used, however, the manager must guard against the spurious credibility sometimes granted their output because of the sophistication of their computational techniques. Proponents of the method argue that the approach helps mediate the production of new knowledge and facilitates the development of action plans (Rouwette, Vennix, et al. 2002). The most useful models often are graphical representations such as trees or flowcharts that capture interactions between factors germane to the forecast. The process of constructing such models can be very helpful in understanding and directing a project, and they require few resources to create. However, iteration is absolutely necessary if a satisfactory model is to be created.

Many forecasting projects will demand more-or-less equal participation by several units of an organization (e.g., R&D, production, and marketing). When this happens, politics usually will dictate a project structure that involves *negotiation among experts*. In this approach, tasks are divided among members of the core group on the basis of expertise or unit responsibility. Thus, this approach endorses limited pairwise negotiation between experts who share some common base of knowledge. Predictably, the initial results will reflect the substantive knowledge as well as the self-interest peculiar to the units (i.e., their turf). These results are integrated by negotiation, and the tasks are redone to reflect that negotiation. This approach tends to preserve depth and expertise but usually at the expense of full integration. However, it can build broad-based support for the forecast or implementation plan that is produced.

An *integration by leader* structure often is employed by strong managers who feel it necessary to maintain tight project control. In this structure, all tasks are assigned by the manager, and he or she becomes the sole integrator of various components of the project. Typically, there is little interaction between team members working on different aspects of the project. This method requires that the manager assimilate and understand each of the contributions before integrating them. Thus, it makes major demands on a single individual. Like common group learning, integration is achieved at the expense of depth, because the manager is unlikely to grasp the details of the variety of areas of substantive knowledge needed for the project (Rossini and Porter 1981). However, tight managerial control increases the probability that the project will be finished on time and within budget. The integration by leader arrangement works best for small, tightly bounded projects.

Collective intelligence involves interaction among different individuals and information systems, often at different physical locations and perhaps operating asynchronously. Interaction is usually facilitated by the Internet; the results are stored and then integrated to produce an information product. This approach leverages the vast array of information and human resources available through the Internet and the nearly instantaneous communication that it provides. However, it usually fails to fully share information and results, and may lead to multiple diverging views that can be very difficult to reconcile. When this occurs, the final result will lack a unified base of support. Further, the rapidity of communication can lead to both confusion and incompletely considered conclusions. An added advantage of this approach is that little overarching structure or coordination is needed. The resulting forecasts may reveal new information not resident with any single actor (Cunningham 2009).

The five management structures are archetypical, and real-world projects seldom rely solely one or another of them. For example, the core group may function in a common learning or negotiation among experts mode while supporting groups use the integration by leader or modeling structures method. In many instances, the manager will inherit a management structure that may benefit from modification. In others, ad hoc project teams formed for a single project may have no established structure and the manager will have to create one. Either way, an understanding of the function and importance of the management scheme will be critical to a successful result. In many instances, the management structure will be inherited by the technology manager and may benefit from modification. In others, ad hoc project teams formed to make a single forecast may have no established structure and the manager will have to create one. Either way, an understanding of the function and importance of the management scheme will be critical to a successful forecast. The manager should organize the project structure to leverage personnel strengths and accommodate established ways of doing business within the firm.

3.3.2 Communications

Unless important decision makers support the forecasting and implementation projects, the time and resources invested in them are wasted. Thus, it is vitally important that team communications are effective in building credibility within the organization and with stakeholders who can influence the technology's development. It is absolutely critical to maintain good contact with these individuals and to convincingly communicate project progress and results. Taking time to identify the ways in which important decision makers and stakeholders prefer to receive information will pay major dividends.

Asking for input from parties outside the team and taking their input seriously is very important. External participation can begin at any stage of the project from definition to evaluation of results. Early involvement can increase commitment and build trust, but it uses more resources than later involvement. The latter uses resources more efficiently but may lead to delays and costs to repeat earlier steps. Waiting too long can alienate stakeholders who believe they have been brought in to endorse conclusions already reached. Susskind (1983) lists two keys to achieving effective external participation: identifying those with a legitimate stake and defining the ground rules for participants who join the process in later stages (e.g., conditions to be set on late joiners, such as whether earlier agreements will be reopened).

Useful vehicles for external participation may include advisory committees (Arnstein 1975) and planning cells. The latter engage small groups, who reflect major stakeholder perspectives and values, to work intensively for a short time (Peters 1986). According to Redelfs and Stanke (1988), such participation can allow parties to provide additional information to decision makers and, if appropriate, it can provide a vehicle for collaborative decision making. In cases in which a major conflict arises that cannot be resolved internally, it may be necessary to bring in a third party as mediator. Susskind, McMahon, et al. (1987) present guidelines for this process:

- Ensure that all parties fully understand the issues and the alternatives.
- Direct the energies of all parties toward achieving a consensus.

- Keep everyone abreast of the progress of negotiations.
- Preempt any escalation of disputes caused by selective perceptions.
- Develop incentives for good faith bargaining.
- Devise ways to bind all parties to agreements they reach.

Following these suggestions will give credibility to those who will make decisions and allocate funds for implementing the technology.

Communication patterns *within* teams are strongly influenced by the management structure that is chosen. If the communication pattern is incompatible with the management scheme, it can produce a de facto structure that functions quite differently from what the manager intends. The old-boy/old-girl network, for example, is independent of organizational charts and impervious to reorganization. Rossini and Porter (1981) list three archetypical communications patterns: all channel, hub-and-spokes, and any channel. Real-world project communication usually is some combination of these three patterns.

In the *all channel* pattern, everyone communicates with everyone else. This arrangement is most compatible with the common group learning and collective intelligence management structures. In the *hub-and-spokes* pattern, individuals communicate with the project manager but not with one another. This pattern is encountered in management structures with a strong manager, centralized responsibility, and team members who are not located near each other. It also is seen in structures in which a single manager controls the input from a group of outside experts. In *any channel* communications patterns, all channels of communication are open but are used only as needed. This pattern is often encountered with negotiation among experts and modeling management structures and likely is the most appropriate for typical forecasting or implementation projects. It provides for sharing of the important knowledge, information, and perspectives necessary for multiskill tasks without the lost effort inherent in the all channel pattern or the isolation produced by the hub-and-spokes arrangement. However, effective use of the pattern does assume that individuals know when they need to communicate with each other.

It is important for the forecast manager to realize that an effective communications pattern does not just happen. Nor does a pattern always evolve in a fashion compatible with the management structure pictured by the organizational chart. Communication patterns must be fostered through meetings and by assigning tasks and responsibilities in a way that forces communication to occur in the desired pattern until it becomes a natural part of daily activity.

3.3.3 Summary Conclusions about Project Management and Organization

The forecasting project can be organized thematically by *system*, *method*, or *process*. For instance, projects organized around a TDS as proposed in this book might explicitly investigate systems and subsystems, recognizing where key uncertainties lie and acting upon this information accordingly. This approach has the advantage of marshaling a comprehensive response to a forecast task. Unfortunately, such projects are difficult to schedule and report since results are often diffuse and organic in character. Projects organized by method can allow easy delegation by task or technical specialty. However, they may not always give clear signals about which methods should be chosen for a specific forecast, and integrating the results may be difficult. Projects organized by process may facilitate integration of forecast results into a project organization. But it may be difficult to iterate the results and to deal comprehensively with uncertainties using this method.

Standard forms of coordination suggest that the machine bureaucracies may be best served by group model building deployed around a process, and divisional forms by collective intelligence activities structured around a system such as the TDS. Simple form organizations and adhocracies make good use of foresight activities but may prefer more loosely structured and informal kinds of activities. The professional bureaucracy effectively conducts foresight by utilizing a highly skilled and mobile workforce. Professional bureaucracies attempting to gain a strategic perspective on their skills might try projects structured around methodology and involving negotiation by experts.

3.4 SUCCESS: THE RIGHT INFORMATION AT THE RIGHT TIME

Recall that Section 3.1.2 noted six types of information the forecaster needs to supply for technology decision making. Forecasting projects that provided the right information at the right time are described in this section. The *Firestone Company*, for instance, projected the substitution of radial for bias ply tires in the U.S. market. *United Technology Corporation* has used substitution forecasts to plan the introduction of new aircraft engines. Historical examples of successful forecasts of the substitution of one technology for another also include fiber-optic cable for copper wire and office facsimile machines for overnight delivery services (Vanston 1985).

Philips Medical Systems is a part of Royal Phillips Electronics that sells medical systems. It develops road maps for different topics, including clinical research and technology road maps. These forecasts are institutionalized to become an integral part of their innovation process, and they help reduce the lead time for innovations. Roadmap construction also has enabled the company to better structure and streamline its innovation process (Van der Duin 2006). Thus, Philips is using forecasting to help manage its R&D processes.

Vanston (1985) notes that it is common for firms to assign monetary values to technologies during their development cycle. This provides a measurement of a technology's present and potential worth that can be used to decide resource allocation. However, it also requires forecasts of likely social, political, environmental, regulatory, and economic parameters over the life of the technology. These are perhaps the most complex and uncertain of all forecasting targets. Examples from industry include *AT&T*, which carried out a program to evaluate the present value of existing and developing technologies. The corporation is particularly interested in the market value of new technologies, a topic that is discussed in greater detail in Chapter 8.

Evaluation of new products and processes that can be supportive of or competitive to the firm's technologies is vital. *General Electric* completed a corporatewide study to examine its competitive technology base (Vanston 1985). *Royal Dutch Shell* famously used a scenario analysis approach to create robust strategies in the event of an oil shortage (Van der Heijden 2005). The resulting strategies greatly improved the competitive position of these companies. General Electric is developing new products, and Shell is developing new processes to manage a disruptive external environment.

Daimler has a dedicated group for carrying out futures research—its Society and Technology Research Group (STRG). The STRG has a broad portfolio of futures research methods and a staff of about 40 employees worldwide. STRG's work spans six areas: developing and analyzing regional perspectives to identify business opportunities in specific regions; assessing the interdependence of its products and services in the societal and technological environments; continuously monitoring the company's broader environment; identifying opportunities for new mobility concepts and manufacturing systems; energy, environment, and resource assessments; and development and deployment of new methodologies (Van der Duin 2006). The STRG uses futures research to look at the development of the company's business and to seek the societal context of Daimler operations. Daimler is particularly interested in the analysis of new technologies that could change its operations and the response of regulators and society to new environmental and technological changes.

3.5 PROJECT SCHEDULING

Scheduling is important to any project to ensure that it is completed on time and within budget. But it is critically important in forecasting and implementation for several reasons. First, timing-forecast results must be available before deadlines for decisions about implementing the technology or they will be of no value. Second, forecasting and implementation processes both often include parties outside of the organization and thus are less time efficient than more conventional projects. Third, iteration of the forecast usually will be required and may alter implementation considerations. Finally, multiple techniques may be needed to increase confidence in the result. All of these factors combine to make scheduling both difficult and vital.

Three scheduling tools are discussed in the following sections: the Program Evaluation and Review Technique (PERT), the Gantt chart, and the Project Accountability Chart (PAC). Many readers with experience in project work will be familiar with the first two. The third combines the concerns of PERT and Gantt methods with information about the responsibility for each project task.

3.5.1 Program Evaluation and Review Technique (PERT)

A PERT diagram depicts the flow of the project and indicates the interdependence of tasks. To construct one, a manager must first list the tasks (or activities) needed to complete the project, taking care to be neither too detailed nor unproductively general. An example is shown in Table 3.3.

Next, these tasks are arrayed in a flowchart displaying the order in which they must be completed. These two steps usually require iteration. The PERT flowchart for a simple forecasting project is shown in Figure 3.2. In that figure, dependence is shown by arrows and tasks by circles. The number in the upper half of each circle uniquely identifies the task. The number at the bottom of the circle is the estimated duration of the task. Tasks that need be only partly completed before subsequent tasks depending on them begin are represented by more than one circle.

Once the flowchart is finished, the manager must estimate the time needed to complete each task (*task duration*). These estimates usually are made in terms

TABLE 3.3 Initial List of Forecasting Project Tasks

Task #	Description	Duration (days)
S	Project start	0
1	Conduct and analyze poll of internal experts	10
2	Conduct and analyze poll of external experts	60
3	Analyze and summarize results	10
4	Reconcile results	7
5	Integrate results	8
6	Initial market penetration analysis	3
7	Adjust results for regulatory factors	2
8	Obtain additional market penetration data	5
9	Review and revise results	3
10	Iterate tasks 6 and 7	6
F	Project finish	0

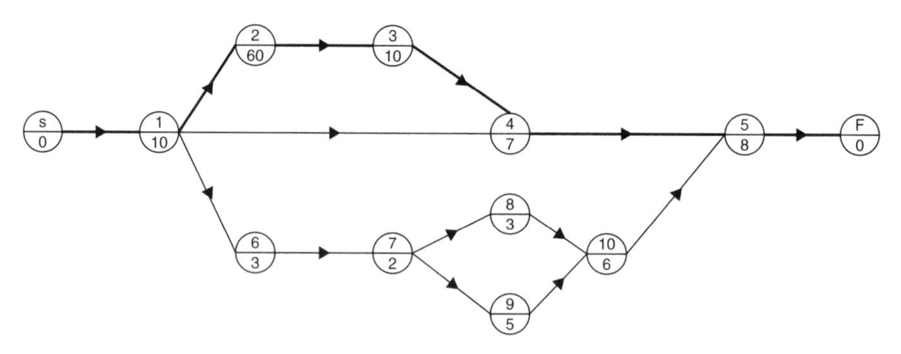

Figure 3.2. Example Project PERT Chart

of both person-hours and working days. Previous experience with the tasks is immensely helpful in making these estimates. However, since it usually is not possible to be precise, a weighting system such as the following one is sometimes used. The estimated durations are recorded in the lower half of the task circles on the PERT chart.

$$T(est) = \frac{T(optimistic) + 4 \cdot T(likely) + T(pessimistic)}{6}$$

The next step is to find the longest path from project start to project finish. This is called the *critical path* (CP), and it defines the shortest time in which the project can be completed as it is currently planned. The CP is shown in bold in Figure 3.2. If the flowchart is complex, computer assistance may be required in this step. If the CP exceeds the time available for the project, either some tasks must be shortened or eliminated and/or more resources must be assigned to complete them more quickly. Even if the PERT process is taken no further than this, the manager will have gained more thorough understanding of the project and its resource requirements. However, more information can be developed.

Completing tasks that are not on the CP sometimes can be delayed (i.e., allowed to *float*) without delaying project completion. To find the length of the delay that can be tolerated, the manager first must find the *earliest start time* (EST) and the *earliest finish time* (EFT) for each task. The EST of a task is defined by the tasks that must be completed before it can begin. The EFT is merely the EST plus the task duration. These times are computed by moving *forward* through the chart. In Figure 3.2, for example, Task 1 can begin immediately (time = 0) and takes 10 days to complete. So, its EST is 0 days and its EFT is 10 days. Since Task 6 cannot begin before Task 1 is complete, its EST is 10 days and its EFT is 10 + 3 days.

Once these computations are complete, the manager can work *backward* through the chart to find the latest time that each task can be finished (*latest finish time* or LFT) without lengthening the project. The *latest start time* (LST) is simply the LFT of the task minus its duration. The difference between the LST and EST of a task is the total amount by which it can float (*total float,* TF) without delaying project completion. Since any delay for tasks on the CP would extend the project, they cannot have float.

Total float (TF) is composed of two types of float. *Free float* (FF) is the float that a task can be allowed before a later task that depends on it is delayed past its EST. *Interfering float* (IF) delays the start of a subsequent task but does not delay project completion. Clearly, IF = TF − FF. Generally, FF is more desirable than IF since IF uses some of the float time available for later tasks, making things more tense. The various start, finish, and float times can be displayed as shown in Figure 3.1.

The PERT chart gives a visual representation of the task sequencing for a forecasting or implementation project and clearly indicates which tasks are most critical to timely completion. It can be used to find tasks that can be delayed, how the project might be shortened, and how lost time might be made up. For example, the project in Figure 3.2 could be completed more quickly if more people were assigned to Task 4 so that it could be completed in less time. However, the PERT chart does not give a clear visual clue to elapsed time. That is provided by the Gantt chart.

3.5.2 Gantt Chart

The Gantt chart is a simple bar chart representation of information generated for the PERT chart. Figure 3.3 is a Gantt chart of the project schedule shown in Figure 3.2. In the Gantt chart, tasks are listed on the vertical axis and time is displayed on the horizontal axis. Floats can be shown by cross-hatching. During the project, completed or partially completed tasks can be indicated by filling in a portion of the task bar. The Gantt chart gives a clearer visual representation of progress and timing than the PERT chart. Unfortunately, this is obtained at the cost of providing a clear representation of the task sequencing. While the two charts supplement each other, neither indicates who is responsible for completing the tasks.

3.5.3 Project Accountability Chart (PAC)

The PAC was suggested by Martin and Trumbly (1987). It combines a visual representation of task responsibility with aspects of the scheduling information provided by the PERT and Gantt charts. A PAC for the example project presented

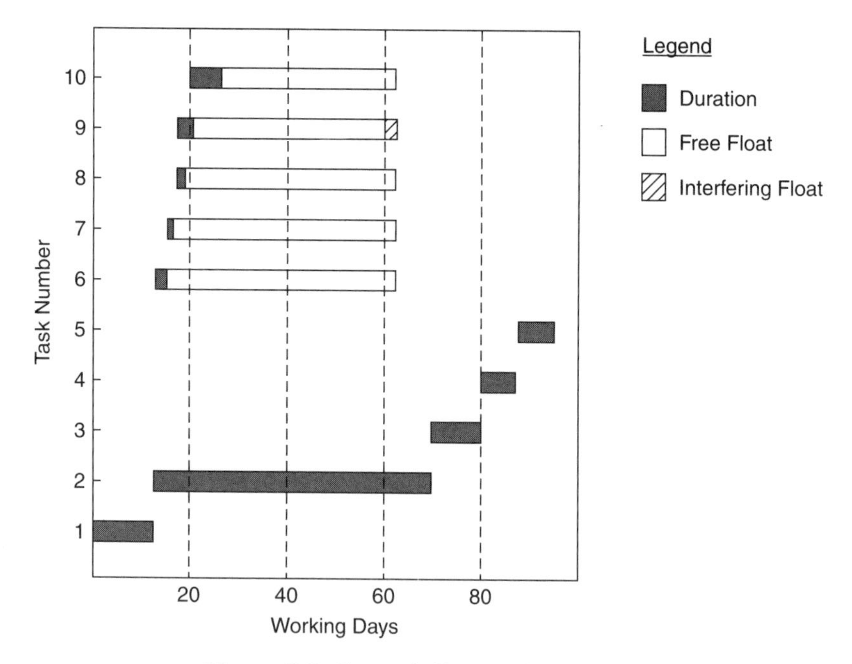

Figure 3.3. Example Project Gantt Chart

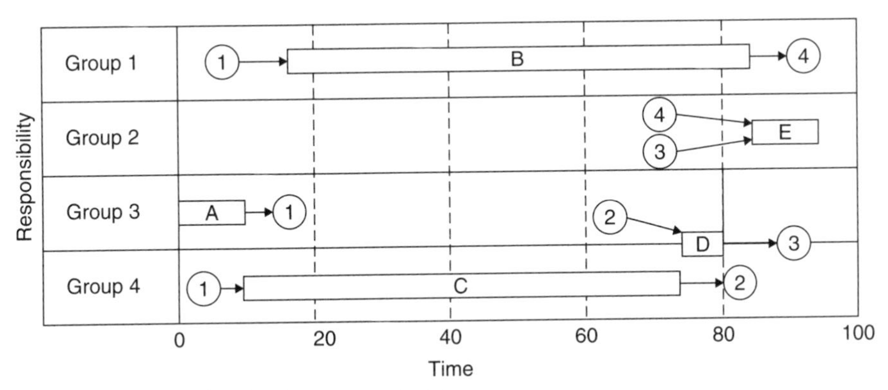

A Obtain internal expert opinion
B Market penetration analysis
C Obtain external expert opinion
D Reconcile opinions of internal and external experts
E Integrate opinion and market penetration results

Figure 3.4. Example Project PAC

in Section 3.5.1 is shown in Figure 3.4. The horizontal axis of the PAC is time. The vertical axis displays those responsible for executing various project components. It usually is best to cluster ones with the largest number of responsibilities near the center. Tasks with shared responsibility can appear twice or, if those sharing them are adjacent, by overlapping the organization boundaries.

Task sequencing is displayed on the PAC using the PERT chart with two significant changes. First, several tasks may be combined into a single identifiable activity if one entity is responsible for them. For instance, Tasks 2 and 3 (Figure 3.2) might be combined as "obtain external expert opinion" and represented by one bar (C in Figure 3.4). Second, the bar for the combined activities extends from the EST of the first task to the LFT of the last. In the example, it extends from the EST of Task 2 (10 days) to the LFT of Task 3 (80 days). The numerals shown in circles on the figure represent input and output dependencies between the three project teams.

The PAC displays responsibility, timing, and the interdependence for groups of project tasks. However, this breadth is obtained at the expense of detail about the tasks themselves. This deficiency can be remedied by exploding individual nodes to show PERT and Gantt diagrams for the tasks they include. Regardless of the approach, however, the intent is to schedule important aspects of the project and to give a quick visual reference for its progress.

3.5.4 Project Scheduling Software

There are many freeware and proprietary software packages for project management. Several can be downloaded from the Internet. Microsoft Project (Microsoft 2011) is a commonly used and readily available proprietary project management program. The program was updated in 2010. Presently, no version of Project is compatible with either MAC OS X or LINUX, nor is one planned.

3.6 CONCLUSIONS

Managing forecasting projects requires many of the management qualities necessary for managing other projects. However, because forecasting differs in essential ways from other projects, there are management differences as well. The manager must carefully bound the forecasting task and determine the information required by the decision the forecast is intended to support. Careful scheduling is required to ensure completion of the project on time and within budget. Dynamics among team members tend to be more complex, and iteration and forecasting by several methods are generally required to produce a satisfactory product. Finally, because the forecast is a multi-skill product, special attention must be given to management structures and communications patterns that acknowledge the need for substantive knowledge and contributions from a variety of fields.

This book is organized according to a staged model of decision making, as discussed in Section 3.2. Techniques are discussed in the phase of decision making where they are most likely to be of assistance. These first three chapters constitute an introduction to technology forecasting. The exploration (cold) phase of technology forecasting begins in Chapter 4. Chapter 5 continues the exploration theme by discussing the range of information available to the forecaster, especially electronic sources of information such as the Internet and science and technology databases. The next phase of decision making, analysis (warm), is discussed in Chapter 6. Chapter 7 turns to the focusing (hot) phase. Chapter 8 continues the discussion of focusing by considering the constraints imposed by institutions and the marketplace and a range of market analysis. Chapter 9 discusses impact analysis, while Chapter 10 focuses more closely on cost-benefit and risk calculations. Chapter 11 considers choosing and implementing the technology development. The book concludes with a vision of putting it all together by "Managing the Present from the Future"—the ultimate intent of technology forecasting. That chapter is followed by a nanotechnology case study (Appendix A) that demonstrates some of the techniques in this book in a real-world technology context.

REFERENCES

Arnstein, S. R. (1975). "A Working Model of Public Participation." *Public Administration Review* **35**: 70–73.

Arthur, W. B. (2009). *The Nature of Technology: What It Is and How It Evolves*. London, Allen Lane.

Brown, S. L. and E. K. M. Eisenhardt. (1998). *Competing on the Edge: Strategy as Structured Chaos*. Boston, Harvard University Press.

Burgelman, R. A., C. M. Christensen, et al. (2009). *Strategic Management of Technology and Innovation*. New York, McGraw-Hill Irwin.

Collins, J. and J. I. Porras. (1997). *Built to Last: Successful Habits of Visionary Companies*. New York, HarperCollins.

CRISP-DM (2010). "Cross-Industry Standard Data Mining Process." Retrieved 21 August 2010 from http://www.crisp-dm.org/.

Cunningham, S. W. (2009). "Analysis for Radical Design." *Technological Forecasting and Social Change* **76**(19): 1138–1149.

De Sanctis, G. and M. S. Poole. (1994). "Capturing the Complexity in Advanced Technology Use: Adaptive Structuration Theory." *Organizational Science* **5**(3): 121–147.

Firat, A. K., W. L. Woon, et al. (2008, September 2008). "Technological Forecasting: A Review." Retrieved 10 May 2010 from http://web.mit.edu/smadnick/www/wp/2008–15.pdf.

Human Metrics. (2010). Retrieved 10 May 2010 from http://www.humanmetrics.com/cgi-win/JTypes2.asp.

Kash, D. E. (1977). "Observations on Interdisciplinary Studies and Government Roles." In *Adapting Science to Social Needs*, ed. R. Scribner and R. Chalk. Washington, DC, American Association for the Advancement of Science: 147–178.

Martin, B. R. (1995). "Foresight in Science and Technology." *Technology Analysis and Strategic Management* **7**(2): 139–168.

Martin, M. and J. E. Trumbly. (1987). "A Project Accountability Chart (PAC)." *Journal of Systems Management* **38**: 21–24.

Microsoft. (2011). "Microsoft Project 2010." Retrieved 12 January 2011 from http://www.microsoft.com/project/en/us/default.aspx.

Mintzberg, H. (1980). "Structure in 5's: A Synthesis of the Research on Organizational Design." *Management Science* **26**(3): 322–341.

Norling, P. M., J. Herring, et al. (2000). "Putting Competitive Technology Intelligence to Work." *Research-Technology Management* **43**(5): 23–28.

Pahl-Wostl, C. and M. Hare. (2004). "Process of Social Learning in Integrated Resources Management." *Journal of Community and Applied Social Psychology* **14**(2): 193–206.

Personality Pathways. (2010). Retrieved 10 May 2010 from http://www.personalitypathways.com/type_inventory.html.

Peters, H. P. (1986). "Social Impact Analysis of Four Energy Scenarios." *Impact Assessment Bulletin* **4**(3/4): 149–167.

Porter, A. L. and F. A. Rossini. (1984). "Interdisciplinary Research Redefined: Multi-Skill Problem-Focused Research in the STRAP Framework." *R&D Management* **14**: 105–111.

Porter, A. L. and F. A. Rossini. (1986). "Multiskill Research." *Knowledge: Creation, Diffusion, Utilization* **7**(3): 219–246.

Porter, M. (1985). *Competitive Advantage: Creating and Sustaining Superior Performance*. New York, The Free Press.

Redelfs, M. and M. Stanke. (1988). "Citizen Participation in Technology Assessment: Practice at the Congressional Office of Technology Assessment." *Impact Assessment Bulletin* **6**(1): 55–70.

Rivette, K. G. and D. Kline. (2000). *Rembrandts in the Attic*. Boston, Harvard Business School.

Rossini, F. A. and A. L. Porter. (1981). "Interdisciplinary Research: Performance and Policy Issues." *SRA Journal* **13** (Fall): 8–24.

Rouwette, E. A. J. A., J. A. M. Vennix, et al. (2002). "Group Model Building Effectiveness: A Review of Assessment Studies." *System Dynamics Review* **18**(1): 5–45.

Ruff, F. (2004). "Society and Technology Foresight in the Context of a Multinational Company." *Proceedings of the EU-US Seminar. New Technology Foresight, Forecasting and Assessment Methods*. Seville, Spain, Institute for Prospective Technology Studies (IPTS). 13-14 May 2004. Institute for Prospective Technology Studies, Seville, pp. 44–70.

Sage, A. P. and J. E. Armstrong. (2000). *Introduction to Systems Engineering*. Hoboken, NJ, Wiley-Interscience.

Susskind, L. E. (1983). "The Uses of Negotiation and Mediation in Environmental Impact Assessment." In *Integrated Impact Assessment*, ed. F. A. Rossini and A. L. Porter. Boulder, CO, Westview Press: 154–167.

Susskind, L. E., G. McMahon, et al. (1987). "Mediating Development Disputes: Some Barriers and Bridges to Successful Negotiation." *Environmental Impact Assessment Review* **7**: 127–138.

Taleb, N. N. (2007). *The Black Swan: The Impact of the Highly Improbable*. New York, Random House.

Van der Duin, P. (2006). *Qualitative Futures Research in Innovation*. Delft, the Netherlands, Eburon.

Van der Heijden, K. (2005). *Scenarios: The Art of Strategic Conversation*. Hoboken, NJ, John Wiley & Sons.

Vanston, J. H. (1985). *Technology Forecasting: An Aid to Effective Technology Management*. Austin, Texas, Technology Futures, Inc.

Vanston, L. (2008, 3 April). "Practical Tips for Forecasting New Technology Adoption." Retrieved 10 May 2010 from http://www.telenor.com/en/resources/images/179–189_PracticalTips-ver1_tcm28–36194.pdf.

von Hippel, E. (1988). *The Sources of Innovation*. New York, Oxford University Press.

von Hippel, E., S. Thomke, et al. (1999). "Creating Breakthroughs at 3M." *Harvard Business Review* **77**(5): 47–57.

Walker, W. E. (2000). "Policy Analysis: A Systematic Approach to Supporting Policymaking in the Public Sector." *Journal of Multi-Criteria Decision Analysis* **9**: 11–27.

Wheatley, K. K. and D. Wilemon. (1999). "From Emerging Technology to Competitive Advantage." *Proceedings of the Portland International Conference on the Management of Engineering and Technology*, 25-29 July 1999. IEEE, Portland, Oregon, vol. 2, pp. 35–41.

4 EXPLORING

Chapter Summary: This chapter lays out the process of initiating a technology forecast in any organizational environment, whether private or public sector, in highly competitive or relatively noncompetitive environments. Recall that Chapter 3 advocated a three-phase approach for any forecast of more than minimal complexity. This chapter is chiefly concerned with the first phase: exploring. To review.

Exploring engages a broad subject matter at a fairly shallow analytical level; identifies basic institutional connections; and makes very few decisions about directions of the technology.

Establishing the context, monitoring, and creativity are the key topics to this chapter: context because it is central to framing the forecast; monitoring because it is the most important method used in exploring; and creativity because all forecasting activities require it.

4.1 ESTABLISHING THE CONTEXT—THE TDS

The process of technology forecasting begins by exploring the broad societal context in which the technology is being developed—the technology delivery system (TDS). The context is progressively narrowed to those institutions directly developing the technology and those impacting and impacted by it. The technology being forecast and its supporting, competing, and related technologies are then considered. Their potential development paths and the barriers and facilitators to their development are explored. All possible information is initially swept into the exploration and, of course, the broader the sweep, the shallower the depth. This information is briefly considered to assess its worth, and a substantial portion is discarded. There is no magic key beyond sound, informed judgment to determine what is to be explored further and what is to be discarded.

As information is taken into the study, it is organized to make a broad but shallow initial map of two things: the technological enterprise components needed to accomplish the innovation and the external forces and factors impinging on them. Boundaries are drawn, and information that is judged irrelevant is bounded out of the study. The detail retained increases in areas of the map close to the technology. It is important to note that the societal context of a technology is not static. It will evolve, sometimes in dramatic ways, with time. Thus, as much as possible, both the technology and its context should be treated dynamically throughout the course of the forecast.

4.1.1 Societal and Institutional Contexts

Technological developments occur within the larger society, its institutions and its values. The relationship between a technology and its societal context is crucial to the nature, effectiveness, and speed of its development. The societal and institutional contexts provide much of the impetus and resources for technological development (impacts *on* the technology). In turn, the technology impacts the societal context, often decisively, as in the cases of the microcomputer and the cell phone (impacts *of* the technology). Technology–society interactions are more of a spiral than a loop, as neither context is static. For instance, the rise of the Internet has substantially altered commerce and the internal communications of organizations of all sorts. Thus, the technology is substantially altering both the organizations that create it and the organizations that use it.

To identify the elements of the societal and institutional context that comprise the TDS, answer the following questions:

1. *Institutions*
   - What government and private institutions are developing the technology and its facilitating technologies? Which ones are pursuing competing technological innovations?
   - What policies and dynamics of these institutions could impact or be impacted by this technological development?
   - Are standards important?
2. *Legal and Regulatory Issues*
   - What legal or regulatory issues may facilitate, inhibit, or modify the development of the technology?
3. *Economic Factors*
   - What are the sources of financial support, and what are the conditions under which they can be obtained?
   - What are the economic situations of potential users and customers (now and over the course of development)?

4. *Environmental Factors*
   - What environmental factors affect the desirability of this technology? How might these factors assist or inhibit its development and use?
   - How do these factors interact with other factors affecting the technology?
5. *Demographic Factors*
   - What demographic factors affect the technology (e.g., the rapidly growing senior population's increased demand for assistive technologies)?
6. *Values and Goals*
   - Is development assisted or inhibited by important social values (e.g., privacy concerns are important in some communications technologies)?
   - Are important social goals involved (e.g., a goal of Mars exploration could encourage robotic and communication developments)?
   - Consideration of these values and goals can be extremely important for novel technologies.

Integration and explanation of the elements of the societal and institutional contexts of the technological enterprise are major parts of mapping the TDS. To continue mapping, critical facets of the technology interfaces with its context must be explored in depth. Interaction with technical and contextual experts is vital to get the stakeholders and relationships right.

4.1.2 Technology Context

The technology context is the heart of the matter. At this point in the forecast, the present state of the technology is the main focus. Yet, even from the present state, it is possible to identify probable directions of the technology's development and issues that may affect its future. A sound approach is to begin with relatively broad, and not overly deep, coverage of the major aspects of the technology. Following is a list of questions (adapted from Porter, Rossini et al. 1980, p. 105) that may prove useful.

1. *Physical and Functional Description of the Technology*
   - What is the technology? What is its present state of development?
   - What technological and scientific areas are involved? What specialized skills are needed to develop and produce the technology?
   - What materials are required to develop and produce the technology?
   - What industrial sectors, specific firms, and organizations are/will be developers and users of the technology? (This question ties institutions to the technology.)
2. *Technological and Scientific Influences*
   - What are the technological and scientific *barriers* to further progress? What breakthroughs are needed to surmount them?

- What are the technological and scientific *drivers* for further progress? How can they be pursued?
- What are the essential users' needs and issues that are driving or inhibiting progress? How can these be addressed?

3. *Supporting and Competing Technologies*
   - What are the complementary and/or supporting technologies? How do they interact with this technology?
   - What are the competing technologies? What are their states of development relative to this technology? How can the effects of their competition on this technology be altered?
4. *Applications*
   - What principal applications are currently perceived? Who are their prospective users? Can these prospective users be involved in the development?
   - Could there be major applications that are not presently anticipated? In what areas? How can these be prepared for?
5. *Future Directions*
   - What is the current thinking about the most promising innovation pathways? What alternatives and paths forward have been identified? Robinson and Propp (2008) provide an appealing methodology to address alternative pathways, effectively illustrated for the case of the "lab on a chip."
   - What are the currently recognized principal positive and negative impacts of development?
   - What organizational or public policy issues most strongly affect future developments?
   - Is there speculation about low-probability, high-impact issues (black swans) relating to its development?

Answering these questions will provide a sound basis for describing the technology, and it's the technological enterprise that is necessary to ground the technological forecasting to follow.

4.1.3 Stakeholders

A stakeholder or party at interest is an organization, group, or individual involved in a technology or an issue. In the development of a technology, stakeholders typically include those involved in developing the technology, funding its development, restricting or promoting its development, using it, regulating it, developing competing or supporting technologies, or aided or harmed by the technology.

Stakeholder analysis involves identifying the major actors in the institutional and technology contexts and studying their interests, goals, values, concerns, perspectives, and resources—their stakes—in the development. Finally, their existing and potential interactions with the development are considered. Much of this information may have been developed in the initial contextual construction activities. However, the institutional and technological dynamics may not have been adequately developed. Understanding these dynamics is helpful in characterizing the linkages among stakeholders that affect development of the technology. Answers to the following questions will help locate each stakeholder in the development process and the TDS.

- What are their interests in the technology? How do these impact its development?
- What are their major organizational/personal goals? How do these impact this development?
- Do they have values that may affect the development? What might be their impact?
- What actions have they taken or will they take relative to the technology? What impacts will these have?
- What resources have they deployed or will they deploy? What will be their impact on development?
- What issues and concerns vis-à-vis the technology are important to them?
- What sorts of interactions with other stakeholders have they had or will they have relative to the technology?

Understanding stakeholder dynamics is quite important in the forecasting process. Answers to these questions should produce a preliminary view of those dynamics.

4.1.4 Understanding the TDS

The information gathered so far provides the elements of a map of the technology development process. This map is the basis from which the forecast can be developed. It portrays the actors, their interactions, and the impacts of their interactions in a broad preliminary view of the technology development. This map is the TDS.

To better understand the TDS, it is useful to consider its operation in a static sense. R&D organizations in universities, industry, private think tanks, and government agencies develop knowledge and capabilities that provide *push* for the delivery of new technologies. Users provide *pull* through demands for goods and services. Push and pull are coupled through the management of the technological organizations that sense demand and capability, gauge external constraints, and assemble and organize the factors of production.

External facilitators/constraints to development can be social, technological, economic, and/or environmental in nature. For societies these include cultural and values factors such as the desire for more and broader social networks and resistance to some forms of genetic engineering. Institutional facilitators/constraints include government needs (e.g., in the health and defense areas) and regulation that inhibits technological development. Technological breakthroughs and bottlenecks also may affect development. Economic factors are exemplified by the availability or lack of development funds. Environmental factors, such as air and water quality standards, can either enhance or inhibit a given technology development.

Government institutions select and prioritize the value preferences of both the general public and organized stakeholder groups. Policies and programs formalize these preferences. The performance of the technological organization and its output are strongly influenced by government through R&D, subsidy, and regulation. In many areas, such as defense and health care, technological directions and outputs have been a shared private and public responsibility. Impediments to the delivery of desired outcomes can also develop within the private or public sector. Factors such as conflicting value preferences, information constraints, and bureaucratic inertia (industrial and public) can constrain the development and use of technologies as well. The forecaster must strive to get at the key influences and gauge their relative strengths.

TDS models provide several critical inputs to the forecaster, including:

Framing considerations—what might be done with the technology to generate attractive innovations for whom (what customers in what sectors)

Arraying the essential "enterprise" components—what is necessary to take R&D advances to market (e.g., a small pharmaceutical business could not bring a new drug to market because of the skill and resource requirements needed to get FDA approval; in the United States, this is likely to cost over $500 million)

Mapping—identifying key contextual institutions and individuals that can affect the development, as well as interactions among them

Spotlighting—locating leverage points

The TDS and its accompanying description (not all the pertinent information can fit in an understandable system sketch) map the technology's developmental context. They provide the basis for the forecast. They may take many forms, with varying levels of complexity, depending on the subject and requirements of the forecast.

Monitoring provides much of the information that makes up the TDS, and it underpins the entire forecast. In any forecast, however, taking advantage of the creativity inherent in the forecaster team can be an important tool in reaching beyond the obvious. Creativity is essential for effective exploring. It can play an equally important role in the subsequent phases of the forecasting process.

4.1.5 An Example TDS Model

Figure 4.1 illustrates a simple TDS model (Ezra 1975; Guo, Huang, et al. 2010). It was originally constructed in the 1970s to help explain why substantial national

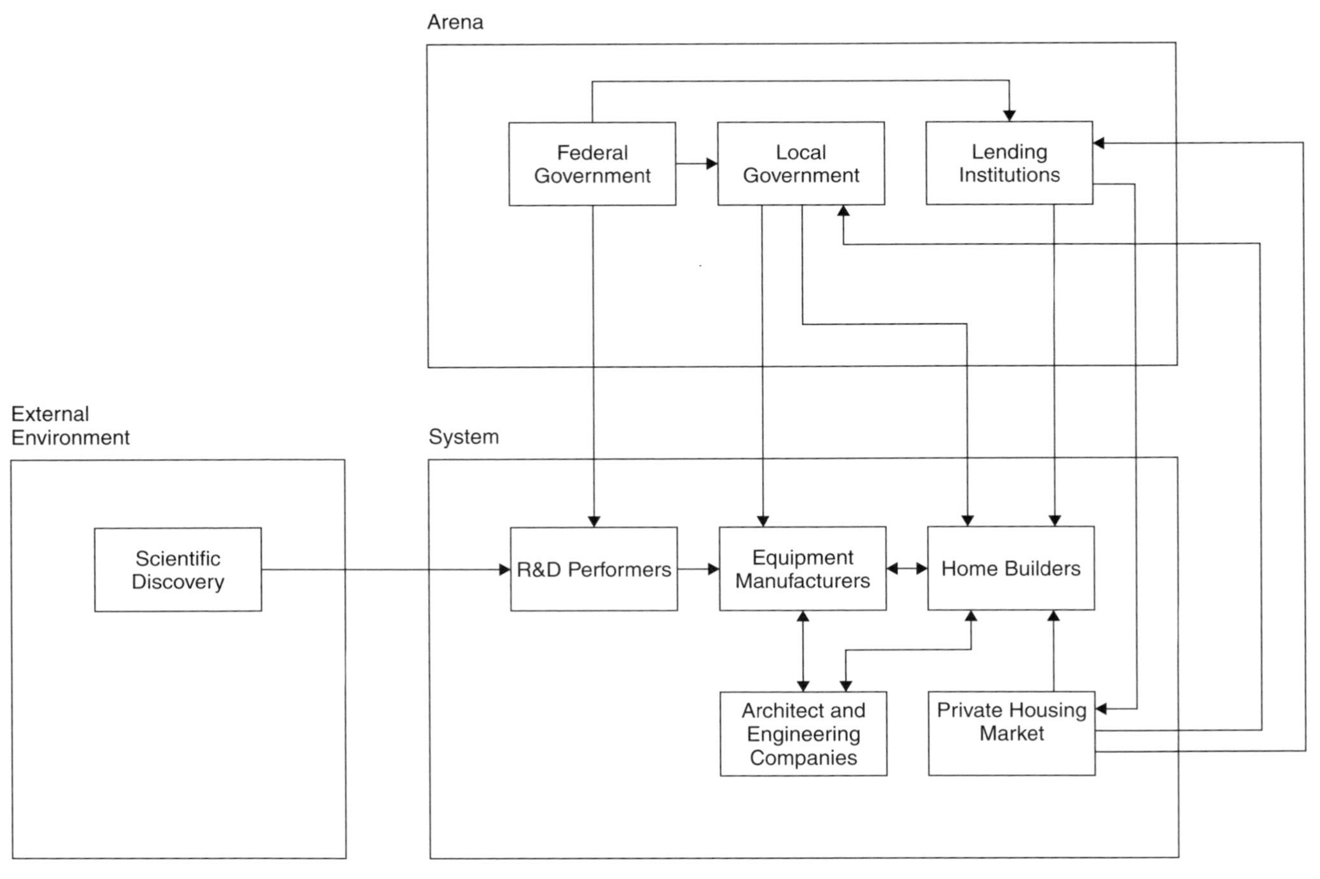

Figure 4.1. A TDS for Innovation in U.S. Residential Solar Energy

laboratory solar energy research had not resulted in more widespread applications in home construction.

In this diagram the chief participants in the *arena* are the federal government, local governments, and financial institutions. A network of interactions connects researchers, equipment manufacturers, construction companies, architects, and consumers. The relationships are relatively fixed, but the participants in the arena have the power to change the rules of the game. The *external environment* is scientific discovery in related fields such as nanotechnology that can deliver surprises and shocks. The two principal feedback loops are profitability and regulatory compliance. The actors in the arena assess the capability of the system to deliver desired outcomes and modify their decisions accordingly.

The TDS in Figure 4.1 helped bring to light several impediments to solar energy use in home construction:

- R&D performers lacked incentives to promote innovation.
- Innovations needed to pass from R&D to manufacturers to home builders, who tended to be risk-adverse and low tech.
- Conservative lending institutions needed to be convinced to provide additional lending for home builders without a confirmed market.
- Building codes required regulatory approval from local and/or regional and state authorities, all of which varied with geographic location.

There are several areas of this TDS for improvement and research. It might be expanded to include an international context, and it might be elaborated with respect to the external environment. Detailed evaluation of the external environment can assist in delivering more robust forecasts. Further improvements in the TDS might stem from including the role of energy generation and transmission companies. De Vries et al. (2007) discuss the role of large rural solar energy plants that sometimes compete with agricultural land use. Government guarantees of higher premiums for solar electricity than electricity generated by conventional means also might add another feedback loop to the diagram.

4.2 MONITORING

Monitoring provides the information upon which forecasting is based. "Monitoring is to watch, observe, check, and keep up with developments, usually in a well-defined area for a specific purpose" (Coates and Coates 1986, p. 31). This description captures the essence of monitoring, the most fundamental and most widely practiced forecasting technique. Monitoring is the backbone of forecasting. It supplies most of the information that is analyzed and structured for the forecast by taking advantage of all relevant information sources. No TDS could be constructed without some form of monitoring, whether simple and informal or highly structured and complex. Moreover, the utility of monitoring in decision making goes well beyond technology forecasting. Monitoring plays a major role in such activities as technology selection, analyses of competitive environments, and following trends in technology development.

The mix of monitoring information sources has changed in recent years. Although the printed word is still important, the Internet has become a principal information source through the widespread use of Google and other search engines to exploit topically focused databases and news sources. The number of databases on scientific, technical, and contextual factors that can be accessed on the Internet is vast. When technology is the focus, Porter and Cunningham (2005) call this process *tech mining*.

If the information sought pertains to a particular technology, or technological monitoring, then historical information may be sought on the development, current information on the state of the art, and/or information about future prospects of the technology. If the primary focus is *contextual*, key elements of the institutional, social, physical, and market environments may be targeted. In monitoring activities, accumulating broadly based information is the first step in the process of producing knowledge to inform decision making.

Table 4.1 presents six dimensions that color how and why one monitors. The focus dimension is discussed later in the chapter, and technological maturity is addressed in terms of the stage of development. The latter dimension is especially worth highlighting because the nature of available information differs, depending on the technology's maturity. Martino (1993) neatly illustrates tracking significant developments over the history of an established technology for the case of plastics use in automobile bodies.

By arraying the progression of a technology's development, "next steps" in the most likely innovation pathways often can be spotted (Robinson and Propp 2008). For instance:

- Emergence of new technological platforms upon which to build alternative innovations
- Inventions that cry out for complementary technological capabilities to enable them
- Potential enhancements in supporting technologies that could provide important system performance gains or cost savings

TABLE 4.1 Monitoring Choices

Dimension	Option 1	Option 2
Focus	Technology	Context
Technological maturity	Established	Emerging
Time frame	Imminent	Unrushed
Purpose	Choosing	Tracking and forecasting
Breadth	Macro	Micro
Structuring	One-time study	Ongoing program

Exhibit 4.1 Macro Monitoring

The Korea Institute of Science and Technology Information (KISTI) and the Technology Policy and Assessment Center (TPAC) at Georgia Tech initiated a project to monitor advances in emerging technologies. KISTI formulated strategies to efficiently monitor areas that presented especially favorable opportunities for Korea. For instance, one strategy that greatly reduced information requirements was to use the most cited papers ("Essential Science Indicators"—www.isiknowledge.com) as the key source instead of the entire Web of Science. Further, the resulting analyses were performed at the level of research categories rather than for specific technologies. In this way KISTI was able to monitor all emerging technologies and achieve their exploring phase objectives.

- Specialty, high-end, early applications that may presage wider applications
- Leading indicators that presage future developments (e.g., Martino 1993 explored the temporal relationship between development of new metal alloys and first aircraft applications)

The time frame dimension is especially important. An imminent time frame puts a premium on finding ready-made reviews and forecasts, using information at hand, and capturing key expert insights on the go. If the monitoring exercise is to feed an explicit decision, then it should focus on the key management of technology needs. In contrast, a longer time frame warrants a broader perspective. *Postdecision monitoring* is another important application. The purpose of the monitoring activity, of course, shifts after a technology is chosen. The breadth dimension (i.e., macro vs. micro focus) colors the types of information sought and how best to digest it to produce effective results (see Exhibit 4.1).

4.2.1 Why Monitor?

Forecasting, and thus monitoring, is especially important in times of rapid social and technological change. Rapid changes in information, biomedical, environmental, and energy technologies, among others, and major shifts in the economic environment require that technology managers have awareness and foresight in order to make sound decisions.

Informal monitoring is a common activity for most people. We routinely check the kitchen to see what we need for next week's meals before going to the grocery, and we check to see if the grass needs mowing. Before we buy a new car, we also do some research and talk to friends who have one. We all monitor.

Professionals tend to follow developments in their areas through colleagues, literature, and the Internet to develop an understanding of present and future developments in their fields. These activities are the basis for structured monitoring programs within many organizations. Basic monitoring skills are already present; they only need to be tuned and focused in a group context. An ongoing monitoring program can be critical before and after a formal forecast. The level of effort and the comprehensiveness of the monitoring program vary with the organization's needs and capabilities.

Monitoring can help all organizations, but it is critical for those operating in highly competitive environments. The need for industries to monitor is obvious. New product and process development and modifications to existing products are driven by information, especially in rapidly changing areas. Government agencies also need to monitor the technological and institutional environments in which they operate. The need for information in defense, energy, intelligence, and the environment is obvious. However, effectively delivering service increasingly depends on understanding new opportunities produced by advances in communications technology. It's amazing that a foreign ATM thousands of miles from home knows that you're broke and that income tax returns are electronically submitted. There are many other areas where awareness of changes in technology drives more effective service delivery and decreases its cost.

4.2.2 Who Should Monitor?

Not all organizations have the resources to monitor. Some large ones have trained staff, data access, and decision processes conducive to using monitoring results. Others (see Exhibit 4.2) lack the need to monitor. Most small and medium-sized organizations don't have the funds and personnel for fully internal monitoring efforts. Options about who should monitor in an organization include:

- *Completely decentralized*: Individuals monitor as necessary for their jobs
- *Centralized*: A staff group or an ad hoc project team
- *Blended*: A combination of centralization and decentralization
- *Outsourced*: Consulting external parties such as consultants or think tanks

Exhibit 4.2 Life in the Slow Lane

If a company's contextual environment is stable, then its product/service mix and markets are apt to be stable as well. Since its well-being probably doesn't depend on early warnings of threats and opportunities, monitoring and forecasting are low priorities. Exploring tech mining in cooperation with a multi-billion-dollar petrochemical company with a stable product mix and established customers revealed that it essentially devoted no staff to monitoring. Instead it relied on reports read by everyone in their business sector. While the company *wanted* to identify new process technologies, new applications, and political-economic trends that could impact demand, it did not *need* to make monitoring a priority.

Given the importance of monitoring to achieve and maintain competitiveness, a completely decentralized or laissez-faire approach is inappropriate for organizations in competitive situations. Centralized operations can be valuable as support for subunit monitoring studies by licensing key information resources, and by gathering software tools and providing training in their use by units of the organization.

Outsourcing sometimes can be a cost-effective way to provide good intelligence. Commercial providers can build up repertoires of data and human sources and maintain up-to-date knowledge of technologies and/or markets. On the other hand, information garnered solely from outside sources is not likely to be tailored to organizational needs. Thus, it will need to be internally supplemented, modified, or interpreted (Brenner 2005). In addition, networking to ensure that information reaches those who need to know clearly benefits from insider involvement. For organizations in highly competitive environments, complete reliance on external sources is potentially dangerous.

Some degree of centralization is usually warranted. A blended approach can work very well. This could combine selective outsourcing and ad hoc monitoring by workgroups supported by a central unit. Allen (1984) showed that self-selecting "gatekeepers" serve well as critical bridges and can funnel a wide sweep of information to a project team or to others within the organization who have a need to know.

4.2.3 Monitoring Strategy

Forecasting and hence monitoring strategies are built around what the forecaster knows, what the forecaster wants or needs to know, and what resources (time, money, people, and techniques) are available. While monitoring is a central component of forecasting, it also is important before and after forecasting. Before a forecast is even contemplated, monitoring can give a rough determination of whether a forecast is appropriate and what its initial parameters might be. After all, monitoring serves one of two purposes. If the forecast showed that no further current action is needed, a modest ongoing program might be appropriate to watch for changes that would alter this situation. If the organization decides to develop the technology, an ongoing monitoring program should be instituted to scan for new conditions that should be factored into the implementation plan.

Depending on the forecaster's current depth and breadth of knowledge about the technology, the appropriate phase (exploring, analyzing, or focusing) at which to begin the forecast can be selected. Following are some approaches that are useful at each phase.

Exploring Phase: The forecaster starts "cold" and is unfamiliar with the subject. Immediate questions include:

- What is the technology? How is it defined and described? What is the state of the art?
- How do other technologies relate to it? What institutional and contextual factors affect it?

- Who are the stakeholders (e.g., individuals, organizations, suppliers, regulators, users) and what are their interests?
- What are commonly understood plausible future development pathways?

To monitor at this phase:

- Use a "shotgun" approach to gather information. Grab anything that is convenient and pertinent, but don't hesitate to eliminate material that doesn't prove useful after considering it.
- Emphasize recent literature and review state-of-the-art articles or books.
- Use online R&D abstract databases and start with a simple search. Use the search interface's instant analyses to help identify important topics (keywords) and key research centers.
- Locate one or two accessible professionals with sufficient expertise to point out information sources and to help ensure that monitoring does not go adrift.
- Prepare a preliminary TDS.

Analyzing Phase: The forecaster either has completed the exploring phase monitoring or is familiar with the subject. Objectives become more tightly targeted, and pertinent questions include:

- What forces are driving this technology?
- Can important interdependencies with other technologies, socioeconomic factors, or interactions among stakeholders be mapped?
- What are the key uncertainties along the technology's development path?

In this phase the sources of information shift:

- Literature searches become more focused; historical searches may become viable ways to identify leading indicators of progress and significant influences.
- Available forecasts for the technology can help answer pivotal questions and ground forecasts.
- Online searches can be fleshed out with more thorough search term phrasing and can yield vital overviews of trends, emerging "hot" technologies, and the names of active researchers (experts).
- Networking can identify experts with a variety of perspectives on the technology.

It now makes sense to begin to synthesize the information by formulating an image of what is happening to the technology and by developing and detailing the TDS. Identify characteristics and interests of the principal stakeholders in greater detail and map their patterns of interaction from forecasts developed or identified in this phase.

Focusing Phase: The forecaster is now very familiar with the subject. The objectives of this phase are closely targeted, and pertinent issues include:

- Specifying and analyzing key factors in the technology's development
- Identifying the most likely development paths for the near future
- Projecting these paths over the longer term and identifying issues
- Developing key recommendations to help manage the technology's development

Actions undertaken in this phase include:

- Extending the information search on the key factors to be as comprehensive as feasible
- Mapping the network of key R&D players and their associations and drawing implications about potential strategic alliances
- Developing the most comprehensive TDS possible within study constraints
- Seeking confirmation of this model and a review of projections from experts
- Generating a credible forecast by integrating the monitoring results with other forecasting techniques and perhaps using scenarios to present the results to users
- Establishing a structure for an ongoing monitoring effort
- Communicating the intelligence to serve organizational decision processes

Clearly, monitoring at each of the three phases implies different requirements. For instance, subject expertise is not essential in the exploring phase but is vital in the focusing phase. Comprehensive access to the premier online databases is not critical in the exploring and analyzing phases, but it is in the focusing phase to ensure thoroughness. Since depth and detail increase with the phase, so do commitments of time, resources, and effort. Therefore, it is sensible to commit only to the exploring phase when investigating the relevance of a technology. Results may be sufficient to conclude that the technology is not critical to the organization. If it does seem sufficiently important, a basis has been established for monitoring in later phases.

As noted earlier, the sources of information and the techniques for monitoring have changed considerably with the increasing importance of the Internet and the enhanced access it gives to many useful databases. Chapter 5 deals with currently available information resources for monitoring and the methods by which these resources can be most effectively accessed and used.

Lastly, the process of structuring of the monitoring project deserves attention. Brenner's description (2005) of the Air Products and Chemicals, Inc., systematic approach is an excellent model of an ongoing monitoring program that also is structured to respond to one-time requests. The project distinguishes two types of alerts (informational and actionable) that are sent out daily to 2400 internal clients. The stage-gate model of decision making involves a series of sequential, limited commitments for pursuing new ideas from initiation to launch. The Air Products monitoring system is an excellent example of careful integration of monitoring with the company's stage-gate business decision processes. Coordinating data resources, standardizing formats, and expediting analyses via programming also are effective features of the Air Products program.

Focusing the monitoring activity on specific aspects of the technology can be useful, and some of these activities are considered in the following sections

4.2.4 Monitoring Focused on Management of Technology Issues

The specific goals to be addressed often derive from a finite set of concerns relating to technology management. A framework and selected framing methods for tech mining can help specify goals for monitoring programs. Table 4.2 presents 13 key issues (based on Porter and Cunningham 2005). Many programs will target one or more of these.

Any forecasting exercise should begin by defining the focal issue and the questions to answer about it. Table 4.3 offers a starter set of questions that can be adjusted to fit decision organizational processes, norms, and priorities (Porter and Cunningham 2005). The list is suggestive, not exhaustive.

Note the prevalence of "what" and "who" questions on the list. These can often be consolidated by asking "who's doing what, where, and when?" Monitoring goes a long way toward answering these questions. Deeper probing of "how?" and "why?" also is vital but requires additional insight to project beyond available data. These questions can be traced to Kipling's inquisitive baby elephant, whose incessant questioning gets him ostracized and launched on the road to perilous adventures.

TABLE 4.2 Management of Technology (MOT) Issues

1. R&D portfolio selection for funding and/or execution
2. R&D project initiation
3. Engineering project initiation
4. New product development and design
5. New market development
6. Mergers and acquisitions
7. External technology target prioritization
8. Intellectual assets and licensing intellectual property
9. Pursuit of collaborative agreements for joint technology development
10. Identification and assessment of competing organizations
11. Identification of breakthrough technologies; assessment of product and market changes
12. Strategic technology planning
13. Technology roadmapping

TABLE 4.3 Framing Questions

1. What emerging technologies merit ongoing attention?
2. What facets of this technology are especially promising?
3. How bright are the prospects for this technology?
4. What are new frontiers for this technology?
5. What are the significant components of this technology? When will they mature?
6. How does this technology fit within the technological landscape?
7. What are the likely development paths for this technology?
8. What is driving this technological development?
9. What are key competing technologies?
10. What form of intellectual property protection relating to this technology should be pursued (e.g., patents, trade secrets, nothing)?
11. When will this technology be ready to apply?
12. How mature are the systems to which this technology applies?
13. What are the technology's commercial prospects?
14. Which aspects of the technology fit our needs?
15. What opportunities does this technology offer locally? Globally?
16. What societal and market needs do this technology and its applications address? Who are its potential users?
17. What is the competitive environment, and how is it changing?
18. What environmental hazards does the technology pose, and what are the appropriate mitigating approaches?
19. Have life cycle assessments been done? If so, what are key sustainability concerns?
20. What stances are government and stakeholders taking toward this technology or its applications and how might they encourage or oppose them?
21. What pertinent standards or regulations are in place or are being considered?
22. Which universities, research labs, or companies lead in developing or applying this technology?
23. What are the pertinent strengths and gaps within our own organization vis-à-vis this technology?
24. What companies are the present leaders in the most important markets for applications?
25. How strong and stable are the leading companies or R&D teams developing the technology?
26. How do their strengths and emphases compare to ours?
27. What strengths does each have in complementary technologies?
28. What organizations or individuals have attractive intellectual property relating to this technology and might any of them make attractive partners or acquisitions?
29. Are there existing partnerships?
30. What are each competitor's related technological and market strengths and weaknesses?
31. Which organizations should be watched?
32. To what organizations might it be possible to license intellectual property?
33. How entrepreneurial is the competitive environment?

4.2.5 Monitoring Focused on the Stage of the Technology Development

Monitoring should be tailored to the developmental stage of the technology, and the current TDS should reflect that stage. Brenner (2005) nicely portrays emphasis shifting from scanning for opportunities at the R&D stage, to focused monitoring of technological progress, to commercial issues as a technology matures. Table 4.4 illustrates the issues, priorities, and information resources for monitoring that are appropriate at various stages of the development process. Note that information resources change as the technology matures.

At the *Fundamental research* stage, "science forecasting" is pursued, but the value of a quick response to major advances in related domains needs to be recognized. As *Applied R&D* drives advances, monitoring can support qualitative and/or quantitative trend analyses. Reducing technological uncertainty is often paramount (i.e., will the technology work?). Detecting potential "show-stoppers" is especially valuable to allow adjustments to innovation strategies. As *Initial applications* emerge, monitoring shifts from technical to socioeconomic parameters. Identifying rates of adoption and spread to additional markets is paramount. As *Widespread adoption* continues, monitors should be alert to next-stage technological advances and/or competing technologies. It is important to recognize that an innovation process is apt to be quite nonlinear. That is, important events can simultaneously occur in more than one stage. The growth of fuel cell technology in the first decade of the twenty-first century exemplifies this, with significant activity occurring concurrently in all stages of development.

4.3 THE STIMULATION OF CREATIVITY

Forecasting requires the capacity to envision what the future might hold. This section describes methods to enhance the forecaster's and technology manager's creativity and to increase their ability to visualize alternative futures. First, methods of stimulating individual creativity are described. These include lateral thinking, suspended judgment, fractionation, reversal, checklists, morphological analysis, and the use of random words. Second, group techniques including brainstorming and Synectics are considered.

4.3.1 Five Elements of Creativity

J. P. Guilford's research into creative behavior established the basis for much of our current understanding of creativity. This research began shortly after World War II as a project funded by the U.S. Navy. Guilford (1959) identified five key elements of creativity: *fluency, flexibility, originality, awareness*, and *drive*. Understanding these elements removes some of the mystery surrounding creativity and paves the way for encouraging its growth.

Fluency usually is thought of as the ability to express thoughts in a flowing, effortless style. In creativity, however, fluency is the ability to provide ideas in volume. A simple test might be to see how many uses of an ordinary item, for

TABLE 4.4 Typical Monitoring Tactics as a Function of Technological Stage of Development

Stage of Development	Key Issues	Monitoring Priorities	Key Information Resources
Fundamental research	Scientific uncertainty	Tracking scientific progress Rapid recognition of relevant breakthroughs	Governmental research awards (NIH Crisp, NSF Awards) Scientific conference and article abstract databases (Web of Science, MEDLINE)
Applied R&D	Technological uncertainty	Technological trends Spotting application potential Detecting any popular or policy concerns	Patents (Derwent World Patent Index) Trade literature (Materials Business File) Environmental Literature (Pollution Abstracts, environmental blogs) Popular and policy literatures (Lexis-Nexis, PAIS)
Initial applications	Market prospects	Assessing markets' uptakes	Market research (marketresearch.com) New products and trade literature (Computer database)
Widespread adoption	Life cycle and replacement	Technological prospects Market prospects	Trade literature Market analysis

instance a used paper cup, can be devised in a limited time. Clearly, fluency is important in forecasting to help ensure that all possibly useful alternatives have been identified.

Flexibility is the ability to bend familiar concepts into new shapes or to jump from old concepts to new ones. *Nimbleness* is an apt synonym. For example, a creative person will consider uses beyond the paper cup's intended purpose, coming up with less conventional ones, such as a seed sprouter. Flexibility can be measured by the number of categories included in a stream of ideas. Many individuals will exhaust one category before moving to another; others will list only a few related ideas before moving on. Practice can increase flexibility.

Originality relates to the unusualness of ideas. An individual with awareness has the imagination to see connections and possibilities beyond the obvious. Throughout the ages, some people have been able to look at one thing and see another: to look at a bird and see an aircraft (Leonardo da Vinci) or to look at a fish but imagine an undersea boat (Jules Verne, Leonardo da Vinci). Recently, many engineers and designers have begun to look at nature for ideas—for example, a hammer with the structural properties of a woodpecker (Vincent, Sahinkaya, et al. 2007).

Awareness is the imagination to perceive connections and possibilities beyond the obvious. The similarities with NASA's Apollo Program are many (three-person crew, launch from Florida, similar cost in constant dollars, and many others). Consider the following examples of awareness, based on the personal experience of the co-author:

- Several friends purchased an abandoned roller skating rink at a very good price and opened a flea market that operated successfully for three years.
- A faculty member who also owned a large material-handling consulting operation purchased a vacant grammar school and converted it into a beautiful set of offices.
- The consultant's father-in-law purchased a large building that had been used for many years for assembling Ford automobiles and converted it into condominiums. Although he was not successful from a financial standpoint, the next owner was very successful.

Instead of saying, "I wish I had thought of that," record your ideas and take action on the ones that show promise!

Individuals with *drive* have "stick-to-itiveness" or motivation. It is a common misconception to equate creativity with instantaneous blinding flashes of inspiration. But like genius in Edison's famous quote, creativity often is 1% inspiration and 99% perspiration. Drive should not be confused with the blind application of brute force. Confronted with a brick wall, the creative person will not attempt to batter it down but will employ fluency, flexibility, originality, and awareness to find another way.

The techniques described in the following sections can enhance the five key elements of creativity. Raising the level at which individuals or groups apply these elements of creative behavior will increase their monitoring, forecasting, and management skills. There are many techniques for enhancing individual creativity. Five are detailed here: lateral thinking, checklists, the use of random words, morphological analysis, and TRIZ. A sixth technique, group creativity, will be discussed in Section 8.3.2.

Lateral Thinking. Our senses provide our minds with a continuous barrage of information without which we could not make decisions. Since we cannot process all of it, we create patterns from which codes are established. The mind needs to process only enough input to recognize the appropriate code to react. Reflex action is one response to the coding and recognition process. For instance, the reflexive response of most men to a warning about pickpockets is to check their billfolds. The pickpocket's response is to note where they check.

Despite the obvious advantages, there is a downside to the pattern/coding/response process. Although our brain readily forms patterns, these can become difficult to restructure, especially if they are repeatedly used. Our minds try to sort information into existing patterns even if it does not fit. Further, the patterns we establish depend on the sequence in which we happen to receive information, and this is unlikely to be optimal. Finally, even though patterns may differ only slightly, one will be selected and the others ignored. This can produce errors and/or missed opportunities.

Established patterns tend to be clustered into groups that eventually grow to become dominant patterns themselves. *Lateral thinking* provides a way to restructure and escape from old patterns and to provoke new ones (de Bono 2010). Thus, it provides a way to increase creativity. Lateral thinking encourages full use of our natural pattern-making capacity without hindering creativity.

Vertical thinking is selective. It seeks the most promising path. Lateral thinking is generative—that is, it generates new paths simply for the sake of finding the range of alternatives. Since it is not a building process, lateral thinking moves by leaps and bounds rather than sequentially. Far from excluding irrelevant information, it welcomes distraction as a stimulus to restructure old patterns and reveal new approaches.

Vertical thinking applies judgment to find the best path or idea. Thus, some approaches are "good" and others are "bad." Lateral thinking, however, does not judge and dictates that all pathways remain open. Instead of following the most likely paths, lateral thinking may follow the least likely paths, seeking new perspectives, perceptions, and patterns. Categories, classifications, and labels are never fixed because new perspectives may reveal different reference frames. Vertical thinking guarantees at least a minimal solution; lateral thinking improves the chances for an optimal solution but makes no guarantee.

Vertical and lateral thinking are complementary processes. Lateral thinking enhances vertical thinking by providing more approaches to a problem. Vertical thinking justifies lateral thinking by developing the ideas it generates. To be more creative, one must not only understand the principles of lateral thinking, but also examine techniques that use these principles. A discussion of several such techniques follows.

Suspended Judgment: The need to be right is sometimes the greatest obstacle to creativity. It inhibits idea generation because we are afraid of being wrong, and it restricts the chances for improved solutions by ruling out ideas that cannot be immediately justified. Thus, applying judgment, whether it is approbation or condemnation, too early in the search for ideas can cripple the creative process. If judgment can be suspended while the search is conducted, the chances for a creative solution are increased.

Judgment may be applied internally by the individual or externally by the group. Suspending both can help ensure that

- Ideas will survive longer to breed more ideas
- Individuals will offer ideas they would have rejected
- Ideas will be accepted for their value as stimulation
- Ideas that are bad in the current frame of reference will survive long enough to be evaluated in any new frames that emerge

There are several guidelines to be followed if the potential benefits of suspended judgment are to be reaped. First, never rush to evaluate an idea; exploration is more important. Second, when an idea is obviously wrong, shift the focus from why it is wrong to how it can be useful. Third, delay discarding any idea as long as possible. Let it provide the maximum possible stimulus to the generation process. Finally, follow behind an idea rather than forcing it in the direction that judgment dictates.

Fractionation: The more unified a fixed pattern, the more difficult it is to visualize in different ways. Fractionation can help escape this inhibiting unity by dismantling the problems into parts or fractions. The object is to look at a problem less complex than the original one and possibly solve it in parts. The fractions are restructured into larger fractions, and the larger structured problem is solved when possible.

There are many examples of restructured problems that were solved by reducing the parts, fractions, or steps required to reach an objective. Some of these are a few years old but nevertheless are quite instructive.

- In 2002 the Sara Lee Bakery Group introduced Iron Kids Crustless Bread to help in preparing a child's sandwich by eliminating the step of cutting off the crust (*St. Louis Business Journal* 2002).
- In that same year IBP's brand, Thomas E. Wilson, introduced fully cooked and precarved pork and beef roasts to eliminate many steps in preparing a meal (*Creative Online Weekly* 2002).
- Since 1995, McDonalds has introduced many innovations to address portions of fractionated processes to reduce preparation time, cost, and the risk of contamination of their products (e.g., precooked hamburger patties, liquefied eggs, precooked frozen pancakes).

- Self-checkout devices, now common in stores, reduce labor costs by allowing customers to ring up, pack, and pay for items on several stations supervised by a single employee.
- Similar reductions in the number of steps that must be performed by employees are embodied in airport self-check-in kiosks.

All of the examples above involve first fractionating a process and then seeking solutions to the parts so as to increase efficiency, effectiveness, or safety.

Reversal: In this method, the problem is turned around, inside out, upside down, or back to front to see what new patterns emerge. The goal, as with all lateral thinking, is to find different perspectives by forcing adoption of a new vantage point. Here are two famous examples. Henry Ford, instead of asking, "How do we get the workers to the material?" asked, "How do we get the material to the workers?" The result was the assembly line, which has not changed significantly since its initiation. As a second example, Alfred Sloan took over General Motors when it was on the verge of bankruptcy. In 1919, with the creation of General Motors Acceptance Corporation, he turned GM around, in part, by reversing the requirement that customers pay *before* they drive by pioneering the concept of installment purchasing, that is, paying for the car *while* driving it (Ellis and Guettler 2010).

By reversing a problem, it's sometimes possible to generate great ideas that otherwise would be overlooked.

Checklists. Checklists are a familiar part of everyday life: grocery lists, things to be done, personal calendars. They also are important parts of many technological tasks: takeoff and emergency checklists for aircraft, checklists for environmental impacts, and so forth. Building checklists can spur creativity, forcing one to think of possibilities and providing a framework that suggests completeness and consistency and that highlights omissions.

Alex Osborn (2007), one of the pioneers of creativity techniques, writing in the 1950s, provided the following checklist for new ideas:

Put to other uses? (New ways to use as is? Other uses if modified?)
Adapt? (What else is like this? What could I copy?)
Modify? (Change meaning, motion, sound, form, shape?)
Magnify? (Stronger, longer, heavier, exaggerated? Add an ingredient?)
Minify? (Shorter, lower, miniaturize? Subtract an ingredient?)
Substitute? (Other materials, processes, power sources, approach?)
Rearrange? (Interchange components? Other patterns, layouts, or sequences?)
Reverse? (Turn it backward, upside down, inside out? Open it? Close it?)
Transpose the positive and negative?
Combine? (Blend, alloy, ensemble? Combine units, purposes, processes?)

This list can be extended into thousands of questions, as Marsh Fisher did when he created the software known as IdeaFisher, which has been redeveloped

Exhibit 4.3 The Power of Checklists

Here are some checklists items—some new, some old—that show the power of the technique:

Can we reverse it? The ketchup world was turned upside down in the summer of 2002 when Heinz introduced squeezable bottles that stand upside down. No shaking, no anticipation, and no knives needed to get the ketchup flowing.

Can we magnify it? In April 1998, Gillette revealed its vision of the future razor, and it had three blades: the Mach3. Not to be outdone, Schick sought an edge with a four-blade razor, the Quattro, in August 2003. If four blades are better than three blades, then five blades must be better than four. Sure enough, in September 2006, Gillette announced the Fusion with five blades.

Can we combine functions? The Schick Intuition razor for women is designed to be used in the shower by combining shaving cream and razor (Schick 2010).

Can we make it smaller? Evolutionary or revolutionary for their time, circa 2004, were Apple's iPod Mini and the BMW Mini Cooper.

Can we change the form? Since the early 1980s, we have been nibbling baby carrots, which are really fully grown carrots cut into small sections.

and is now marketed by ThoughtOffice (Thoughtoffice 2010). Checklists are simple but very powerful devices for freeing creativity; however, they must be carefully constructed to allow the user to exercise latitude and imagination (see Exhibit 4.3).

Random Words. Everyone has had conversations in which a random word sparked a completely unrelated discussion. Random words often bring about a fresh association of ideas and trigger new concepts or new perspectives of familiar ones. In a way, they provide verbal links that help us look at one thing and see another. Table 4.5 provides a list of "link-rich" words similar to a table devised by von Oech (1986). These words are familiar, and many connections and similar concepts can be generated by using them.

The procedure is to select a word at random from the table and then try to force a connection between it and the problem being considered. To use the table, one could generate a random number between 1 and 400 and select the associated word or simply choose one. For example, suppose the random number is 301, *camera*, and the problem is how to limit graffiti in public places. An obvious connection this suggests is to use Internet cameras in high-risk areas. Other less obvious connections might include interesting graffiti artists in photographic art,

TABLE 4.5 Random Trigger Words

001–010	Knife	Insect	Robot	Pan	Crown	Banana	Accent	Bottle	Violin	Computer
011–020	Pants	Dress	Grill	Tree	Peach	Motor	Buffalo	Floor	Plastic	Leopard
021–030	Barn	Town	Bingo	Club	Class	String	Lot	Gold	Trailer	Butterfly
031–040	Pine	Nose	House	Spice	Button	Key	Auto	Oven	Jungle	Picture
041–050	Staff	Bat	Paper	Lock	Brain	Face	Mask	Nail	Sight	License
051–060	Boat	Board	Cellar	Purse	Lime	Copy	Border	Vein	Milk	Window
061–070	Rose	Muscle	Mirror	Stove	Bed	Park	Fire	Line	Bone	Alligator
071–080	Wasp	Pail	Tribe	Nap	Court	Child	Stomach	Glass	Ring	Attorney
081–090	Photo	Lion	Magnet	Bow	Iron	Suit	Emblem	Car	Train	Stadium
091–100	Arrow	Water	Gym	Race	Voter	Pitcher	Chair	Ice	Razor	Highway
101–110	Ankle	Tower	River	Torch	Elm	Hawk	Circle	Test	Tie	Mountain
111–120	Twins	Snow	Flag	Factory	Track	Joker	Ghost	Play	Network	Building
121–130	Vault	Monkey	Bank	Skates	Rock	Cook	Pearl	Cover	Sling	Battery
131–140	Exit	Hotel	Street	Road	Alley	Sheriff	Top	Meter	Bottom	Jaguar
141–150	Apron	Fox	Fork	Clamps	Blender	Basket	Book	Peak	Union	Station
151–160	Guitar	Grass	Scale	Brush	Shell	Coach	Radar	Branch	Melon	Soda
161–170	Pole	Roll	Star	Oil	Cement	Torpedo	Piano	Smoke	Paint	Escalator
171–180	Mail	Zoo	Needle	Yard	Watch	Belt	Point	Badge	Gorilla	Handle
181–190	Bus	Candle	Comet	Fan	Knee	Spider	Role	Oval	Anchor	Stereo
191–200	Axe	Fiddle	Desk	Door	Back	Fist	Tent	Apple	Moose	Machine

201–210	Mat	Message	Officer	Port	Jockey	Sea	Ship	Seal	Trap	Weight
211–220	Bucket	Chariot	Agent	Garlic	Plate	Gate	Home	Ink	Helmet	Kitten
221–230	Chorus	Laser	Lungs	Pizza	Moon	Worm	Cream	Sink	Cloud	Magazine
231–240	Glove	Winter	Dance	Drum	Friend	Rug	Shoe	Radio	Zebra	Elevator
241–250	Roof	Knot	Folder	Fund	Bride	Glue	Grade	Hammer	Horse	Teacher
251–260	Ruffle	Artery	Wall	Tray	Rodeo	Vase	Ruler	Salad	Cup	Envelope
261–270	Camp	Dice	Bell	Cord	Escape	Judge	School	Present	Song	Football
271–280	Sailor	Puppet	Whale	Wheel	Flash	Colt	Turkey	Coupon	Deer	Flower
281–290	Plant	Crane	Record	Temple	Boxer	Team	Saddle	Athlete	Stunt	Telescope
291–300	Booth	Candy	Party	Organ	Tub	Diamond	Mouse	Jazz	Ocean	Hospital
301–310	Camera	Saloon	Rake	Flute	Ticker	Gas	Halo	Waiter	Hay	Calendar
311–320	Horn	Sole	Script	Energy	Garden	Pantry	Light	Jacket	Lodge	Canteen
321–330	Goose	Marble	Level	Noose	Elbow	Stamp	Memory	Boot	Farm	Elephant
331–340	Tiger	Storm	Kite	Ladder	Fawn	Globe	Spear	Turtle	Rope	Sweater
341–350	Peanut	Shrimp	Oak	Sand	Money	Maze	Cactus	Orange	Swan	Periscope
351–360	Map	Cake	X-ray	Dock	Goat	Chip	Perfume	Chain	Pipe	Cookie
361–370	Bear	Tooth	Polish	Lantern	Skull	Lap	Shark	Sugar	Label	Knuckle
371–380	Snake	Heater	Stage	Eagle	Wolf	Lash	Fly	Potato	Camel	Rooster
381–390	Menu	Pool	Cobra	Towel	Sky	Stool	Table	Eye	Lemon	Armadillo
391–400	Sponge	Rocket	Soap	Scarf	Statue	Poodle	Tack	Police	Pencil	Telephone

offering digital cameras as a reward for the capture and conviction of graffiti artists, or funding merchants to photograph anyone buying spray paint. Alternatively, the word *camera* could be played with to see if novel ideas arise—came ra . . . come rah; perhaps a rally with community leaders could be organized to support the end of graffiti.

The point is not whether the sample ideas are good, practical, economic, or even legal. Rather, it is that they are different paths to solving the problem, paths that would never emerge from preconceived notions. Other ideas may be better, but the purpose of creativity is to multiply the paths for reaching a solution.

Morphological Analysis. Morphological analysis combines features of fractionation and checklists and expands them in a powerful new direction. It was developed by Fritz Zwicky, a Swiss astrophysicist and aerospace scientist at the California Institute of Technology in the 1940s and 1950s (Zwicky 1962, 1969; Swedish Morphological Analysis 2010).

In morphological analysis, first, fractionation is applied to choose the parameters of importance to a concept, and then the alternative possibilities for each are defined. Next, a checklist is created by making an exhaustive list of all combinations of the possibilities. Each of these combinations is examined in turn. Although some will be meaningless, some may already exist, and some may be eliminated for other reasons (such as impracticality or expense), others may merit serious consideration.

When there are two parameters, the possibilities form a plane. Three parameters form a cube. Each is relatively easy to represent and visualize. If there are four parameters, visualization is trickier, but there are several approaches that can be taken. For instance, any parameter could be chosen and a cube built for each alternative possibility. While a computer could be used to generate all possible combinations, an obvious limitation is that the combinations increase rapidly with the number of parameters and alternatives generated. Suppose, for instance, that there are three parameters, each with five possibilities. The number of combinations is $5 \times 5 \times 5 = 125$. Adding a fourth parameter with five possibilities raises the number of combinations to 625.

Suppose one is exploring the possibility of new mass transit technologies. Maybe parameters such as the power source, the transport medium, and the guidance mechanism are selected. The list of alternative possibilities under each parameter might appear as in Table 4.6.

This brief list produces 252 combinations. One is an electrical-underground-guided path that already exists. Another combination, diesel-underground-driver, would probably be rejected unless a method could be developed to eliminate the effects of engine emissions. Each of the combinations is examined in an analogous manner to complete the morphological analysis. The power of the technique stems from considering which combinations were eliminated and why. Inevitably, one realizes that this functional decomposition could be extended or sharpened. If novel new combinations are discovered that cannot be functionally eliminated, then the technique has delivered!

TABLE 4.6 Example Morphological Analysis

Power Source	Transport Medium	Guidance Mechanism
Hydrogen	Roadway	Driver
Gasoline	Air	Towed
Diesel	Water	Guided path
Electrical	Underground	Electronic map
Steam	Conveyor	Collision avoidance
Battery	Rail	None
	Magnetic levitation	

The Swedish Morphological Society's website provides methodologies and numerous examples (Swedish Morphological Analysis 2010). It lists 80 projects on which morphological analysis was used from 1995 to late 2009. One of the featured articles is about using morphological analysis to conduct futures studies (Ritchey 2009), an important component of this book.

Morphological analysis is intended to provide a disciplined framework for creativity. Simply, it provides a kind of accounting system for an array of possibilities that are too extensive for the mind to track. Like other creativity-enhancing techniques, morphological analysis encourages abandoning preconceived patterns. Through it, one is forced to develop possibilities that might otherwise be overlooked or rejected and to consider ways to implement possibilities that might be eliminated.

TRIZ. *TRIZ* (pronounced "trees") is an acronym for four Russian words meaning "the theory of inventor's problem solving." TRIZ is a mixed methodology, combining elements of creativity, matrixed systems analysis, morphology, and patent analysis. Its development was begun by the Soviet engineer and researcher Genrich Altshuller and his colleagues in 1946. It has continued to evolve since. The brief discussion provided here can be extended by examining Barry et al. (2011).

Genrich Altshuller worked in the Inventions Inspection department of the Soviet Navy in the late 1940s. His job was to examine and help document proposals and to assist others to invent. By 1969 he had reviewed some 40,000 patent abstracts to determine how innovation occurred. From his studies, he developed "40 principles of invention," several "laws of technical systems evolution," the concepts of technical and physical "contradictions," the concept of system "ideality," and many other theoretical and practical approaches. By scanning a large number of inventions to identify the underlying contradictions and to formulate the principle the inventor used to remove them, he concluded that *inventing is the removal of a technical contradiction with the help of certain principles*.

TRIZ is an algorithmic approach to the invention of new systems and the refinement of old ones. It is based on Altshuller's system of creativity as a set of connections:

Specific Problem → Typical Problem → Typical Solution → Specific Solution

By 1985, the basic structures were established and proponents have continued to improve and add to the methodology.

According to TRIZ proponents, the technique has been used by large companies worldwide to solve manufacturing problems and to create new products. They cite BAE Systems, Procter & Gamble, Ford Motor Company, Boeing, and others as companies that have used TRIZ to systematically solve complex technical and organizational problems. TRIZ has been successfully applied to biomedical research, medicine, and many other areas.

A browser pointed to TRIZ consulting will reveal many in the TRIZ business offering training and services. One software package is Invention Machine (Invention Machine 2010). I-TRIZ is another (I-TRIZ 2011). Relatively simple techniques, simply, robustly, and transparently applied, are probably best.

4.3.2 Group Creativity

Since technology managers and forecasters often work as members of a group, they must be concerned with ways to increase group creativity. Many of the techniques that stimulate individual creativity also can contribute to group creativity. However, the concerns of individual creativity are intertwined with concerns about the dynamics of group interaction. The group techniques presented in this section address these additional concerns.

Brainstorming: Brainstorming is an old concept. Its formalization as a group creativity process is largely the work of Alex Osborn, who coined the term in 1939 (SkyMark 2010). The members of a brainstorming group are asked to respond to a central problem or theme. Emphasis is on generating a large number of ideas (fluency), and criticism or evaluation is deferred (suspended judgment). Thus, brainstorming is a group implementation of the concepts of lateral thinking and, as such, the results of brainstorming eventually must be treated with vertical thinking.

The brainstorming session is consciously unstructured. Four general guidelines are observed:

1. Criticism is ruled out.
2. "Freewheeling" and wild ideas are welcome.
3. A large number of ideas are sought.
4. Participants are encouraged to combine ideas into new or better ideas.

The setting for the process should be relaxed and isolated, and participants should be encouraged to verbalize their responses as quickly as they come to mind. The session should involve at least six but not more than twelve participants, with perhaps one-third of them directly involved in the topic under consideration. It may be important not to have both superiors and their subordinates in attendance. Subordinates and their superiors often do not feel free to generate "far-out" ideas for fear of appearing foolish in front of each other. A broad range of backgrounds and interests should be represented in the group to enrich ideas and perspectives. Sessions should not last too long. Some suggest that one to six hours is ideal, but, we have had better success with shorter sessions, say 30 minutes to one hour.

If participants are not familiar with the technique, a warm-up session dealing with a familiar but unrelated problem can be useful. In a productive brainstorming session, the ideas may flow so rapidly that it is difficult to keep track of them. Thus, some means of recording ideas must be provided. The interactive electronic whiteboard is a modern method, but older approaches such as whiteboards or flip charts are quite workable. The interactive electronic whiteboard allows ideas to be saved as a file, played back, and revised into a more effective format.

The chairperson must keep ideas flowing smoothly and control traffic so that only one person talks at a time and everyone has an opportunity to speak. He or she also must ensure that no evaluation takes place during the session. The chairperson occasionally may need to control the pace, slowing things down for the note taker or jumping in with ideas if the session slows prematurely. Finally, as with any meeting, the chairperson is responsible for organizing the session, reserving space, issuing invitations, preparing the problem definition, making sure that the recording device (flip chart, interactive electronic whiteboard, or other mechanism) is there and functioning, and naming a note taker.

Problem definition is an extremely important part of the brainstorming process that often is given too little attention. The problem should be stated clearly but not too narrowly: narrow statements invite a narrow range of ideas and may inadvertently eliminate the richness that is sought. For example, a firm concerned that too many trips are being taken to branch offices might state the problem in positive terms: "How can all forms of communication with branch offices be enhanced?"

Brainstorming is a useful process, but it is not without problems. For example:

- Delayed evaluation may cause some participants to lose focus.
- Dominant individuals may influence other participants and try to monopolize the floor.
- Bandwagon and other "groupthink" phenomena can undermine creativity.
- It is difficult to prepare reference material in advance because the ideas that will be generated are unpredictable.
- Some participants become emotionally involved, thus stifling the participation of others.

In contrast, brainstorming offers the positive benefits of suspended judgment, lateral thinking, and the use of random key words in a group setting. Furthermore, since the group members "own" the ideas that are generated during the process, their support may be greater for implementing the solutions derived from them.

A few variations on brainstorming are given here without much comment. They can be used to provide an interesting or varied experience for the participants.

Brainwriting: Ideas (some select number, such as three) are written individually then passed to another person, who adds more, possibly triggered by those. A hybrid technique begins by individually writing a few ideas and then asking each participant to offer one idea orally. Mycoted (2011) offers many interesting creativity techniques with clever brainwriting variations.

- *Stop-and-go:* Participants ideate for three minutes, reflect for three minutes, ideate for three minutes, and reflect for three minutes. This gives them a chance to think, make notes, and build on the ideas of others; it also reduces pressure.
- *Sequenced:* Participants take turns in a set routine, say a round robin. If one person has nothing to say, the next person is called upon. The process stops when there are no more ideas. This gives everyone a chance to offer ideas, and the structure encourages reticent members to participate.
- *Computer-assisted:* Many firms provide computer-assisted brainstorming services. (Enter "computer-assisted brainstorming" into a browser to see what is available.) The service can vary from software for a single computer to sessions for networked computers in one location.

Synectics: There are many other group techniques for creativity stimulation. One, developed by Synectics, Inc., of Boston, for instance, is a technique with both contrasts and parallels to brainstorming (Synecticsworld 2010). It is much more involved, however, and can require several days and perhaps the help of a consultant. The technique is based on the concept that, since only the individual or group with the problem (client) can implement a solution, the goal of the process should be to inspire the client. Thus, Synectics is not so much designed to produce ideas as to provide effective interaction so that ideas will be implemented.

Synectics groups and brainstorming groups are about the same size, but Synectics sessions typically are calmer and shorter (45 minutes) than most brainstorming sessions. The goal is not fluency but the generation of a few ideas at a time. The thought process is more vertical than lateral and seeks to expand and improve one of a few original ideas. Thus, judgment (evaluation) *cannot* be suspended. It can be softened, however, by asking for two positive comments before a negative or critical one (reservation) is allowed. The process continues until the client is satisfied with the solution.

One principle of Synectics is to look for a solution provided by nature. There have been many successes using Synectics. One is Pringles (Procter & Gamble 2010), the potato chips in a can. Pringles were born in a Synectics session when a participant remembered how much easier it is to rake leaves when they are damp because they stack on top of each other. Voila! Pringles resulted.

Two other principles are springboarding and excursions (Creating Minds.org 2010). *Springboarding* is a method of triggering new ideas by rewording old ones. This is accomplished by prefixing the statement with "I wish" or "How to." "I wish" tends to be used for more speculative ideas and "How to" for more specific problems, although people tend to have their own preferences. For example, using "IW," shorthand for "I wish," it is easier to say things like "IW e-mails answered themselves" than saying, "The e-mails should answer themselves." "H2" is used as an abbreviation for "How to," as in "H2 find new product ideas in a mature business that sells disposable containers and other packaging for restaurant and other food-service operators."

Excursions are exercises that use various techniques to find ideas that might be unusual, weird, or nonstandard but that can be brought back and used, perhaps after some interpretation. Elsewhere, in creativity, the technique is called the *use of metaphors and analogies*. Suppose that the subject is insertion of components into a circuit board. The moderator could ask, "What else is like this?" A response that it is like inserting a sword into a scabbard leads to the statement "The sword has a groove for blood to drain, which acts as a lubricant to insert the sword even deeper." This leads to the suggestion that a lubricant might make the insertion of computer components easier.

4.4 CONCLUSION

This chapter began exposition of the forecasting process by describing its first phase, exploring. It introduced TDS and monitoring, which respectively are the primary model and the premier technique for technology forecasting. Building the context for the TDS was considered in some detail, as were the needs, techniques, and strategies for monitoring. Lastly, methods of enhancing the creativity of the forecast team were presented.

All of these topics are important for the remaining two phases of the forecasting project: analyzing and focusing. The TDS is used as an integrating theme throughout the book. Monitoring is emphasized as central to all phases of the forecast. In fact, the central question about monitoring is whether it will be formally done with an organizational commitment or informally done as part of the daily work routine. The answer to this question is pertinent to every phase of technology development in the corporate, government, and university environments. Finally, creativity is indispensable to envision the future.

REFERENCES

Allen, T. J. (1984). *Managing the Flow of Technology*. Cambridge, MA, MIT Press.

Barry, D., E. Domb, and M. S. Slocum (2011),"What is TRIZ?" Accessed 24 January 2011 from http://www.triz-journal.com/archives/what_is_triz/.

Brenner, M. (2005). "Technology Intelligence at Air Products: Leveraging Analysis and Collection Techniques." *Competitive Intelligence Magazine* **8**(3): 6–19.

Coates, J. F., V. T. Coates, et al. (1986). *Issues Management*. Mt. Airy, MD, Lomond.

Creating Minds.org. (2010). "Synectics." Retrieved 14 May 2010 from http://creatingminds.org/articles/synectics.htm.

Creative Online Weekly. (2002, 18 March). "New Thomas E. Wilson Dinner Roasts Are Fully Cooked and Ready-to-Serve." Retrieved 13 May 2010 from http://www.creativemag.com/onlineweekly.html.

de Bono, E. (2010). "Lateral Thinking and Parallel Thinking." Retrieved 13 May 2010 from http://www.edwdebono.com/debono/lateral.htm.

de Vries, B. J. M., D. P. van Vuuren, et al. (2007). "Renewable Energy Sources: Their Global Potential for the First-Half of the 21st Century at a Global Level: An Integrated Approach." *Energy Policy* **35**: 2590–2610.

Ellis, K. and E. Guettler. (2010). "A Better Life: Creating the American Dream." Retrieved 13 May 2010 from http://americanradioworks.publicradio.org/features/americandream/b1.html.

Ezra, A. A. (1975). "Technology Utilization: Incentives and Solar Energy." *Science* **187**: 707–713.

Guilford, J. P. (1959). "Traits of Creativity." In *Creativity and Its Cultivation*, ed. H. Anderson. New York, Harper and Brothers: 142–161.

Guo, Y., L. Huang, et al. (2010). "Research Profiling: Nano-Enhanced, Thin-Film Solar Cells." *R&D Management* **40**(2): 195–208.

I-TRIZ. (2011). "Welcome to Ideation TRIZ," Retrieved 12 January 2011 from http://www.ideationtriz.com.

Invention Machine (2010). "The Innovation Software Company," Retrieved 14 May 2010 from http://www.invention-machine.com.

Martino, J. P. (1993). *Technological Forecasting for Decision Making*. New York, McGraw-Hill.

Mycoted. (2011). "Main Page," Retrieved 12 January 2011 from http://www.mycoted.com/Main_Page.

Osborn, A. (1948). *Your Creative Power: How to Use Your Imagination*. New York: Charles Scribner's Sons.

Porter, A. L. and S. W. Cunningham. (2005). *Tech Mining: Exploiting New Technologies for Competitive Advantage*. Hoboken, NJ, John Wiley & Sons.

Porter, A. L., Rossini, F. A., Carpenter, S. R. and Roper A. T. (1980). *A Guidebook for Technology Assessment and Impact Analysis*. New York: North Holland.

Procter & Gamble. (2010). "Pringles." Retrieved 14 May 2010 from http://www.pringles.com/en_US/Pages/Home.aspx.

Ritchey, T. (2009). "Futures Studies Using Morphological Analysis." Retrieved 14 May 2010 from http://www.swemorph.com/pdf/futures.pdf.

Robinson, D. K. R. and T. Propp. (2008). "Multi-Path Mapping for Alignment Strategies in Emerging Science and Technology." *Technological Forecasting & Social Change* **75**: 517–538.

Schick. (2010). "Schick Intution." Retrieved 18 May 2010 from http://www.schickintuition.com/.

SkyMark. (2010). "Alex F. Osborne: Father of the Brainstorm." Retrieved 15 August 2010 from http://www.skymark.com/resources/leaders/osborne.asp.

St. Louis Business Journal. (2002). "Sara Lee to Roll Out Crustless White Bread." Retrieved 13 May 2010 from http://www.bizjournals.com/stlouis/stories/2002/04/29/daily84.html.

Swedish Morphological Society. (2010), "Swedish Morphological Society," Retrieved 14 May 2010 from http://www.swemorph.com/.

Synecticsworld. (2010). "Synecticsworld." Retrieved 14 May 2010 from http://www.synecticsworld.com/.

Thoughtoffice. (2010). "Whatever Happened to IdeaFisher Software?" Retrieved 13 May 2010 from http://www.thoughtoffice.com/?page_id=148.

Vincent, J. F. V., M. N. Sahinkaya, et al. (2007). "A Woodpecker Hammer." *Proceedings of the Institution of Mechanical Engineers, Part C: Journal of Mechanical Engineering Science* **221**(10): 1141–1147.

Von Oech, R. (1986). *A Kick in the Seat of the Pants*. New York, Harper & Row.

Zwicky, F. (1962). *Morphology of Propulsive Power.* Pasadena, CA, Society for Morphological Research.

Zwicky, F. (1969). *Discovery, Invention, Research: Through Morphological Approach*. Toronto, Macmillan.

5

GATHERING AND USING INFORMATION

Chapter Summary: This chapter explores two major ways of gaining data for the technology forecast. One technique is to consult experts. Another is to sift content from sources such as the Internet. The chapter concludes with strategies for presenting and communicating forecast-relevant information.

Technology forecasts are successful when the forecaster is able to integrate diverse sources of information to produce convincing and holistic portraits of possible futures. Technology forecasters need two types of information—information about the technology and information about its context (i.e., the TDS). In addition to conventional written materials, they seek this information from experts and the Internet. Since approaches to accessing conventional sources are well known, this chapter emphasizes acquiring information from the latter two sources.

Technology experts can contribute in-depth knowledge in specific fields of the environmental and physical sciences and technology. They can provide expertise in every step of the development process from discovery to commercialization. Societal experts can give insight into governments, public and private organizations, and society at large, as well as provide knowledge about world regions and cultures. Technologies are demanded and supplied by society. Thus, the forecaster must not overlook the complex web of societal rules, incentives, and obligations that characterize the TDS.

"The Internet" is shorthand for a variety of information sources that are networked and sometimes integrated by the World Wide Web. Technology databases include a profusion of scientific and engineering information as well as patent databases that give indicators of the stage of development and commercialization. Besides the usual databases dealing with societal measures (e.g., population, employment, economic activity), numerous and varied news sources and blogs provide unique insights into social issues and perceptions. Some of these are free; others are not. All aggregate a variety of opinions.

5.1 EXPERT OPINION

Rarely do forecasters or managers have sufficient technical depth and a full-spectrum understanding of the business and human dimensions of the technology being forecast. "Expert opinion" is shorthand for a spectrum of factual and tacit knowledge that the forecaster needs but cannot personally supply. Much of this knowledge derives from the superb human ability to synthesize information, an ability that empirical data mining technologies cannot replace. Tacit knowledge can help to interpret existing situations and anticipate future ones, albeit uncertainly. This section considers who to engage and how to engage them to obtain worthwhile input.

5.1.1 Selecting Experts

The techniques used to select experts influence the choice of experts and vice versa. Start by identifying the range of vital perspectives that are needed. Usually, this will yield a diverse cross section of knowledgeable people, stakeholders, and publics. Some concerns about each are:

- Knowledgeable persons
  - *Area of expertise:* Area of knowledge may be technological or contextual.
  - *Depth of knowledge:* Depth may range from world class, which is rarely needed, to that of a graduate student who understands the essentials of the technology.
- Stakeholders
  - *Who they are:* The TDS identifies the actors engaged in implementing, using, and dealing with the consequences of the technology
  - *Perspective:* The analyst must distinguish different perspectives, recognizing, for instance, that the technology users may not have been the ones to actually make the purchase decision.
- Publics: What different interests should be tapped?

Review the management of technology issues and questions (Section 4.2.4, Tables 4.2 and 4.3) to prioritize the most salient knowledge needs. Perhaps secondary sources can be substituted for expert opinion to cover less vital viewpoints.

Should expert opinion come from inside or outside the organization—or both? In-house experts offer strong advantages. For instance, they likely understand the decision context and the organizational culture, and they may have built-in incentives to participate. Moreover, involving them may boost their commitment to the resulting decisions. Outsiders bring fresh perspectives uncolored by the organizational culture. If they have been part of similar innovation processes, they can provide valuable insights about system requirements, pitfalls, and parameters that need to be tracked. Vendors, technology developers, and established users also can be valuable. They often can offer hard data that are not easily accessible to the forecaster, as well as opinions about industry stability, directions of change, and how best to introduce new products. Personal links or network introductions may help gain cooperation, but beware of sampling only the like-minded. Remember, *no information source is without bias*. However, knowing the source of the information helps you to understand what the bias may be.

Lipinski and Loveridge (1982) list some personal attributes to consider when selecting experts. These include substantive knowledge of the desired aspects; the breadth of perspective to relate their specialized knowledge to the innovation; the mental agility to extrapolate from their knowledge to address future prospects and uncertainties; and imagination. Asking prospects for a self-assessment of their substantive expertise can help screen suitable participants. Consider phrasing this question to address the question "How expert are you?" on the key dimensions. Doing so forces you to be clear about what facts or opinions you want.

How many experts do you need? The answer could range from one to help "get the technology right" to enough to sample diverse population segments and address multiple issues. When it is important to document the views of different populations, sampling issues become paramount. These concern strategy (e.g., a random, stratified, or multistage survey) and size. In such cases, consult a suitable text (c.f. Rea and Parker 2005).

Some of the experts you need can be found by networking with experts you know. Others can be located through professional society databases, patent databases, citations in books and papers, and academic department rosters. Bryson (2004) provides an excellent overview of techniques for identifying relevant stakeholders. Beware of sampling from among conveniently available individuals who may share biases—"prevailing wisdom" or the zeitgeist. These biases may be the result of shared cultural norms and backgrounds or of something as simple as reading the same news sources.

5.1.2 Selecting Expert Opinion Techniques

This text defines expert opinion inclusively to cover a gamut of expertise and of sought-after information and methods, including "participatory methods" central to some technology foresight and constructive technology assessment approaches. Glenn and Gordon (2010) have devoted 39 chapters to future-oriented technology analysis (FTA) methods, only a few of which can be discussed here.

The first edition of this book emphasized formal gathering of information from fairly large samples (e.g., surveys). This is still a sound approach. This second edition, however, stresses interactive expertise. A few experts who are willing to engage with you on multiple facets of the forecast can add great value. Although the Internet can be an excellent facilitator, this approach puts a premium on local talent willing to participate on an ongoing basis to:

- Guide your understanding of the technology and possibly help formulate effective database searches (e.g., review early search results, help tune the search algorithm)
- Help depict the TDS by pointing out key roles, actors, and relationships.
- Relate R&D to technological capabilities that can increase functionality, leading to improved products or processes and potential new users
- Review and critique draft results, interpretations, reports, and recommendations
- Become bona fide collaborators

You've identified whose input you want; now you just have to get it. Think hard about how to get busy people to participate (what's in it for them?) and make it easy for them. You can't tell them too often why their input is so valuable. Leave out "nice to know" questions and ask only what you need to know. Go for the "Goldilocks" solution—the "just right" degree of detail.

Know what you need to learn and formulate your queries to get it. With technology experts, you will find that you are asking mainly about facts or estimates. With societal experts, you may be relying more on preferences and opinions. Even when dealing with qualitative information, however, you will need to develop measurement scales.

Gustafson et al. (1973) distinguished three expert opinion processes: feedback, or one-way communication; interaction; and estimates. Nelms and Porter (1985) considered schemes for choosing, and possibly combining, these processes and suggested weighing six factors in tailoring expert opinion techniques:

1. *Logistics*: Resources and time available may constrain your options (e.g., preclude multiple feedback techniques).
2. *Feedback*: Beware long delays in multistage processes; seek to balance consideration of outlier positions.
3. *Communication medium*: In-person or electronic; real-time or "whenever" participation.
4. *Sample size*: Particularly for interactive modes. Group processes are valuable, but their value added declines as groups grow beyond five or six.
5. *Stopping rule*: In multistage processes, time and tolerance favor two, or at most three, rounds.
6. *Interaction*: More is needed for complex tasks and where serious disagreements arise.

Consider location, availability, and time commitments when choosing how to gather opinion. Consider whether target group members speak or even know the same technical vocabularies. Factor in whether their worldviews are compatible. Weigh requirements for detailed input against the time available. Each technique

TABLE 5.1 Classification of Expert Opinion Techniques

Technique	Talk (T)	Feedback (F)	Estimate (E)	Process Summary
Committees	X		X	TE
Brainstorming	X			T
Nominal group process	X	X	X	EFTE
Survey			X	E
Delphi		X	X	EFE
Shang Inquiry		X	X	EFE
EFTE	X	X	X	EFTE
POSTURE	X		X	ETE

has different strengths, weaknesses, and costs; each delivers different outputs. Think like a craftsman—don't use a sledgehammer to go after that fly on the plate glass window. Consider using multiple expert opinion methods.

Table 5.1 shows a few of the many available expert opinion-gathering techniques. These vary according to their allowance for talk, feedback, and estimation. Some allow for repeated iteration. Dividing lines blur, but the key consideration is whether experts interact among themselves. Individual techniques, such as interviews or surveys, don't provide for this; group processes do.

You can moderate the degree of interaction. The purpose affects the amount of interaction and the choice of technique. For instance, in group foresight processes, the purpose might be fostering consensus on one or more of multiple options. In the Delphi process described below, it might be broadening individuals' consideration of factors in estimating the likelihood of an event. Slight interaction processes often use anonymous feedback, as in Delphi techniques. Full interaction might involve designing group decision processes.

Whatever techniques you employ, follow sound survey principles. When the Delphi method became popular, some technical professionals ignored the basic principles of sound sampling and questioning. Bad questions mean bad results. Consult a basic text on survey methods (e.g., Fowler 2008). Until then, here's a list of things to avoid:

- Insufficient background information (but don't say too much either; it may color responses)
- Leading questions
- Ambiguous questions
- Compound statement questions
- Technical jargon not understood by some participants
- Random question sequencing (e.g., general to specific; basic to advanced)

Interviews. You can obtain individual input in various ways;

- In person, by phone, or semi-real-time on the Internet (e.g., back-and-forth messaging).

- Structured (set questions) versus nonstructured (elicit the respondent's attitudes by following leads as they emerge). A middle-ground—the focused interview—is often best. It follows a preset interview guide but opportunistically explores leads.
- Collaboratively—interacting in multiple ways to obtain ideas and feedback.

Surveys. Surveys solicit input from multiple individuals without interaction among them. Its advantages include its relative ease and inexpensiveness. Its disadvantages (shared to varying degrees with other techniques) include sampling difficulties, especially nonresponses. The best advice is to review a basic text on how to construct a sound survey (Fowler 2008). For technology forecasts, consider:

- Closed-form questions with a limited set of choices, enabling tabulation of averages and distributions
- Open-ended questions that allow responses to explain meaning and rationale.

Consider which mode will enable you to reach the most participants, obtain the best response rate, and fit your time/budget. There are several choices:

- In-person (ensures understanding but is very expensive and time-consuming)
- Mail (cheap and slow)
- E-mail (cheapest and fastest; allows easy follow-up and response but is easy to ignore)

Finally, construct the survey instrument. Begin at the end: consider what you need to get from the responses, their expected format, and how to process the responses. Decide if you want to quantify responses and the level of detail you want. You may consider ranking (i.e., ordering from least to most on some scale) versus rating (on a fixed scale). Draft the survey and pilot test it first on colleagues, then on some target respondents. Besides having them answer the questions, explicitly ask for their feedback on the survey itself.

Delphi Process. The Delphi process is a form of survey intended to provide participant anonymity, controlled feedback, iterative responses, estimates of the likelihood/timing of technological developments, and statistical response measures. It was developed in the early 1950s by the Rand Corporation and has become a staple of technology forecasting. Since the time of Sackman's (1974) critique, an ongoing discussion of issues, underlying assumptions, requirements, and modifications has ensued. A huge number and variety of published articles underscore Delphi's wide range of applications and variations. Linstone and Turoff's (1975) survey is one of the most interesting.

Delphi entails more than one round of questioning. Respondents are asked a set of questions and then are asked those same questions again after being provided with statistical feedback from the first round. A popular format is to present the median and upper- and lower-quartile responses (i.e., the 25th, 50th, and 75th

Round 3

How important is the operating system in your selection of a smart phone?

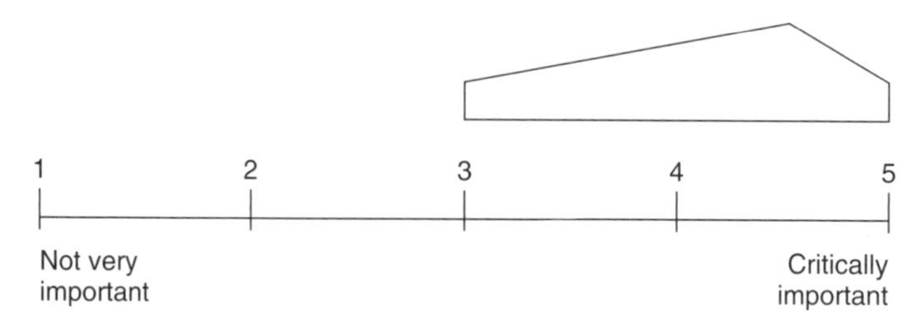

Figure 5.1. Delphi Respondent Summaries

percentiles) superimposed for each item on its response scale. Selective reasons for outlier responses may also be provided, usually anonymously. Figure 5.1 illustrates a response summary for a question concerning smart phones.

Advantages of Delphi may include less contentious exchange of perspectives than other group methods. Anonymous feedback and absence of interpersonal exchanges in which certain individuals may dominate favor Delphi in this regard. Delphi can serve both empirical and normative forecasting objectives and can include multiple stakeholder factions. Of special concern is the typical falloff in response rate over additional survey rounds that may undermine sampling. Delphi administration entails double effort, given multiple rounds, frequent requests for a response from participants, and interim round tabulations needed to generate the feedback.

Focus Groups and Workshops. Traditionally, these group techniques required participants to be in the same location. However, electronic communications such as e-mail, video conferencing and Web-based conferences can be substituted. These processes vary widely in terms of the

- Number of individuals involved, from a few (committees) to very many (conferences)
- Formality and degree of interactivity
- Frequency—one time to ongoing, organized in committees or other such group decision formats
- Agendas—single focus to broad or open agendas
- Composition—homogeneous to heterogeneous

For expert opinion purposes, these activities need to be carefully guided. Focus groups are popular activities for eliciting consumer preferences in one-time meetings, typically with some remuneration. They can be adapted to obtain expert input. Larger-scale participatory approaches, such as national foresight studies, can engage large numbers of different groups via a sequence of activities. Germany twice stepped through its "FUTUR" process to engage stakeholders and publics in helping to prioritize national R&D (Cuhl 2004). Such processes can involve expert-generated content presented to publics, stakeholders, and policy-makers in different forums. For instance, Constructive Technology Assessment (CTA) strives to engage such diverse groups in valuing potential impacts of technological innovations (c.f. Genus and Coles 2005).

The Center for Nanotechnology in Society at Arizona State University is evolving a Real Time Technology Assessment (RTTA) technique that can be used while the technology is still largely in the R&D stage. They also have conducted a National Citizens Forum on Nanotechnology (National Citizens Forum on Nanotechnology 2010).

Small-scale participatory approaches also present special appeal. Working with a few experts on an ongoing interactive basis can greatly enrich a forecasting project. Georgia Tech has had success in getting professors engaged in nanotechnology R&D to help explain the technology and to participate in two-hour workshops to explore potential innovation pathways.

Small-Group Processes. Brainstorming, a small-group process to generate creative ideas, was presented in Section 4.3.7. The Nominal Group Process (NGP) offers an appealing alternative format to brainstorming (Delbecq and Van de Ven 1971). Like brainstorming, it seeks to overcome certain deficiencies of unstructured meetings. In NGP, participants first generate ideas individually without interacting. Then they share these ideas communally. In one variation, each participant, in turn, is asked to contribute one idea that is written or posted for all to see, without comment. Discussion to clarify and consolidate ideas follows. Then individuals may be asked to select the best ideas, and votes are tabulated. Discussion can follow to assess these results and often leads to sharpening and reformulating the ideas. This may be followed by a final individual voting (estimate) round. Experience suggests limiting NGP groups to ten or fewer persons (Roper 1988). For larger numbers, separate subgroup sessions followed by a consolidated group session can work. Variations such as a single sequence of idea generation, talk, and estimate can expedite NGP, but at the cost of reduced commitment to the final selections.

Compiling expert opinion poses several challenges. For instance, collapsing information to its central tendency may lose rich counterviews and outlier viewpoints. Compiled tacit knowledge must be integrated with other (empirical) information and with one's local knowledge. Don't hide behind expert opinion. Forecasters must generate and stand behind forecast conclusions or recommendations.

5.2 GATHERING INFORMATION ON THE INTERNET

This section shifts the discussion from getting information "live" to more diffuse, asynchronous techniques associated with mining the Internet. Historically, the techniques discussed here might have been called *content analysis* or *archival analysis* (Krippendorf 2004). This is not to say that content analytic approaches are not useful. They are! Only the means of accessing the information online has changed.

Despite the Internet, forecasters still must consult a wide variety of sources. This section gives some guidance about sources that are "the best of the best," but ultimately, source selection depends upon the specific technologies and topics being addressed. Subsections addressing science and technology and societal sources of information follow.

5.2.1 Science and Technology on the Internet

"On the one hand information wants to be expensive, because it's so valuable. The right information in the right place just changes your life. On the other hand, information wants to be free, because the cost of getting it out is getting lower and lower all the time. So you have these two fighting against each other" (Brand and Henron 1985, p. 49).

As suggested by Brand, the primary trade-off in acquiring that information is its cost. Information is valuable, so database providers often charge a hefty fee to access it. That is why universities, large research institutions, and government labs often bear the cost of licensing science and technology databases for their researchers. Technology forecasters need database information, but rather than one or two articles, they are interested in finding patterns across whole literatures—often a thousand articles or more.

There also are electronic journals and databases on the Internet that can be accessed free. In an era where information wants to be free, the primary added value of pay databases is the extra data structuring that may be provided. Structuring may involve machine-readable text, additional information not otherwise available on the Internet, or a guarantee of the quality of the information provided.

A conceptual model of the R&D process, Figure 5.2, helps define the types of information sought on the Internet. R&D occurs in an institutional context. It involves decision making about funding, hiring, and market entry, as well as choices of strategic collaboration, and policies or regulations that can change the rules of the game. Input indicators include information about personnel, financial capital, and facilities and their locations. Output indicators involve published papers, patents, new product announcements, and sales. In between is an elaborate, only partly understood process of R&D. Because the R&D process is incompletely measured by papers and patents, the forecaster must be careful to correctly interpret these indicators correctly.

Technology forecasters strive to produce a complete and unified picture by filling in the details of Figure 5.2. While a comprehensive model for all new technology would be useful, building the model is a customized task that varies among technologies and across societies. The material that follows provides more information about how scientific and technical databases can help.

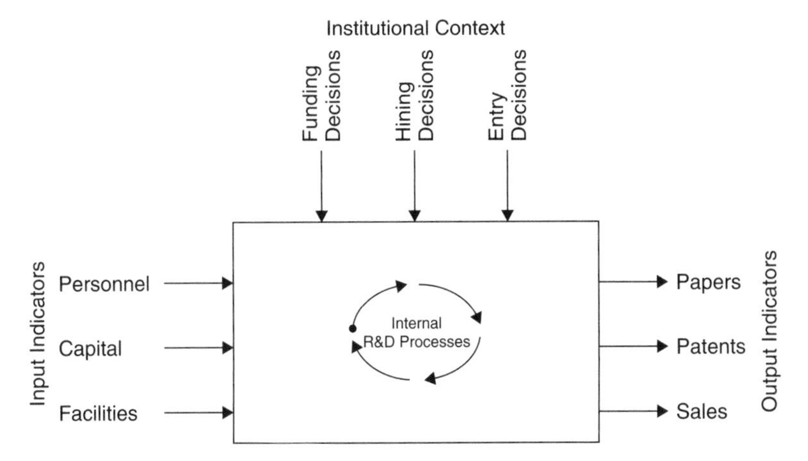

Figure 5.2. An Input-Output Model of R&D

The forecasting question drives the search for information. Use scientific databases for more fundamental or basic research. Use engineering or medical databases for applied research. Use patent databases for information about technologies that are close to market. The following material highlights a few sources, the best of the best, for science and technology information.

Paid sources include the Web of Knowledge provided by Thomson Reuters and Scopus by Elsevier. These databases are subscription only. Thus, they must be accessed from an institution with a subscription. Ready access to the data via a web browser is then provided. These sources are worth paying for if you

- Value access to peer-reviewed research
- Want to disambiguate authors' names
- Need at least one good institutional address associated with a given piece of research
- Desire machine-readable text
- Prefer to have preset classifications of content
- Must track the intellectual history of ideas

In contrast, Google Scholar and CiteSeer are free. They provide a richer variety of sources but may require additional effort in structuring the data for analysis. There is also a trend toward *open-access* scientific articles for which a research institution pays but allows wide distribution and access (see Table 5.2).

While you can find plenty of information about engineering and medicine in science databases, more specialized databases dealing with advances in applied science often are useful. The best of the best are given in Table 5.3. Engineering Village (Elsevier) offers two of the best databases for engineering and computer science—Inspec and Compendex. These have many of the advantages of the paid

TABLE 5.2 Best of the Best in Science

Paid	Free
Web of Knowledge (Thomson Reuters 2011c)	Google Scholar (Google 2010a)
Scopus (Elsevier 2010c)	CiteSeer (Penn State 2010)

TABLE 5.3 Best of the Best in Engineering

Paid	Free
Engineering Village (Elsevier 2010a)	MEDLINE (U.S. National Library of Medicine and National Institutes of Health 2010)
	NTIS (U.S. Department of Commerce 2011)

TABLE 5.4 Best of the Best in Technology and Patenting

Paid	Free
Delphion (Thomson Reuters 2011a)	U.S. Patent Office (U. S. Patent and Trademark Office 2011)
MicroPatent (Thomson Reuters 2011b)	European Patent Office (European Patent Office 2010)
PatentCafe.com (Pantros IP 2011)	JPO Japanese Patent Office (Japanese Patent Office 2010)
Questel-Orbit (Questel 2010)	WIPO World Intellectual Property Organization (World Intellectual Property Organization 2011)
WIPS (WIPS 2010)	
Derwent World Patent Index (STN International 2010)	
IFI CLAIMS (IFI CLAIMS 2011)	

databases noted above. There are many other specialized sources of engineering information, such as chemical research and paper research. Free sources also include MEDLINE and NTIS. The articles accessed there are of the highest quality and provide useful insight into the science funding priorities of specific U.S. government agencies. Other national government sites and the Organization for Economic Cooperation and Development (OECD) also provide worthwhile information.

Patent databases provide information about discoveries that companies and inventors seek to protect by patents. Thus, they often provide information about technologies that are nearing market. The best of the best are given in Table 5.4.

There are a great many pay databases for patents, all competing to provide superior service. The principal value-added that these databases provide is machine-readable records. A new trend is for them to bundle the patents with prepackaged analytics and indicators. The costs for these databases are high, and the transparency of the analysis services is low. However, the convenience of analysis on a Web browser is unquestioned! In contrast, all major patent offices provide searchable databases. Unfortunately, they do not invest as much effort to make the records structured and easy to analyze as commercial services.

5.2.2 Society and Culture on the Internet

As emphasized thus far, institutions and society as a whole play critical roles in implementing new technologies. Thus, the forecaster must identify important institutions and devise a strategy for collecting relevant information. When that information is available on the Internet at all, it is free. The distinction is between information available only in written form and electronic publications available on the Internet.

The first step in searching for societal information is to consider the time frame of the forecast. Long-term forecasts will necessarily require more information about societal processes, while short-term forecasts will be more concerned with microeconomic data of sales and demand.

Nobel Prize winner Olivier Williamson alerted people to the great variety of institutions of which they should be aware (Williamson 2000; Table 5.5). His framework involves different levels of society, each operating with a characteristic time frame and each occupied with its own societal questions. The framework is a reminder that institutions are nested and interdependent. Longer-term trends shape even the most immediate questions of product sales and new product announcements. The following paragraphs highlight some of the most valuable sources available at each level.

Level One. This level involves the long, slow sweep of history. Questions concerning customs, traditions, norms, and religion occupy time frames ranging from 100 to 1000 years. These informal institutions often provide the societal impetus for developing new technologies and for accepting or rejecting them when they are first proposed. The "clash" of civilizations also occurs at this level, driving the military impetus to develop new technologies. Social theories are most helpful here (see Table 5.6).

There are many sources of information on social theory. However, the nature of the topic lends itself to expert input. The leading thinkers concerning the political economy of science and technology are Schumpeter (1937) and Marx (1887). Marx provides useful insights into the means of production, as well as the social relations of production. The conflict between means and relations serves as a major source of societal disruption. One can find Marx's insights into technology valuable even without endorsing his prescriptions for change. Schumpeter provides a vivid account of how waves of innovation periodically

TABLE 5.5 Levels of Institutional Analysis

Time Frames	Unit of Analysis	Relevant Knowledge
100–1000 years	Civilizations, states, nations	Social theory
10–100 years	Bureaucracies and polities	Political theory
1–10 years	Networks, markets, alliances	Institutional economics
Continuous	Product sales and placement	Microeconomics, marketing

Source: Adapted from Williamson (2000).

TABLE 5.6 Sources for Cultural Analyses

Books	Online
Social theory (Marx 1887; Schumpeter 1937; Schumpeter 1942)	Social technical theories (Freeman 1984; Perez 1985)
Systems views of history (Toynbee 1987; Tainter 1990; Diamond 2005)	Evidence of waves of innovation (Marchetti 1980)
Empirical writings on culture (Hofstede 2001)	Macrotechnology data (U.S. Energy Information Administration 2010)
Writers with the long view (Friedman 2010)	Social critique and commentary on foreign affairs (Council on Foreign Relationships 2010)

reshape society. His work has been extremely influential in much of the thought that follows. While it is always helpful to read the original sources on the matter, Rossini and Porter (1982) offer a pointed application of political economic theories for social and institutional forecasting.

Freeman (1984) and Perez (1985) postulated five waves of inventive activity since the dawn of the industrial era. Opinions vary as to whether these waves are structural or simply a post hoc description of what has happened. Marchetti (1980) argues that they are regular, even predictable. Using long-term historical statistics of infrastructural changes, for instance in the energy and transport sector, Marchetti argues that society should be prepared for future waves of change. Sites such as that of the U.S. Energy Information Administration (U.S. Energy Information Administration 2010) provide useful information at the national as well as international levels that can give advice about potential changes at level two.

Diamond (2005) offers the widest sweep of history, linking the fate of societies to their biogeographical endowment. In another masterful work about societal change, Toynbee (1987) embraces an amazing sweep of history in categorizing the rise and fall of civilizations. He provides a systematic view of history while denying the historical necessity of societal collapse. Tainter (1990), although surveying a more select sweep of history, provides a more detailed discussion of how overinvestment in complexity ultimately undermines civilizations, exposing them to external shocks that they can no longer manage. He then offers some interesting insights for anyone studying technologies as societal artifacts.

Friedman (2010) writes his near-future forecasts with a strong awareness of the interplay of political forces. Lempert also discusses forecasting changes over the longer term for government policy analysis (Lempert, Popper, et al. 2003). The journal *Foreign Affairs* (Council on Foreign Relationships 2011) provides a variety of commentary at the intersection of the news and academic scholarship. Hofstede (2001) describes how to empirically measure and verify meaningful differences in cultures. His results can be used to categorize culture differences between nations and organizations.

TABLE 5.7 Best of the Best in Political Economy of Technology

Books	Online
Political theories (Bueno de Mesquita 2009) Social and political change (Naisbitt 1982; Dunnigan and Bay 1985; Naisbitt and Aburdene 1989; Howe and Strauss 1992) Empirical evidence of democratic differences (Lijphart 1999)	U.S. politics (*Congressional Quarterly* 2011) International politics and economic affairs (*Economist* 2010)

Level Two. Time frames at this level are from 10 to 100 years. Level two involves the formal rules of the game, particularly those concerning taxation and property. Technology forecasters should be aware of how changes in governance can enhance or deny the possibilities for technological change. Legislative forecasting occurs at this level and is an important concern for companies trying to anticipate medium- to long-term futures. Political theories are useful here to give frameworks for organizing and modeling political change (see Table 5.7).

Level two involves the evolution of politics and policy. Bueno de Mesquita is an academic practitioner and consultant in the area of "positive" political theory. He creates explicit models of political change in countries to reveal mechanisms and offer recommendations for change. His book *The Predictioneer's Game* (Bueno de Mesquita 2009) is useful to anyone specifically interested in the cases that are discussed. The author reveals little of methodology, but the book nonetheless is a useful gateway to a more scholarly literature. Another highly recommended book is *Generations* (Howe and Strauss 1992). The authors describe how a repeated generational dynamic seems to drive social and political change in the United States. Citizens in other nations, such as the Netherlands and England, also seem to identify with this generational dynamic.

Naisbitt's *Megatrends* (1982), although dated, is interesting because of the way the author gathers short-term news and market trends and organizes them into a framework encompassing a wider span of history. The book has been revised and updated for the year 2000 (Naisbitt and Aburdene 1989). Dunnigan and Bay (1985) offer *A Quick and Dirty Guide to War*. The book gives a thorough overview of the forces leading to war, as well as other events of violent social change including insurrection and political upheaval. This book has aged well, and the authors have proven remarkably foresighted. Another use for the book is to provide methodological insight into simulation games—the primary analytical procedures utilized in this book.

Science fiction also can provide insight into emerging social changes. In retrospect, *Earth* (Brin 1990) has shown remarkable foresight. Some have credited Brin with recognizing the future impacts of the World Wide Web, micro-blogging, and cloud computing. Finally, Neal Stephenson's (1996) *The Diamond Age* has been very influential in imagining a future where nanotechnology has reshaped society.

Many online and news sources also are available at this level. The journal *Futures* offers a useful mix of case study work and methodological discussion often emphasizing the social aspects of technological change (*Futures* 2010). For the United States, the *Congressional Quarterly* provides both free and paid news services that track bills and other political issues in Congress (*Congressional Quarterly* 2011). The *Economist* provides coverage of U.S., British, and international political issues (*Economist* 2011). It also provides excellent coverage of economic events and sharp, if limited, coverage of technology. Other sources include *Business Week*, the *New York Times, Time,* the *Wall Street Journal,* and the *Washington Post* (*Business Week* 2010; *New York Times* 2010; *Time* 2011; *Wall Street Journal* 2010; *Washington Post* 2010).

Level Three. This level involves new product development and announcements and the formation of strategic alliances with other organizations. Governments must be concerned with the creation of new industries and the delivery of new research discoveries from the laboratory to the market at this level. Transaction cost economics are useful here, as these approaches help consideration about the development of new markets and the creation and maintenance of strategic networks (see Table 5.8).

If you want to consult only one book, it probably should be Michael Porter's (1980) *Competitive Strategy*. Some academic reviewers consider the book incomplete or incorrect. Nevertheless, it undoubtedly sets the agenda for research in technology strategy. More importantly, it seems to guide actual decision making in companies. Online resources include the *Strategic Management Journal*, which emphasizes private sector questions, and *Research Policy*, which often emphasizes more public sector questions. The database *ABI Inform* is very useful for tracking strategic alliances, market entry, and new product announcements.

There is a vast literature on the relationship between technology and the firm. Teece, Pisano, et al. (1997) and Henderson and Clark (1990b) offer an excellent entry point. They remind us that the pursuit of technological opportunity (and not mere market opportunism) provides the greatest value to technology-based firms. Henderson and Clark (1990b) discuss the coevolution of technologies and the firm, as well as the reconfiguration of technologies out of existing components.

TABLE 5.8 Best of the Best in Strategic Management

Books	Online
Technology strategy (Porter 2008)	Articles (Henderson and Clark 1990a; Teece, Pisano, et al. 1997) Journals: *Strategic Management Journal* (John Wiley & Sons 2010); *Research Policy*(Elsevier 2010b) Databases: ABI Inform (ProQuest LLC 2010)

TABLE 5.9 Best of the Best in Technology Trade and Marketing

Books	Online
International trade (Krugman and Obstfeld 2008)	Databases: Forrester Research (Forrester 2010); Gartner Dataquest (Gartner 2010)
Microeconomics (Mankiw 2006)	Encyclopedias: Wikipedia (Wikipedia 2010)
	Search engines: Google Trends (Google 2010b)
	Technology news: Slashdot (Slashdot 2011)

Level Four. This level involves the immediate and continuous process of balancing technology supply and demand (see Table 5.9). This is the realm of microeconomics and marketing. Forecasters need to know how many units of a technology have been sold and the potential for new technologies to penetrate the marketplace. There are many choices in this area. Two excellent ones are Krugman and Obstfeld (2008) for international trade and Mankiw (2006) for microeconomics.

A range of online resources for evaluating product sales and diffusion are available; unfortunately, they are all priced at a premium. The technology forecaster could begin with Forrester Research or Gartner Dataquest. Affordable alternatives are open source encyclopedias like Wikipedia. While these may not provide much information about product sales, open source communities contain a lot of latent information about technological configuration. See Cunningham (2009) for an example of this approach. Moreover, the future of detailed numerical data online may be changing, and efforts to create the semantic web may provide new tools for users. Those seeking to monitor the almost daily changes in technology trends and culture could benefit by monitoring technology blogs such as Slashdot (2011).

Amanatidou, Cachia, et al. (2008) suggest the World Wide Web as a networking medium that can be used to obtain, not just disseminate, information. Web 2.0 points future Internet users toward the use of websites as interactive mechanisms for engaging a wider range of knowledgeable minds. Online social networks can contribute by providing evidence, expertise, creativity, and collective intelligence.

5.3 STRUCTURING THE SEARCH

Gathering forecasting data is an iterative process. At each step, a little more about the topic under investigation is uncovered. Sometimes the right approach is to press ahead, complete the original search for information, and then consider a follow-up. At other times, it is best to restart the process with a new query that more accurately captures the topic. Framing the search query is very important. An experienced technical librarian suggested the following general process:

- Describe the information you seek in general terms (e.g., list the subject areas).

- Nominate terms (words or phrases) that seem to capture that subject information.
- Translate the terms into search logic (i.e., Boolean phrasing).
- Determine the types of sources desired (e.g., patents applied for or granted; journal articles or conference presentations; books or popular articles).
- Consider which sources to search (which databases, websites, etc.).
- Try a small-scale search (e.g., the most recent year or so); assess the results; and refine.

The search for data can be based on a variety of different strategies including

- Substantive terms
- Names (people, institutions, regions, or countries)
- Indices or classifications
- Citations or hyperlinks

Consider the trade-offs between broad and narrow searches (Table 5.10). The choice depends on your topic and target tech mining uses. If an aim is to spot unusual, nonmainstream R&D, you want to capture items with the barest threads of association to your topic—reach out broadly. Conversely, the more you know about what you're looking for—say, what research group X is publishing on subtopic Y—the narrower the query can be.

A consideration in framing the query is the choice of *natural* versus *scientific* language. Natural languages are spoken by scientists and laypeople alike. They are rich, expressive, and deeply and often deliberately ambiguous. Computer scientists contrast natural with *machine* languages—languages constructed for use by computers. Machine language is precise, structured, and unambiguous, but rather limited in its expressive power. Scientific language exists because scientists and engineers desire precision. To express scientific or technical ideas in an exclusive way, scientists either appropriate words or create entirely new words (often rooted in Latin or Greek). These new scientific words occur within a specific theoretical (and often disciplinary) framework.

Many tech mining studies concern a specific person, institution, region, or country. For these studies, it is appropriate to form a query that searches fields

TABLE 5.10 Breadth of Queries

Nature of the Query	Consequences
Broad	Broad queries capture many articles across multiple disciplines or fields Scientific progress in the form of transfer of ideas is best captured in these queries
Narrow	Narrow queries capture fewer articles of a more specific nature Scientific progress in the form of research concentrations is best captured in these queries

such as the "Author" or "Institution" for names. Such studies can be particularly important when the goal is to obtain competitive technological intelligence—for instance, if General Motors wants to discover the R&D emphases of Ford or Toyota. These searches also can contribute to monitoring, foresight, and technology management.

Another searching option could be to take advantage of the database's categorization. MEDLINE's "MeSH" index is a prominent example of a multilayer, hierarchical index. Its structure helps associate variations in categorization; for instance, "asthma" nests under lung diseases. Use of indexes is particularly critical in patent searches. Searches might make use of:

- Classification codes (e.g., INSPEC codes such as "A4255N" for fiber lasers and amplifiers)
- Database-controlled terms (keywords; e.g., EI Compendex indexes articles using "Fuel cells" or "Fuel cells—electrolytes")
- "CODENs" (designators of specific journals; likewise, Conference Codes in EI Compendex)
- Document type (e.g., restrict to journal articles)

Patent search nuances include the importance of combining index-based searching (e.g., using patent classes) with term searching.

Many different professionals use data searches. Scientists investigate the literature and reference-related research as part of the discovery process. Likewise, inventors (or patent attorneys) cite other patents to delineate their intellectual property (IP) from prior work. Patents also increasingly cite scientific work. The elaborate network of citations among papers and/or patents in an R&D domain can sometimes be mapped to ascertain intellectual and social ties.

You may want to examine the body of literature that references a specific scientific paper, author, or institution. Scientific authors exist in a competitive marketplace where the reputation of ideas and research is at stake. Most papers are rarely read (the most common number of citations of a paper is zero). However, a few papers and authors in any specialty are cited repeatedly. Papers that are already cited are easier to find and more attractive to scientists looking for key references. As a result, "the rich get richer."

Tech miners can exploit this *reputational market*. For instance, Porter and Cunningham (1995) examined nanotechnology by the use of citations. The authors probed who did and did not cite Eric Drexler. Drexler has done a great deal to introduce the concept of nanotechnology to a wider audience; it is interesting, therefore, to study nanotechnology as it is understood by specialists as well as a wider audience.

How do you know if your query is working well? There are five approaches:

1. Use multiple redundant search terms and search sources and compare the results.
2. Use indicators to determine how well your query is faring.

3. Read a small fraction of the articles yourself.
4. Ask an expert to review the results.
5. Utilize "queries by example."

What happens when the query doesn't meet your needs? Try cleaning and augmenting existing data before deciding to download entirely new data sets. Sometimes compounding several discrete queries can help to filter results so that they more closely meet your needs. Consider reframing the query. Tech mining is an iterative process.

5.4 PREPARING SEARCH RESULTS

Many of the applications of search results use quantitative measures. Indeed, even when the information is qualitative, some scale or rating scheme often needs to be applied. For instance, names or a nominal scale might be attached to expert data. Or decision makers might be asked to represent their preferences from most to least preferred—this is an ordinal scale. If there are established units of measurement (e.g., numbers of publications or patents), an interval scale that measures technological development might be constructed. Finally, you might be able to assign a ratio scale to the measurement—for instance, by asking how profitable a new technology will be.

In what would become one of the most famous papers in the annals of social science, Stevens (1946) reported the conclusions of a committee tasked to provide "quantitative estimates of sensory events." He concluded that the problem of measurement involves finding rules for assigning quantitative indicators and considering their mathematical properties. The classic alternatives for measurement scales, as developed by the committee, are shown in Table 5.11.

Stevens recommended four basic measurement scales, increasing in detail. For Stevens, the more permissible the operations that can be conducted on a measurement without loss of meaning, the finer the degree of measurement. He further argued that the scale of measurement dictates the kinds of modeling techniques to which the data can be usefully applied.

Subsequent writers have been less confident that the appropriate analysis technique is determined by the available scale of measurement. Ultimately, measurement turns on the question of meaning. What does the scale mean? How will it be used in decision making? Analysts must consider the meaning of what they

TABLE 5.11 Units of Measurement

Scale	Nature of Operations	Statistical Equivalent
Nominal	Determination of equality	Number of cases
Ordinal	Determination of greater than or less than	Percentiles
Interval	Determination of equality of intervals	Means Standard deviations
Ratio	Determination of equality of ratios	Correlations

are trying to convey. Overly simple rules for selecting analysis techniques will not suffice (Velleman and Wilkinson 1993).

5.5 USING SEARCH RESULTS

Results of increasing complexity can be derived from search results. These include:

- Lists—the "top 10" organizations patenting the technology
- Matrices—the distribution of technology patents on the technology by year
- Maps—social network of researchers within an organization
- Profiles—top researchers to provide selected fields of information

Depending on the type of search information retrieved and the forecast focus and intent, many different uses can be made of the results. For instance, a rich network of relationships found in a single scientific article may provide the names of the authors, the year of publication, the journal, a content classification and keywords, actual content, institutional names and addresses, and associated nations. Similarly complex relations can be extracted from patents or web pages (Figure 5.3). Experts can provide complex chains of data as well.

These networks can be used to construct elaborate chains and relationships. For instance, one could try to:

- Determine which nations are pursuing which fields of knowledge (e.g., Youtie, Shapira, et al. 2008)

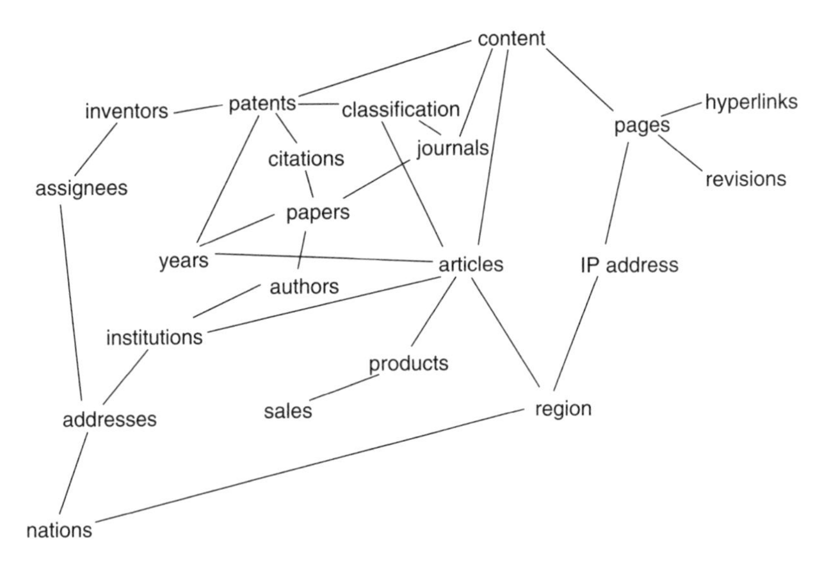

Figure 5.3. Relationships in the Data

- Identify authoritative web pages based on hyperlink patterns—this is what Google does
- Identify high-quality science through citation patterns (Porter and Youtie 2009)
- Evaluate the interdisciplinary character of teams and individuals (Porter, Roessner, et al. 2008).

Many of the forecasting methods overviewed in Chapter 2 depend on quantifiable data found during database searches. The validity of these methods strongly depends on the quality of available data. Techniques vary from very simple to highly complicated statistical models that require powerful computers to apply. A sophisticated method, however, does not ensure a valid forecast. Changing conditions and data limitations are two reasons why any forecaster should apply multiple forecasting tools.

Trend extrapolation uses the past to anticipate the future. It applies mathematical and statistical techniques to extend quantitative time series data into the future. Trend extrapolation results are of use only if the future proves to be like the past, at least in some important respects.

Observations of natural phenomena can lead to the recognition of scientific laws that describe growth processes. Humans develop patterns of behavior based upon their experiences, and these sometimes allow social scientists to anticipate events in a manner analogous to the predictions of physical scientists. Although social relationships are more complex and uncertain than physical ones, the past still is the richest source of information about the future.

Events such as scientific breakthroughs, political upheavals, economic recessions, or natural disasters can affect the reliability of the past as a guide to forecasting the future. Such factors imply that the forecaster should avoid making single point predictions, provide confidence intervals, and perform an explicit sensitivity analysis of the results.

A critical assumption for the use of growth patterns and trend extrapolation in technology forecasting is that technical attributes advance in a relatively orderly and predictable manner. This is no trivial assumption; exceptions abound. Indeed, discrete events often can cause discontinuities in orderly growth, as when development of the U.S. supersonic transport was derailed by the government's decision to cease major subsidies. Fortunately, the complex mix of influences that operate on technology growth tends to moderate discontinuities. Like the rationale statisticians use to test observations against a normal distribution, the cumulation of many small contributions results in a predictable pattern of deviations about a central value. For technological change, this condition is best met when progress reflects a series of ongoing engineering improvements that do not require scientific breakthroughs or major inventions.

Regression among two or more variables can begin from various standpoints. In the strongest case, a solid theoretical basis is reflected in the regression equation. In the weakest, a database is explored to discover correlation among certain variables. In this case, it should not be assumed that causation has been "proven" in any sense. Correlation does not imply causation. The forecaster must assess the causal strength of any regression model before using it to forecast.

5.6 DEVELOPING SCIENCE, TECHNOLOGY, AND SOCIAL INDICATORS

Indicators are aggregate measures of various phenomena. Collectively, they indicate the state of a system. Indicators can address both technology and societal factors. Each is considered in the sections that follow.

5.6.1 Science and Technology Indicators

There are restrictions on the choice of variables to be used as technology indicators. First, the variable must measure the level of functionality of the technology. Thus, an understanding of the technology and its application is required for choosing the correct variable. Second, the variable must be applicable both to the new technology and to any older technology it replaces. Third, and often most limiting, sufficient data must be available to compute historical values.

If the ideal indicator is not available or is less complete than alternative measures, a compromise may be necessary. For instance, in devising indicators of high-technology development among nations, Roessner and Porter (1990) used technology sales as surrogate production measures.

The tech mining framework posits a list of some 200 "innovation indicators" (Porter and Cunningham 2005). The list offers ideas about how to draw on science and technology information resources to help understand particular factors, such as the effect of a rising trend in research activity. These innovation indicators can be used to derive knowledge from available data in less obvious ways that are attuned to sensitivities in the innovation process.

Innovation indicators also can be used to address the management of technology issues (Table 4.2) and questions that drive monitoring (Table 4.4). In this regard, there are three general indicator categories:

- *Technology Life Cycle Indicators:* Locating a technology's progression along a growth curve and possibly projecting likely future development
- *Innovation Context Indicators:* Tallying the presence or absence of particular success factors, such as R&D funding, suitable platforms, standards, and the fit with existing applications
- *Market Indicators:* Addressing prospects for successful innovation

Two examples of the use of technology indicators are presented in Exhibit 5.1. The first involves the analysis of ceramic engines. The technology forecaster in this example focused on time, sources of publication, and types and numbers of keywords. A simple bar chart helped communicate the findings to decision makers, demonstrating the trends and maturities in an emerging engine technology.

Exhibit 5.1 Ceramic Coating of Engine Parts

In the mid-1990s, Watts did a technology opportunities analysis for the U.S. Army's Tank-Automotive Research, Development & Engineering Center (TARDEC). He began to explore the potential for innovative applications of ceramics in automotive engines (especially tank engines). A review of local history revealed that TARDEC had invested in ceramics R&D in the 1980s without achieving a significant return and subsequently discontinued funding. Nonetheless, Watts saw promise and pursued his investigation.

Figure 5.4 provided the key innovation indicator that pointed to an important acceleration in the maturation of this technology. The message in the figure has several parts. The back row shows the "Number of Publications." In the 1987–1988 time slice, this number reached 200 or more and then crashed in 1989–1990. During the most recent period (1993–1995), the number began to recover slightly. The second row, "Number of Distinct Sources," shows a similar pattern. But the front row, "Number of Keywords" tells a markedly different story. It indicates that the richness of R&D discourse about ceramic engine applications had sharply accelerated recently. That was the signal for action.

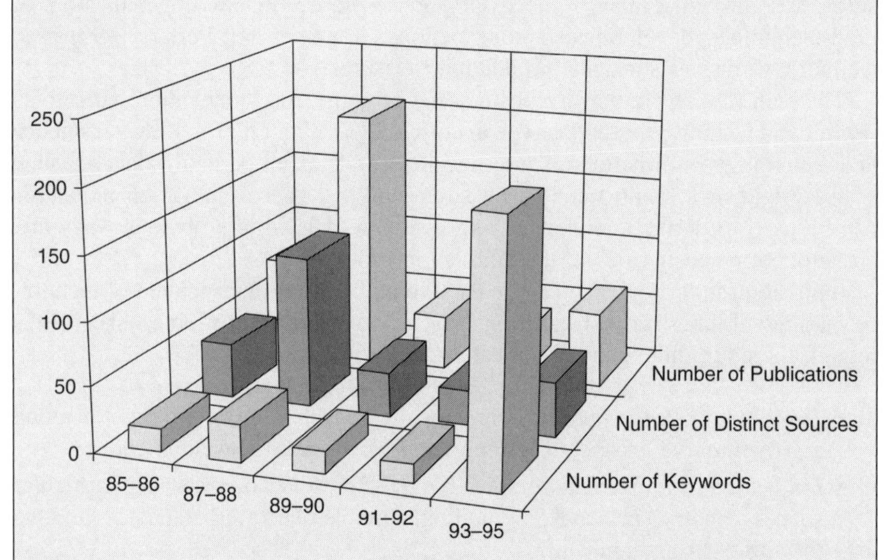

Figure 5.4. Keyword Behavior Change as an Innovation Indicator
Source: Watts and Porter (1997)

Ceramics experts confirmed that the research arena had gotten much more specialized. In the 1980s there was much general "hand-waving" about the prospects for ceramic applications. Key terms included "materials," "alloys," and "automotive engineering." More recently, key terms implied that serious research was underway on topics such as "silicon nitride," "braided ceramic fiber seals," "microstructure," "fatigue testing," "math models," and "thin films." This proved compelling to TARDEC senior management, and Watts pushed on.

The second forecasting phase found that prospects for thin film ceramic coating applications in the automotive context were especially promising. Surprisingly, the research leaders were not from the structural ceramics community. Instead, the semiconductors sector was doing the cutting-edge R&D on thin ceramic films for chip design and development. Technology forecasting work was able to span these traditional bounds by searching in Engineering Index (EI Compendex—a major database for engineering and selected science areas). Watts identified particularly exciting R&D activities at a national lab and a company. Those researchers had not considered applying their coating technology in the engine environment!

To wrap up the story, TARDEC made contact with those research organizations and funded two large projects to explore coating applications. In 2004, a major production plant opened to coat used Abrams tank turbine blades to extend their life. The ceramic coating offered additional advantages in terms of higher operating temperatures and reduced air pollution. A success!

The second example is drawn from the field of nanotechnology. It uses geographical data, coupled with publication measures, to show the concentrated nature of research. The results have particular relevance for those assembling world industrial regions (see Figure 5.5).

Source: Based on Watts and Porter (1997); Porter and Newman (2011).

Figure 5.5. Nanotechnology Districts

5.6.2 Social Indicators

Social indicators collectively represent the state of society or some subset of it, such as a community. They can be used to measure economic, demographic, educational, welfare, employment, health, and criminal justice factors, as well as other factors important to sociopolitical description and forecasting. Social indicators are analogous to measures of the functional capacity of a technology. They possess three important characteristics: they are numerical and thus can be used with quantitative methods; they lend themselves to geographic disaggregation and analysis; and they are widely collected (Ferriss 1988).

A 1997 study in *Nature* demonstrated the use of expert opinion, simulation, and trend extrapolation to provide probabilistic forecasts of world population (Lutz, Sanderson, et al. 1997). Other studies have examined the causal links between wage inequality and violent crime (Fowles and Merva 1996) and the relationship between lead exposure and a range of socially undesirable outcomes (Nevin 2000).

Since indicators are frequently projected singly, the forecaster must assess their dependence on context. As always, when using forecasts produced by regression analysis, the forecaster must assess the continued relevance of the independent variables or indicators employed, and the continued irrelevance of those omitted, over the time horizon of the forecast.

Social indicators are collected by a wide variety of organizations (Ferriss 1988). Data can sometimes be disaggregated by taking successively smaller geographic areas: nation, region, state, district, county or even census tract, and block. The forecaster should ensure that social indicators used as surrogate measures are used validly. In this regard, surrogate social indicators are no different than surrogate technological or economic measures. Forecasts based solely on social indicator trends do not provide as rich or as integrated a picture of a changing societal context as scenarios, but they can add credibility to scenarios.

Richard Florida's book *The Rise of the Creative Class* provides a powerful example of what can be done with social indicators (Florida 2002). Rigorously analyzing extensive data sources, Florida was able to draw conclusions about why some areas have prospered in the knowledge economy, while others have declined in wealth and even population. His work shows that social characteristics such as diversity and tolerance lead to more creative development, and that this environment attracts the best minds and leads to prosperity producing innovation.

5.7 COMMUNICATING SEARCH RESULTS

Decision makers may well have concerns about what data were used and how analyses were done so that they can assess the validity of monitoring results. To address these concerns, explicit explanations of the questions posed in Table 5.12 should be given in the report.

Different users prefer different information presentation modes. Most people do best with visual representations, but others favor numeric and tabular presentations. Still others want to read or be presented with detailed explanations.

TABLE 5.12 Informing the Customer

Issue	Considerations
Bounding and Assumptions	How was the tech mining effort framed?
Information Resources Used	Which databases or other Internet resources were searched? Which were not searched and why? Were there coverage limitations of note?
Searching Algorithms Used	How well targeted are the resulting data and subsequent analyses? Can you estimate what was excluded?
Data Cleaning	How dirty are the data?
Analytics	What analytical approaches were used and why?

Know your audience's preferences. In general, providing multiple representations works best; it allows users to key on their favored modality.

Users get used to familiar information forms. Established business decision processes, such as stage gate decision making, can reinforce this tendency by calling for standard forms. Many organizations use stage gate processes, wherein a development progresses through a series of decisions, each requiring that certain information be provided in a structured manner. For instance, at stage gate zero, one expects a column trend chart that compares the company's patenting in the target domain to that of two key competitors. Technology forecasting can support standardization through the use of macros that generate particular charts and graphs in a repeatable way.

How one delivers the information can and should vary, depending on the audience. Reports usually should be provided at multiple levels: a succinct statement about the analysis and its punch line; a brief summary highlighting key findings for various users (possibly with different levels of technical familiarity); the main report; and appendices (documentation and analytical details). Visual presentations, when combined with oral briefings, are attractive. The Internet can provide wide access within an organization as well as archival benefits. Such Web-based sharing can be augmented by opportunities for comment, feedback, and requests for elaboration.

5.8 CONCLUSIONS

This chapter discussed gathering data from experts and various techniques for doing so. Special attention also was devoted to the collection of information from the Internet. The need to evaluate both technical and social sources of information in order to assemble a complete picture of possible future developments was emphasized. Thus, both technical and societal database sources were considered, and the "best of the best" were nominated in each area.

The search strategy and framing of the search query were considered, and preparing, using, and communicating the results were discussed. Exploiting the information obtained requires the careful design of indicators, scales of measurement, and the fusion of various kinds of data. The concepts and construction of science, technology, and social indicators were introduced. The next chapter advances the discussion further by closely examining the role of analytical models and trend extrapolation in technology forecasting.

REFERENCES

Amanatidou, E., R. Cachia, et al. (2008). "Web 2.0 Can Support the Way of Doing Foresight." *Conference on Future-Oriented Technology Analysis.* Seville, Spain: Institute for Prospective Technological Studies: 159–160.

Brand, S., Herron, M. (1985). "1984 Ad," *Whole Earth Review* **46**: 49.

Brin, D. (1990). *Earth*. New York, Bantam Books.

Bryson, J. M. (2004). "What to Do When Stakeholders Matter." *Public Management Review* **6**(1): 21–53.

Bueno de Mesquita, B. (2009). *The Predictioneer's Game: Using the Logic of Brazen Self-Interest to See and Shape the Future*. New York, Random House.

Business Week. (2010). From http://www.businessweek.com/.

Congressional Quarterly. (2011). "Roll Call." Retrieved 12 January 2011 from http://www.rollcall.com/politics/index.html?cqp=1.

Council on Foreign Relationships. (2011). "Foreign Affairs." Retrieved 12 January 2011 from http://www.foreignaffairs.com/.

Cuhl, K. (2004). "Future—Foresight for Priority-Setting in Germany." *International Journal of Foresight and Innovation Policy* **1**(3–4).

Cunningham, S. W. (2009). "Analysis for Radical Design." *Technological Forecasting and Social Change* **76**(19): 1138–1149.

Delbecq, A. L. and A. H. Van de Ven. (1971). "A Group Process Model for Problem Identification and Program Planning." *Journal of Applied Behavioral Science* **7**(4): 466–498.

Diamond, J. (2005). *Guns, Germs and Steel: The Fates of Human Societies*. New York, W. W. Norton.

Dunnigan, J. F. and A. Bay. (1985). *A Quick and Dirty Guide to War: Briefing on Present and Potential Wars.* New York, Quill/William Morrow.

Economist. (2010). "The Economist." Retrieved 19 May 2010 from http://www.economist.com/.

Elsevier. (2010a). "Engineering Village." Retrieved 18 May 2010 from http://www.engineeringvillage2.com/controller/servlet/Controller.

Elsevier. (2010b). "Research Policy." Retrieved 19 May 2010 from http://www.elsevier.com/locate/respol/. http://www.scopus.com/home.url.

Elsevier. (2010c). "Scopus." Retrieved 18 May 2010 from http://www.scopus.com/home.url.

European Patent Office. (2010). "European Patent Office." Retrieved 19 May from http://www.epo.org/.

Ferriss, A. L. (1988). "The Uses of Social Indicators." *Social Forces* **66**(3): 601–617.

Florida, R. (2002). *The Rise of the Creative Class: and How It Is Transforming Work, Leisure, Community and Everyday Life*. New York, Basic Books.

Forrester. (2010). "Making Leaders Succesful Every Day." Retrieved 19 May 2010 from http://www.forrester.com/.

Fowler, F. J., Jr. (2008). *Survey Research Methods*. Thousand Oaks, CA, Sage Publications.

Fowles, R. and M. Merva. (1996). "Wage Inequality and Criminal Activity: An Extreme Bounds Analysis for the United States, 1975–1990." *Criminology* **34**(2): 163–182.

Freeman, C., ed. (1984). *Long Waves in the World Economy*. London, Frances Pinter.

Friedman, G. (2010). *The Next 100 Years: A Forecast for the 21st Century*. New York, Anchor Books.

Futures. (2010). From http://www.elsevier.com/wps/find/journaldescription.cws_home/30422/description#description.

Gartner. (2010). "Gartner Research." Retrieved 19 May 2010 from http://www.gartner.com/technology/home.jsp.

Genus, A. and A. Coles. (2005). "On Constructive Technology Assessment and Limitations on Public Participation in Technology Assessment." *Technology Analysis and Strategic Management* **17**(4): 433–443.

Glenn, J. C. and T. J. Gordon. (2010). "Futures Research Methodology Version 3.0." Retrieved 16 May 2010 from http://www.millennium-project.org/millennium/FRM-v3.html.

Google. (2010a). "Google Scholar." Retrieved 18 May 2010 from http://scholar.google.com/.

Google. (2010b). "Google Trends." Retrieved 19 May 2010 from http://www.google.com/trends.

Gustafson, D. H., R. K. Shukla, et al. (1973). "A Comparative Study of Differences in Subjective Likelihood Estimates Made by Individuals, Interacting Groups, Delphi Groups, and Nominal Groups." *Organizational Behavior and Human Performance* **9**: 280–291.

Henderson, R. M. and K. B. Clark. (1990a). "Architectural Innovation: The Reconfiguration of Existing Product Technologies and the Failure of Established Firms." *Administrative Science Quarterly* **35**: 9–30.

Henderson, R. M. and K. B. Clark. (1990b). "Architectural Innovation: The Reconfiguration of Existing Product Technologies and the Failure of Established Firms." *Administrative Science Quarterly* **35**: 9–30.

Hofstede, G. (2001). *Culture's Consequences: Comparing Values, Behaviors, Institutions, and Organizations across Nations*. Thousand Oaks, CA, Sage Publications.

Howe, N. and W. Strauss. (1992). *Generations: The History of America's Future, 1584–2069*. New York, Harper Perennial.

IFI CLAIMS. (2011). "IFI CLAIMS Patent Services: The Patent Research Experts." Retrieved 12 January 2011 from http://www.ificlaims.com/.

Japanese Patent Office. (2010). "Japan Patent Office." Retrieved 19 May 2010 from http://www.jpo.go.jp/.

John Wiley & Sons. (2010). "Strategic Management Journal." Retrieved 19 May 2010 from http://www3.interscience.wiley.com/journal/2144/home.

Krippendorf, K. (2004). *Content Analysis: An Introduction to Its Methodology*. Thousand Oaks, CA, Sage.

Krugman, P. and M. Obstfeld. (2006). *International Economics: Theory and Policy*. New York, Addison Wesley.

Krugman, P. R. and M. Obstfeld. (2008). *International Economics: Theory and Policy*. Upper Saddle River, NJ, Pearson Education Limited.

Lempert, R. J., S. W. Popper, et al. (2003). "Shaping the Next One Hundred Years: New Methods for Quantitative, Long-Term Policy Analysis." *Monograph MR-1626-RPC*. Retrieved 19 May 2010 from http://www.rand.org/pubs/monograph_reports/MR1626/.

Lijphart, A. (1999). *Patterns of Democracy: Government Forms and Performance in Thirty-Six Countries*. New Haven, CT, Yale University Press.

Linstone, H. A. and M. Turoff. (1975). *The Delphi Method: Techniques and Applications*. Reading, MA, Addison-Wesley.

Lipinksi, A. and D. Loveridge. (1982). "Institute for the Future's Study of the UK, 1978–1995." *Futures* **14**: 205–239.

Lutz, W., W. Sanderson, et al. (1997). "Doubling of World Population Unlikely." *Nature* **387**(6635): 803–805.

Mankiw, N. G. (2006). *Principles of Microeconomics,* 4th ed., Mason, OH: Thomson South-Western.

Marchetti, C. (1980). "Society as a Learning System: Discovery, Invention, and Innovation Cycles Revisited." *Technological Forecasting and Social Change* **18**: 267–282.

Marx, K. (1887). *Capital: A Critique of Political Economy*. Moscow, Progress Publishers.

Naisbitt, J. (1982). *Megatrends: Ten New Directions Shaping Our Lives*. New York, Grand Central Publishing.

Naisbitt, J. and P. Aburdene. (1989). *Megatrends 2000*. New York, Morrow.

National Citizens Forum on Nanotechnology. (2011). "Center for Nanotechnology in Society," Retrieved 12 January 2011 from http://cns.asu.edu/cns-library/highlights.

Nelms, K. R. and A. L. Porter. (1985). "EFTE: An Interactive Delphi Methods." *Technological Forecasting and Social Change* **28**: 43–61.

Nevin, R. (2000). "How Lead Exposure Relates to Temporal Changes in IQ, Violent Crime, and Unwed Pregnancy." *Environmental Research* **83**(1): 1–22.

New York Times. (2010). From http://www.nytimes.com/.

Pantros IP. (2011). "PatentCafe." Retrieved 12 January 2011, from http://www.patentcafe.com/.

Penn State. (2010). "CiteSeer: Scientific Literature Digital Library and Search Engine." Retrieved 18 May 2010 from http://citeseerx.ist.psu.edu/.

Perez, C. (1985). "Microelectronics, Long Waves and World Structural Change: New Perspectives of Developing Countries." *World Development* **13**: 441–463.

Porter, A. L. and S. W. Cunningham. (1995). "Whither Nanotechnology? A Bibliometric Study." *Foresight Update* **21**(4): 12–15.

Porter, A. L. and S. W. Cunningham. (2005). *Tech Mining: Exploiting New Technologies for Competitive Advantage*. Hoboken, NJ, John Wiley & Sons.

Porter, A.L., and Newman, N.C. (2011), "Mining External R&D," *Technovation*, **31**: 171–176.

Porter, A. L., D. J. Roessner, et al. (2008). "How Interdisciplinary Is a Given Body of Research." *Research Evaluation* **17**(4): 273–282.

Porter, A. L. and J. Youtie. (2009). "Where Does Nanotechnology Belong in the Map of Science." *Nature Nanotechnology* **4**: 534–536.

Porter, M. (1980). *Competitive Strategy: Techniques for Analyzing Industries and Competitors*. New York, The Free Press.

Porter, M. (2008). *Competitive Strategy: Techniques for Analyzing Industries and Competitors*. New York, The Free Press.

ProQuest LLC. (2010). "ABI Inform." Retrieved 19 May 2010 from http://www.proquest.com/.

Questel. (2010). "Questel, Freedom to Operate." Retrieved 19 May 2010 from http://www.questel.com/.

Rea, L. M. and R. A. Parker. (2005). *Designing and Conducting Survey Research: A Comprehensive Guide*. San Francisco, Jossey-Bass.

Roessner, D. J. and A. L. Porter. (1990). "High Technology Capacity and Competition." In *Management of Technology*, ed. I. T. M. Khalil, B. A. Bayraktar, and J. A. Edosomwan. Geneva, Interscience Enterprises Ltd: 779–790.

Roper, A. T. (1988). "A Technique for the Early Stages of an Assessment." *Proceedings of the International Workshop on Impact Assessment for International Development*. Barbados, West Indies, International Association for Impact Assessment.

Rossini, F. A. and A. L. Porter. (1982). "Forecasting the Social and Institutional Context of Technological Developments." *Proceedings of the International Conference on Cybernetics and Society*: 486–490.

Sackman, H. (1974). *Delphi Assessment, Expert Opinion, Forecasting, and Group Process*. Santa Monica, CA, RAND.

Schumpeter, J. (1937). *Economic Development*. Boston, Harvard University Press.

Schumpeter, J. A. (1942). *Capitalism, Socialism and Democracy.* New York, Harper.

Slashdot. (2011). "News for Nerds, Stuff that Matters." Retrieved 12 January 2011 from http://slashdot.org/.

Stephensen, N. (1996). *The Diamond Age*. New York, Spectra.

Stevens, S. S. (1946). "On the Theory of Scales of Measurement." *Science* **103**(2684): 677–680.

STN International. (2010). "Derwent World Patents Index (DWPI)." Retrieved 19 May 2010 from http://www.stn-international.com/6449.html.

Tainter, J. (1990). *The Collapse of Complex Societies*. Cambridge, Cambridge University Press.

Teece, D., G. Pisano, et al. (1997). "Dynamic Capabilities and Strategic Management." *Journal of Strategic Management* **18**(7): 509–533.

The Economist. (2011). "World News, Politics, Economics, Business & Finance," Retrieved 12 January 2011 from http://www.economist.com/.

Thomson Reuters. (2011a). "Delphion." Retrieved 12 January 2011 from http://www.delphion.com/.

Thomson Reuters. (2011b). "MicroPatent." Retrieved 12 January 2011 from http://www.micropat.com/static/index.htm.

Thomson Reuters. (2011c). "Web of Knowledge." Retrieved 12 January 2011 from http://www.isiwebofknowledge.com/

Time (2011). "Breaking News, Analysis, Politics, Blogs, News Photos, Video, Tech Reviews," Retrieved 12 January 2011 from http://www.time.com/time/.

Toynbee, A. (1987). *A Study of History, Abridged*. Oxford, Oxford University Press.

U.S. Department of Commerce. (2011). "National Technical Information Service." Retrieved 12 January 2011 from http://www.ntis.gov/.

U.S. Energy Information Administration. (2010). "EIA U.S. Energy Information Administration: Independent Statistics and Analysis." Retrieved 19 May 2010 from http://www.eia.doe.gov/.

U.S. National Library of Medicine and National Institutes of Health. (2010). "PubMed." Retrieved 18 May 2010 from http://www.ncbi.nlm.nih.gov/pubmed/.

U.S. Patent and Trademark Office. (2011). "United States Patent and Trademark Office." Retrieved 12 January 2011 from http://www.uspto.gov/.

Velleman, P. F. and L. Wilkinson. (1993). "Ordinal, Interval, and Ratio Typologies are Misleading." *The American Statistician* **47**(1): 65–72.

Wall Street Journal. (2010). From http://europe.wsj.com/home-page.

Washington Post. (2010). From http://www.washingtonpost.com/.

Watts, R. J. and A. L. Porter. (1997). "Innovation Forecasting." *Technological Forecasting and Social Change* **56**: 25–47.

Wikipedia. (2010)."Wikipedia," Retrieved 19 May 2010 from http://www.wikipedia.org/.

Williamson, O. E. (2000). "The New Institutional Economics: Taking Stock, Looking Ahead." *Journal of Economic Literature* **38**(3): 595–613.

WIPS. (2010). "WIPSGLOBAL Version 4." Retrieved 19 May 2010 from http://www.wipsglobal.com/.

World Intellectual Property Organization. (2011). "WIPO: Encouraging Creativity and Innovation." Retrieved 12 January 2011 from http://www.wipo.int/portal/index.html.en.

Youtie, J., P. Shapira, et al. (2008). "Nanotechnology Publications and Citations by Leading Countries and Blocs." *Journal of Nanoparticle Research* **10**(6): 981–986.

6 ANALYZING PHASE

Chapter Summary: Modeling and simulation are major techniques for anticipating possible new technology futures. This chapter begins with an introduction to basic data analysis techniques such as regression and trend extrapolation. It then discusses the use of continuous and discrete simulation techniques in the context of technology forecasting. The chapter introduces the second major phase of forecasting, which entails analysis.

Recall that in Chapter 3, a three-phase approach to technology forecasting was presented. Material in this chapter deals with the second phase, analyzing. Analyzing narrows the range of subject matter, develops a deeper and more detailed insight, and makes significant decisions about the directions of the technology.

The future is uncertain, but forecast we must. This chapter carries forward the notion of "using the data" begun in Chapter 5. Here the emphasis is on extrapolating historical data into the future. Obtaining trend analysis information from Internet sources is described first. Then some quantitative methods for short-term forecasting are considered. Special emphasis is given to growth models. Next, simulation methods including quantitative and qualitative cross-impact and Monte Carlo techniques are described. The chapter concludes with a consideration of system dynamics. While many of the approaches are technically sophisticated, the forecaster must remember the mantra on every investment advisement: "Past results may not be indicative of future ones."

6.1 PERSPECTIVE ON DATA AND METHODS

By its very nature, extrapolation depends on past behavior. Thus, any method the forecaster can use must employ information about an established trend of the technology being forecast or about the behavior of similar technologies or analogous processes (e.g., market penetration). Thus, the acquisition of reliable data is central to all extrapolation. The following sections describe Internet opportunities for gathering the requisite data.

6.1.1 Overview and Caveats

As the saying goes, "Get a life." It's now quite possible to have a "Second Life" (Linden Research 2010). Through your avatar, you can take on a different persona and live another life, a "virtual life," on the Internet. Your avatar can travel, attend lectures, visit museums, and even have a virtual sex life! This isn't free. If you want your avatar to dress nicely, you have to pay real money. Businesses also can have a virtual life on the Internet. For example, a watch company could offer different designs to see which might sell best to avatars in Second Life.

The growth in the number of persons signing up for Second Life from January 2006 to March 2007 was phenomenal, as shown in Figure 6.1. But Figure 6.2, which presents the growth in premium accounts, those with a higher level of service and a higher cost, shows a very different behavior.

The point here is that you must look below the surface when using data sources to make forecasts. The number of committed users of Second Life is quite different from the number of people who merely sign up, perhaps because signups are free, while premium accounts are not. Moreover, extrapolating the early trend in Premium Account Growth does not produce a good forecast of the trend for the final months (Figure 6.2).

Technology forecasters must always remember that:

1. The future is not only unknown, it is unknowable. (If we could predict the future, we would all be rich.)
2. The further into the future one attempts to predict, the worse the forecast is likely to be. (The variance is additive.)
3. It is easier to forecast the general than the specific. (e.g., to forecast the number of pairs of shoes that will be sold than the number of men's brown loafers, size 10½, with tassel).
4. Extrapolation depends on continuity of the future with the past.

Forecasters must also be wary of making causal presumptions. For instance, while time is a natural choice of independent variable, rarely will time, per se, be the driver of the trend.

This said, it is necessary to forecast technology—that is, to make and assess time series projections, identify plausible development pathways, estimate their likelihoods, and think through the effects of the many potentially important interactions (e.g., how the 2007–2009 economic slump might alter trends).

The following sections introduce a range of methods to forecast changes in technology and its context. All begin with data—often historical time series—but go beyond them to incorporate human judgment. Discussions begin with some useful Internet data and trend resources. Then ways to extrapolate data by using

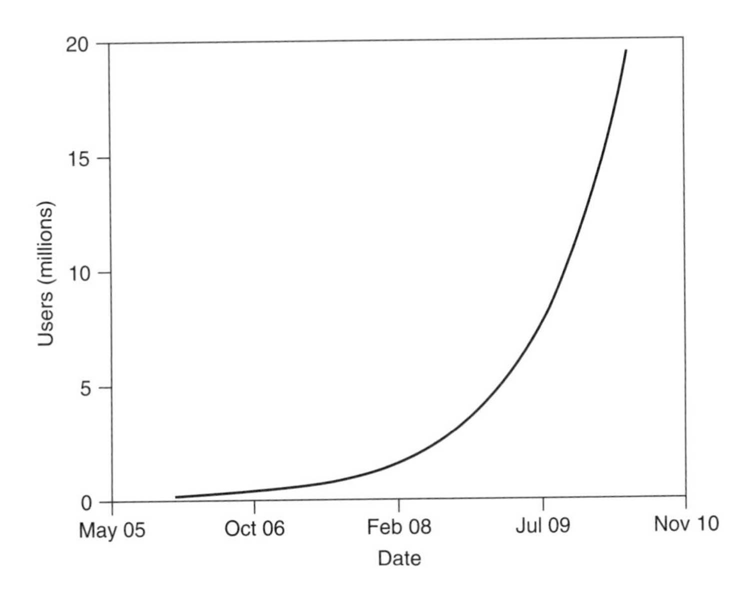

Figure 6.1. Total Sign-ups for Second Life
Source: Based on data from Linden Research (2010)

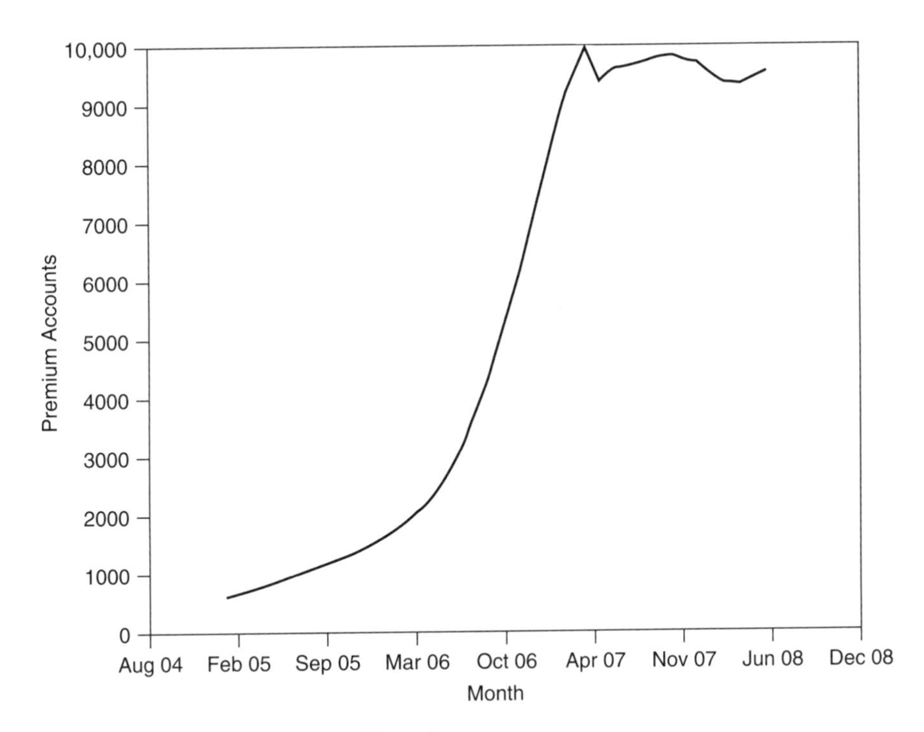

Figure 6.2. Premium Account Growth
Source: Based on data from Linden Research (2010)

analytical tools are considered. Various statistical tools are brought to bear, beginning with linear regression and extensions of it.

6.1.2 Internet Time Series Data and Trends

The Internet provides a rich source of information. The first search uses Google Trends (Google 2010b). Suppose that you are trying to forecast the future of Radio Frequency Identification (RFID) by looking at the past. RFID is a system by which a unique identity, within a predefined protocol, is transferred from a device to a reader via radio-frequency waves. The fast lane on a toll road, where your passage is detected and a fee is subtracted from your account, is an example of RFID technology. For more information on the topic, see Banks, Hanny, et al. (2007).

Enter <RFID> at www.google.com/trends and then ask Google to "Search Trends." The output includes the information shown in Figure 6.3. First, the good news: the history of search volume over a period of more than five years is presented. Now for the bad news: there is no information about the specific number of searches. Instead, the scale is normalized between 0 and 100. The upper graph shows that the search volume in 2009 is about half that in 2004–2005 and that there has been a steady decline in interest since 2004.

The letters within the boxes that are displayed in Figure 6.3 refer to news events that may have impacted the trend. To the right of the graph (but not shown in Figure 6.3), Google provides hypertext that leads to each news item.

The lower graph shows the news reference volume, that is, the amount of new material appearing on the Internet about RFID. It's fairly consistent over the 2005–2008 time periods, but recently, the volume may be declining. This Google site also adds a world map that shows where the interest in the topic arises. One can track that concentration over time by moving a slide. For example, during the period May–August 2009, the highest regional interest in RFID tags was from the United States. The site also lists the top search terms (e.g., "active tags") and cities and regions with the highest search volume.

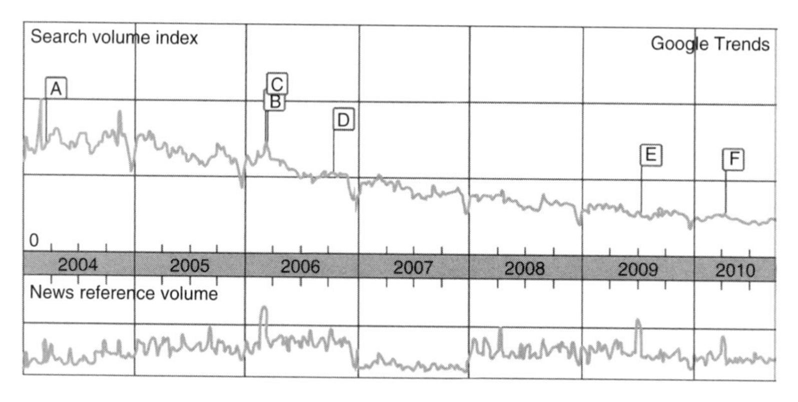

Figure 6.3. Initial Information Using Google Trends

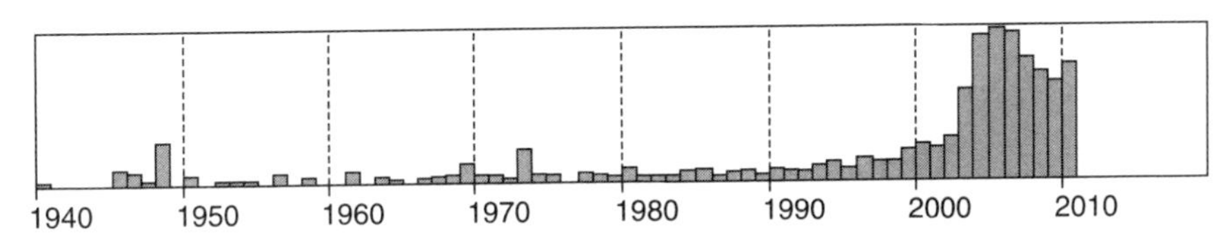

Figure 6.4. Google Insights Search for RFID

Alternatively, Google Insights can be used to mount a search (Google 2010c). First, enter the search term: "RFID tags." This produces a graph very similar to Figure 6.4. An axis with the number of searches scaled to a maximum of 100 is provided as well.

A quick search of U.S. patents can be accomplished using Google as well (Google 2010a) (http://www.google.com/patents). A search for the number of RFID patents issued, year by year, since 2003 yielded the results in Table 6.1. See Chapter 5 for other research and patent sources.

According to the *Performance and Accountability Report Fiscal Year 2008* issued by the U.S. Patent Office (USPTO; U.S. Patent and Trademark Office 2010), the average time from patent application to patent issuance has been 32.2 months. So, patents issued in 2008 were filed two or three years previously. Thus, 2004 probably was a high-water mark for patent filings for RFID innovations given the peak in issuance three years later.

Another useful website tool is Keyword Discovery (Trellian 2010), which compiles search statistics from multiple search engines. Checking select RFID terms produces the searches and search volumes shown in Table 6.2.

Clearly, there is an abundance of information on the Internet. But forecasters must use judgment in deciding what is suitable and how it might help answer the management of technology issues and framing questions (see Section 4.2.4) that have been posed. If it does the job, great; if not, consider other information resources noted in Chapter 5.

6.1.3 Analytical Modeling

Several analytical modeling techniques are described in the following pages. First, simple linear regression is examined from the standpoint of understanding the

TABLE 6.1 RFID Patent Issuance

Year	Issuances
2003	235
2004	275
2005	260
2006	312
2007	337
2008	262

TABLE 6.2 RFID Keyword Hits

Keyword	Occurrences
RFID	1089
RFID chip	128
Review of literature in RFID	69
RFID technology	66
RFID door access diagram	56
RFID Japan	43
RFID software	43
RFID system history	42
RFID chips	41

process rather than detailing the computational steps. The MS Excel spreadsheet and statistical software can take care of the manipulations. Next, three very different methods that involve computer simulation to mimic reality are described: cross-impact analysis, the Monte Carlo method, and system dynamics. Then gaming is discussed. Decision trees are recommended as a way to attach revenues, costs, and probabilities to technology decisions and Bayesian estimation as a way to think about how given or observed information changes decisions. Finally, the value of perfect or imperfect information is discussed and real options analysis is considered.

6.2 LINEAR REGRESSION AND EXTENSIONS

Suppose that you want to find if the relationship between a single independent variable, X (e.g., Year), and a dependent variable, Y (e.g., Sales), can be closely represented by a straight line. The method of least squares can be used to compute an intercept and a slope to give an equation of the form

$$Y = b + kX + e \tag{6.1}$$

where b is the estimated intercept, k is the estimated slope, and e is the error between the true value of Y and the value predicted by $(b + kX)$. It *is important to understand that no cause-and-effect relationship has been posited between X and Y*. You are merely studying an ongoing association between the two. This is sometimes called *naive modeling*, as opposed to causal modeling.

The Excel macro referenced in Equation 6.1 provides other functions that are useful in technology forecasting:

- Cumulating time series values (e.g., the sum of annual sales for all years)
- Inverting a series; models sometimes better fit the inverse of the data series
- Projecting the year in which a trend will reach a certain level

An essential statistical tool to help gauge future growth is the Prediction Interval (P.I.). The P.I. is in the nature of a tolerance applied to projections that extend beyond the observed data range. Since underlying error distribution is assumed to be normal, the equation is:

$$\text{P.I.} = \pm(t_{\alpha/2,d.f.=n-2})\sqrt{s_e^2\left[1+\frac{1}{n}+\frac{(X_*-\overline{X})^2}{\sum_{i=1}^{n}x_i^2}\right]} \tag{6.2}$$

where

n = the number of data points

t = a statistical value from a t-table determined for $\alpha/2 = (1 -$ the desired P.I.)/2 and $n - 2$ degrees of freedom. In Exhibit 6.1, for example, if a 90% P.I. is desired, look up t for $\alpha/2 = .05$ and if $(n - 2) = 4$ and find that $t = 2.312$.

s_e^2 = the standard error

where

$$s_e^2 = \frac{\sum_{i=1}^{n}(\text{Y}_i - \overline{\text{Y}})^2}{n-2} \tag{6.3}$$

and the numerator is the sum of the difference between the observed and predicted values of Y (the deviations) squared. For example, in the data used for deriving Figure 6.4, in 2006 the observed value is 150 sales, while predicted sales are 205, so the deviation is 55.

The [ ] contains three terms that reflect sources of error:

- 1 reflects error about the regression line
- $1/n$ reflects error about the mean of the data
- The final term reflects the error in the extended slope where $\overline{X}$ = the mean value of X, X_* = the target value of X, and $x_i = X_i - \overline{X}$.

Exhibit 6.1 Applying the Excel Trend Macro to RFID Sales

MS Excel spreadsheet software has been used for this case study. For example, suppose you have the number of passive RFID tags sold over the past several years (Table 6.3). These are not actual data; they are used only as an example.

As Figure 6.5 shows, a straight line provides a pretty good fit to those data. How should you fit such a line? You could use a straightedge and then calculate the slope and intercept. Or you could use MS Excel (for instance)

TABLE 6.3

Year	Passive Tags Sold (millions)
2003	18
2004	40
2005	91
2006	150
2007	275
2008	413

to perform a linear regression (the "LINEST" linear estimation function fits a straight line using the least squares method). Even more easily, you could use the macro provided for MS Excel, which lets you fit several different types of growth curves (Technology Policy and Assessment Center 2010).

A standard indicator of the goodness of fit of the curve to the data is R^2, which represents the fraction of the variance between the observed and estimated values explained by the regression line. The square root of R^2 is called the *correlation*. The regression coefficient (k in Equation 6.1) is the slope of the straight line fit; the correlation coefficient is a measure of how closely the data points fall to that line. While the R^2 for this case of 0.95 is

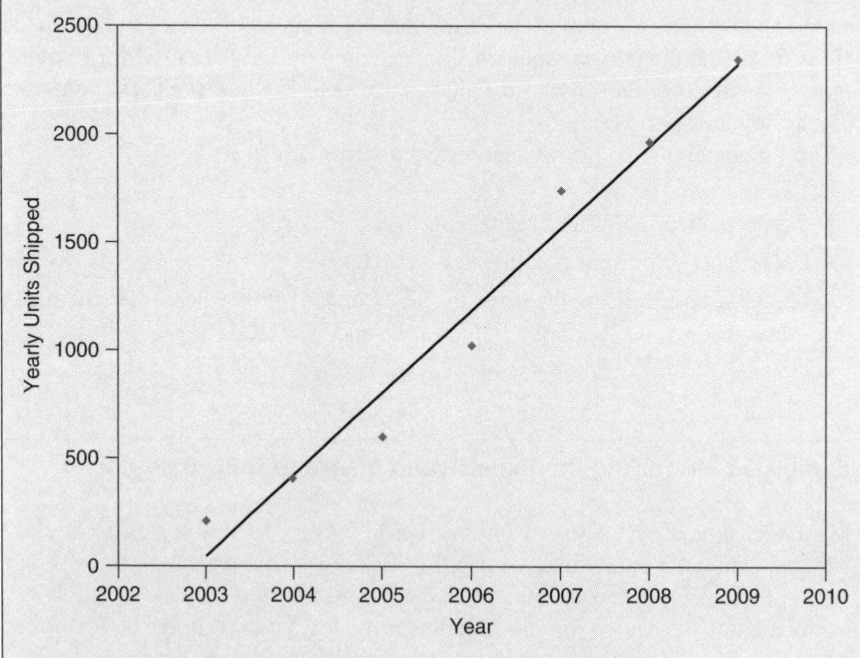

Figure 6.5. Plot of Passive RFID Tags Sold with Fit

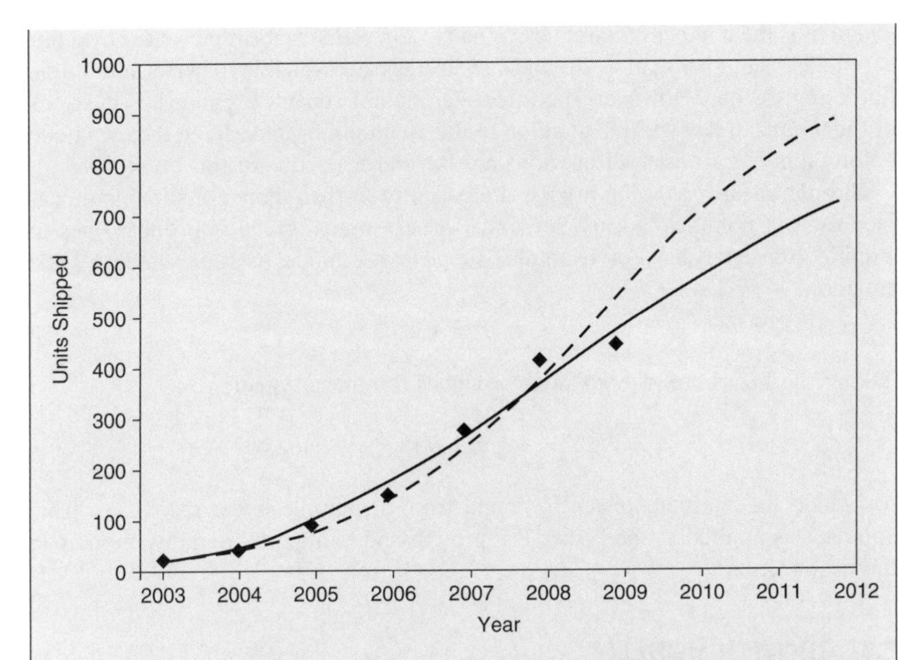

Figure 6.6. Alternative Growth Models for RFID: Fisher-Pry and Gompertz

very high for technical parameters or sales data, Exhibit 6.2 will show that other growth models fit these data even better.

For example, see the fitted lines in Figure 6.6. They are based on the Fisher-Pry (dashes) and Gompertz (line) growth models, and are discussed later in Exhibit 6.2. Should you rush out and invest in RFID enterprises and expect exactly \$681 million in sales in 2012? Of course not! Trend extrapolation is just that—a projection of what would happen if the recent growth trend were to continue unchanged. However, many factors could alter this growth—competing technologies or a changed socioeconomic context, for instance.

The P.I. narrows if you desire a lower prediction level or have more data. It also narrows if the data are closer to the mean or fall closer to the regression line. Finally, the P.I. will narrow if the target of the extrapolation is closer to the data set (i.e., short extrapolations of known trends can be made with more confidence than longer ones).

It is not unusual for a parameter to depend on more than a single independent variable. For instance, a corn yield depends on more than the amount of rainfall. It also depends on fertilizer, hours of sunlight, the type of seed, and other factors. To model such a case, the forecaster might attempt a *multiple linear regression*

$$Y = B_0 + B_1X_1 + B_2X_2 + \cdots + B_mX_m + e \tag{6.4}$$

where m is the number of variables. Whether the yield can be represented by a linear model is another matter. Computer packages are available to perform multiple linear regressions. However, the forecaster should consult expanded discussions of the topic and pay special attention to the assumptions underlying the technique before applying it. Such discussions are beyond the scope of this book.

Simple linear regression models are easier to fit than more complex nonlinear models. But nonlinear forms can sometimes be transformed into linear ones to simplify the analysis. For example, suppose the model is represented by the equation

$$Y = ax_1^b x_2^c$$

Taking the logarithm of both sides produces the linear equation

$$Z = \log Y = \log a + b \log x_1 + c \log x_2$$

for which the coefficients can be found from a multiple linear regression. This approach is applied to the Fisher-Pry growth and Gompertz mortality models in the next section.

6.3 GROWTH MODELS

Growth follows many different patterns. Some are linear, but more often they are not. For instance, Moore's Law describes the roughly exponential growth of semiconductor capacity (e.g., integrated circuit components per unit size or memory per dollar), a trend that has persisted for over half a century. Industry has used this trend to "roadmap" the timing of future capabilities (Semiconductor Industries Association 2010). Roadmapping is briefly treated in Chapter 11. More typically, exponential growth patterns do not hold for such extended periods; rather, growth confronts some form of limit (e.g., physical, technological, economic, or social).

More commonly, technological capabilities follow S-shaped growth curves. These are characterized by slow initial growth, followed by a period of rapid (almost exponential) growth, and finally by an asymptotic tapering to a limit. Market penetration curves also often follow similar trajectories, as do many biological system characteristics (e.g., the weight of a corn plant as it grows). There are several important mathematical formulations for these curves.

6.3.1 The Models

While almost any conceivable growth curve might fit a special case, the focus here is on those most often found to describe technological innovation—the S-curves. The popular concept of exponential growth usually is not sustainable, although Moore's Law has hitherto been an important exception.

S-shaped or logistic curves are loosely modeled on biological system growth patterns. They have been found to accurately model the rate at which a new technology penetrates the market as well as the rate at which it substitutes for an older technology (e.g., water-based paints for oil-based ones). Fisher and Pry (1971) presented 17 cases that fit such a model, and Lenz (1985) expanded the number to nearly 100.

Observed data must exist to apply a growth model. For market penetration, some suggest that the data sample should reflect 10% market penetration, while others maintain that the curve's slope can be identified with as little as 2 to 3% penetration by a disruptive technology, as discussed by Christensen, Horn, et al. (2008). There is no reason, however, that the forecaster cannot fit and refit the model as more data become available. That is a prudent course of action in any case.

Some versions of the growth models take the dependent variable to be

$$f = \frac{Y}{L} \tag{6.5}$$

where

L = the upper bound for the growth of Y

Other versions of the model require the user to specify the year in which the new technology takes half of the market (also called the *inflection point*). Still others require the user to input some different constants; the bottom line is that some assumptions must be made. Table 6.4 shows various growth models in forms to facilitate ordinary linear regression fitting of the data.

This table includes both cumulative sales of the technology (Y) and yearly sales of the technology (dY/dt). In some models, the market fraction (f) is used instead, as defined above in Equation 6.5. The variable Z is a transformation of the data, which linearizes the S-curve and thereby eases further regression analyses.

The following parameters are shown in the table: b, c, k, m, p and q. The parameters shown in the Bass model (m, p, q) have special significance and are further discussed below. The variable e is error, which is a normally distributed random variable as assumed for linear regression. The model parameters are all estimated using linear regression. Growth projections and prediction intervals on growth can then be calculated, as described previously (Section 6.2).

TABLE 6.4 Families of Growth Curves

Linear	$Y = b + kt + e$	(6.6)
Exponential	$Z = \ln[Y] = b + kt + e$	(6.7)
Gompertz (mortality)	$Z = \ln[\ln[f]] = \ln(c) - bt + e$	(6.8)
Fisher-Pry (substitution)	$Z = \ln\left[\frac{f}{1-f}\right] = \ln(c) - bt + e$	(6.9)
Bass	$\frac{dY}{dt} = pm + (q-p)Y - \frac{q}{m}Y^2 + e$	(6.10)

While the equations in the table are posed in terms of time as the independent variable, other choices could be made (e.g., cumulative number of units manufactured). These models provide ways to get more precise pictures of the paths of technology growth. However, *precise* does not mean *accurate*. As noted earlier, the further into the future one looks, the more speculative the view. However, it does mean that these forecasts produce repeatable results. With limited data, they can be used to extrapolate a forecast of the growth of technology that has the S-shape, the pattern of technology adoption and diffusion.

The Fisher-Pry (Fisher and Pry 1971) model is a *growth model* that also is referred to as a *substitution model,* because it is used to forecast the rate at which one technology will replace another. The fundamental assumption of the Fisher-Pry model is that the rate of change of f is proportional both to f and to the fraction of the market remaining, $(1-f)$. This is a good assumption when initial sales make the sale of subsequent units easier—for example, when sales lead to growth in distribution, service, and/or repair networks that, in turn, encourage additional buyers. It also applies well to products like cell phones or Facebook that have network externalities, that is, when the value to each buyer increases when the number of users on the network increases. As indicated above, the fit of the equation improves substantially as data are taken from years beyond the 10% penetration level.

The underpinnings of the Gompertz model are quite different from those of the Fisher-Pry model. The Gompertz model is often referred to as a *mortality model.* It is most appropriate in cases in which existing products are replaced because they are worn out rather than because of advantages of the new technology. The Gompertz model also produces an S-shaped curve, but one that usually rises more sharply and tapers off earlier than the Fisher-Pry curve. Like the Fisher-Pry curve, the Gompertz curve can be adjusted by the forecaster by adjusting the empirically determined constants b (position) and c (shape).

The Bass model (Bass 1969) is intended to describe the first adoption of a new technology. Thus, once the technology is purchased by all interested initial consumers, there are no further sales. Some customers unconditionally adopt the technology without the advice of others; these are represented by the *coefficient of internal innovation*, p. Others await news about the experiences of early customers; this fraction is represent by the *coefficient of external innovation*, q. Together, p and q determine the dynamics of adoption, peak sales, and decline. The constant m is an arbitrary scaling constant that is related to the cumulative shipments of technology at the start of the data series. This quantity may be nonzero if the forecaster is working with an established technology but a limited time series history.

Some forecasters use published data about previous technologies to evaluate estimates of p and q in the Bass model. Such estimates can be used to produce growth scenarios when a technology is still very new and data remain limited. Bass's model has seen renewed interest—in part because it is a superset of other models, including the exponential, logistic and Gompertz curves. Another reason for enduring interest in the Bass model is its more recent use in analyzing social networks (see, for instance, Hu and Wang 2009).

The coefficients p and q have explanatory significance given Bass's hypothesized dynamic of technology adoption (see Bass 1969). In reality, as Bass himself argues, they may also be slowly varying over time (Bass's Basement Institute, 2011). Despite the theoretical significance attached to p and q, these coefficients are still estimated as regression parameters within a linear regression procedure. This may result in parameter estimates that are not very robust, since yearly sales may be highly noisy. A robust alternative is to use nonlinear regression (e.g., see Srinivasan and Mason 1986). Bass offers a useful Excel macro for accomplishing a nonlinear regression of the model (Bass's Basement Institute 2011).

Miranda and Lima (2010) offer a multilogistic function model that fits several S-curves to the time series in question. This model recognizes that several processes can be at play in a system. Miranda and Lima analyze long growth series (U.S. corn production since 1866), showing how residual analyses can help sort out causal forces.

The growth models described above require that the forecaster specify a value for L, the limit or upper bound for the growth of Y. This is usually not cut-and-dried. Consider the passive RFID tags case (Exhibit 6.1):

- Is there an ultimate limit? None is apparent that can be strongly justified. It might be useful to think in terms of the market size (based on U.S. or global sales data).
- Recasting f as the percentage of market penetration can help, but it is not foolproof since most technologies never capture 100% of the market. For instance, cable television subscribers failed to reach a 100% limit because of competition from broadcast and satellite TV. On the other hand, a computer sales limit of one per person has been wildly exceeded for some special groups.

Experience suggests that it is often more useful to think in terms of a reasonable limit over the time horizon of interest. So, if the forecast concerns RFID sales through 2012, there is no need to worry about ultimate limits—just a feasible upper bound for sales by then. While this approach may make one intellectually uneasy, it seems to work pretty well.

After considering the rationale underlying the models, it is perhaps best to try several models with different limits to see how well each fits the data. Exhibit 6.2 illustrates this process. Providing results from alternative models is a reasonable form of sensitivity analysis for plausible ranges of projections. The macro provided for MS Excel at the Technology Policy and Assessment Center (2010) makes this easy. As always, it is important to revisit the model as more data become available.

Time is not the only independent variable that might be chosen in a model. The key notion behind learning curves, for instance, is to use a measure of "effort" or experience instead of time as the independent variable.

Exhibit 6.2 Modeling RFID

To illustrate the process of choosing a model, return to the RFID data in Exhibit 6.1. Recall that a linear projection (Figure 6.5) seemed to fit the observed data reasonably well. Figure 6.6 shows Fisher-Pry (upper curve) and Gompertz (lower curve) models and the observed data (diamonds). The models have been fit with $L = 1000$ using the MS Excel macro cited earlier. Both project faster than linear growth through 2012. For that year the linear estimate is 681 sales, while the Gompertz model projects 729 sales and the Fisher-Pry model projects 892 sales.

How well do the models fit the data? From a cursory examination of Figures 6.5 and 6.6, the linear model is the least satisfactory. The statistical measure of goodness of fit, R^2, was 0.95 for the linear model, and it is 0.98 and 0.99, respectively, for the Fisher-Pry and Gompertz models. But look further.

Data show that sales did not grow much in 2009, when there was an unusually difficult economic climate. If sales for 2009 are omitted as an outlier, the resulting R^2 values are 0.92 (linear), 0.996 (Fisher-Pry), and 0.989 (Gompertz). Visually, the Fisher-Pry model looks best; it is structurally suitable (this is a growth situation, not the mortality situation modeled by the Gompertz equation); and statistically, it is at least as good as the Gompertz model. Therefore, it seems the best choice.

Table 6.5 is a table from the Excel macro that gives 90% prediction interval (P.I.) values for the Fisher-Pry model, with $L = 1000$ and the 2009 sales removed. Note that the P.I. is not constant. For 2005, it ranges from 64 to 98, bracketing the regression value of 79 and including the observed value of 91. For 2012, the P.I. is wider because the model is extrapolated further from the observed data. Note also that the P.I. is not symmetric. The high side is closer to the regression value (933 sales) than the low side. This reflects the constraint exerted on the S-curve by the L of 1000.

TABLE 6.5

Year	Prediction Interval Low	Forecast	Prediction Interval High
2003	15	20	25
2004	32	40	50
2005	64	79	98
2006	124	151	183
2007	225	269	319
2008	370	432	497
2009	542	612	677
2010	703	765	818
2011	825	871	906
2012	904	933	954

The *technical progress function* models the notion that progress will be greatest for a period after initial impediments are overcome and before maturation sets in (the S-curve in another guise). For example, the trend of RFID patenting by a corporate lab might show that patents continue to accumulate, while a plot of the annual rate of patenting might show a decline in R&D productivity that is coincident with the leveling of the S-curve.

The *productivity curve* uses experience as the independent variable. A classic example (Argote and Epple 1990) shows the hours needed to assemble an aircraft versus the cumulative number of aircraft produced. The curve suggests a 20% decline in unit production cost with each doubling of cumulative output. Such decreasing costs would be important in assessing the threat of a competing technology.

How should the forecaster choose from among the available models? Perhaps the best approach is to first examine the underlying characteristics of the technology, and those of the model, and discard those that clearly are inappropriate. Then fit the data to the remaining models to see which is statistically best. Begin by asking questions like these:

- Is there a strong model already (e.g., Moore's Law)?
- Is there a strong analogy for which a particular growth curve has been shown to be a good model?
- Are there known limits to growth?
- Are there general principles, a theory, or an explanation that favors one form of growth?
- Are there particular events that could have affected the observed trends?

6.3.2 Dealing with the Data

There seldom are as much data as the forecaster would like, and the data that exist are often erratic. What to do? If the underlying structure of the phenomena can be related to a growth curve, then the model's inherent structure greatly reduces the data requirements. If it cannot, a simple approximate rule of thumb for extrapolating time series data is

$$P \sim 4\sqrt{F}$$

where P is the number of past time periods for which data exist and F is the time periods to be forecast. Thus, if the forecaster had eight years of data, an extrapolation of four years into the future could be made.

The rule of thumb given by the equation is very aggressive compared to much economic or engineering time series extrapolation. There, as many as 100 points might be desired to formulate a robust statistical model from which to project one or two time periods ahead. There's no magic here. There seldom are much data available at the time that technology projections must be made. But the forecaster must recognize that such projections are highly uncertain. The best approach is to analyze the technology attribute being forecast to determine if a structural model such as those in Table 6.4 is appropriate; gather as much data as possible; do and redo the forecast as more data emerge; and hedge your bets on the results.

Erratic time series data are not uncommon. There are many ways to deal with such data. Two that are especially useful are:

- Use a moving average (e.g., replace individual values by the average value over, say, three years).
- Smooth the data by weighting recent values more heavily than earlier ones. Here again, there are many schemes. An exponential approach sets the weight of the ith term as

$$w_i = \frac{a(1-a)^i}{\sum_i^n a(1-a)^i}$$

where

$0 < a < 1$ (assigned by the analyst to adjust the weighting)
n is the total number of data points

An approach to highlight real growth is to cast monetary values in terms of constant dollars to eliminate the effect of general inflation.

6.3.3 Regression and Growth Modeling: What Can Go Wrong?

The strategy suggested above is a blend. First select the model structure that best fits innovation system behavior and then investigate projections based on alternative formulations and sensitivities. Think carefully about a model if its value of R^2 seems low.

What can go wrong? Plenty! First, don't confuse correlation and relation. Just because an input and an output variable are statistically related doesn't mean that they are, in fact, structurally related. Examples of this mistake appear in newspapers almost daily. For example, an article was posted on the Web on March 11, 2007, entitled "Young Black Americans at Higher Risk for Colon Cancer" (Smith 2007). This would lead one to believe that blacks are more susceptible to colon cancer than members of other races. But more than two years later, that misperception was corrected. The real difference was income, not race.

Second, correlation does not necessarily mean causation. While there are many, many examples of this mistake, one will suffice. *Freakonomics* (Levitt and Dubner 2005) states that the presence of books in a home is a good predictor of how children will perform in school. There may be a relationship between the two, but don't rush out and buy a lot of books. The linkage doesn't work that way. It might be that parents who have books are more likely to read to their children and take an interest in their learning, thus increasing their performance, rather than the mere presence of books.

Other concerns include such things as outliers. In Exhibit 6.2, the low value of observed sales for 2009 was attributed to the economic recession. Should that value be kept or discarded? In quality control, outliers are discarded if they can be explained, for instance, by an inspector's inexperience. That's usually a good guideline for forecasting as well. However, it's a good idea to analyze with and without the outlier to establish its effect. If the outlier cannot be explained, it may or may not be discarded. The decision is up to the forecaster.

Be concerned if forecast errors are not random. The premise underlying regression modeling is that errors are random. If there is a discernible pattern to them repeated over time, then autocorrelation (i.e., correlation of the errors with themselves) is at work in the dependent variable series. For example, autocorrelation might be present if observed data systematically alternate above and below the forecast or if a series of points (say eight) is consistently above it. This is a major concern when developing control charts in quality control. See the *Engineering Statistics Handbook* (2010) for more information about control charts.

Many errors may arise in multiple regression analyses. The fundamental assumption is that the independent variables are truly independent of one another. Multicolinearity may be present if there is a nontrivial correlation between pairs of independent variables in the model. This could be very obvious, such as when two of the variables are personal height and weight. Since taller people are likely to weigh more, the variables are not independent and only one is needed. Statisticians suggest trying the model with one of the correlated variables at a time and then selecting the model that best fits the data.

To close this section, what should the forecaster report? The cornerstone of the answer is sensitivity analysis: that is, how sensitive are results to the models and limits chosen? This information must be conveyed to decision makers along with the prediction intervals. In the RFID exhibit, Exhibit 6.2, the forecaster could indicate that the trend analyses show good consistency for an S-shaped growth curve that is beginning to saturate.

6.4 SIMULATION

Simulations portray some aspect of a real-world system that can be used to study its behavior. This section introduces simulation and several simulation methods useful in forecasting and technology management. The first method discussed is cross-impact analysis, which is useful for understanding the forces surrounding an evolving technology. It can also serve as a vehicle for launching discussions about the impacts of the technology. Next, Monte Carlo simulation is presented and applied to economic decision making. RFID is used as an example of both cross-impact and Monte Carlo analyses. System dynamics methods also are described as a philosophy for analyzing and understanding complex real-world systems. Lastly, gaming is introduced as a means to study the behavior of decision makers in pursuit of goals or in competition with each other.

Simulation means different things to different people. To an airline pilot, for instance, it means physical emulation of the cockpit and analog or digital emulation of aircraft flight behavior. Operations researchers, however, generally think of simulation in terms of discrete-event computer models that imitate the system they wish to study. Once such models have been verified and validated, they can be used to study the effect of changes in the real system. A military analyst may see simulations as war games with stochastic behavior and may use them to study strategy and tactics, the effects of new weapon systems, or other battlefield possibilities. The common notion among these different perspectives is that simulation simplifies a real-world system, yet captures and portrays its major characteristics in ways that are useful for training and learning system behaviors.

6.4.1 Quantitative Cross-Impact Analysis

A basic limitation of many forecasting techniques is that they project events and/or trends independently (Gordon 2009); thus, they fail to account for the impact of events or trends on each other. For example, hydrogen fuel technology could have a major effect on petroleum exploration. Likewise, the scarcity of petroleum resources holds great economic implications for the development of hydrogen fuel cells. These two technologies do not exist in isolation. Each has a history; each is affected by developments in the other.

One approach to capturing interactions among events is to construct a model, that is, a formal representation of interactions among significant variables. There are several types that can be employed. A *mathematical model* uses equations to represent the system in which the events occur. Such models often require major time and money investments to construct. Even with these investments, model coverage usually is limited (e.g., mathematical models of inventory systems, of the economy, or of resource allocation systems). There are, however, special models that cut across disciplines and account for the effect of one event upon another. In the technology forecasting area, one such model is *cross-impact analysis* (CI). Basic CI concepts are widely used and have applications in many areas, including natural resource depletion, institutional change, organizational goals, communication capability, societal planning, regional planning, and others.

Since CI deals with the future, it involves uncertainty. Therefore, it is a stochastic rather than a deterministic model. Traditional CI is focused on the effects that interactions among events have on their probability of occurrence. Thus, it deals with discrete events and incorporates no dynamic (time) dimension. While still discrete, the dynamic dimension can be added to CI using the concepts of Markov chains. The topic is beyond the scope of this book.

The concept of CI arose from a game called "Future" that Gordon and Helmer devised for Kaiser Aluminum in 1966. The method was first documented in Gordon and Hayward (1968). In the game, a future world was constructed in which some or all of 60 events might have taken place (technological breakthroughs, passage of laws, natural occurrences, international treaties, etc.). Each event was assigned an initial probability of occurring, and as play progressed, these probabilities changed. Part of the change was due to actions of the players; the remainder was determined by the occurrence or nonoccurrence of other events. Changes of the latter type gave rise to the concept of CI.

A specific example is useful to understand how traditional CI works. Suppose that your organization is considering adopting Radio Frequency Identification (RFID). You aren't sure if you should do this. Some colleagues tell you of events that may influence your decision. For example, one says that you could spend a lot of money and a replacement strategy, even better than RFID, could come along within two years. Another says that you should just improve the barcode system that you have because it can do everything that RFID can do. Yet another says that much cheaper tags are coming that use organic ink rather than copper. She says that no two RFID installations are the same and that implementing one is a "black art." To top it off, one colleague has heard of a new system that is as easy to set up and use as a TV.

Suppose the events that impact your decision are identified as $E_1, E_2, E_3, \ldots E_m$. These represent entirely *external determinants*—that is, natural or man-made events over which you have no control (e.g., *you* aren't going to develop tags that use organic ink). *Events completely under your control are not included* and must be treated differently. While only four possible external events were introduced above, many more are possible. If the number grows too large, you may need to retain only those that are most important. This could be accomplished in several ways (e.g., having knowledgeable parties rank their importance).

For this example, suppose you have identified the four events mentioned above and shown in Table 6.6. For convenience, this *occurrence matrix* is arranged with the events E_1, through E_4 ordered both across the top and down the left-hand side of the array. The next step is to estimate the probability that each event will occur. These are called the *marginal probabilities.* (They also are sometimes referred to as *ceteris paribus* [*all-else-equal*] *probabilities* to indicate that they are estimated without considering any of the other events.) These probabilities are subjective and might be estimated by consulting experts. Table 6.6 shows that you have estimated the probability that there will be a replacement technology within two years as 0.35.

TABLE 6.6 Occurrence Matrix

If This Event Occurs (Column)				
The Probability of This Event (Row) Becomes:	E_1	E_2	E_3	E_4
E_1 Replacement technology (0.35)*	1.0 $P(1\|1)$	0.40 $P(1\|2)$	0.44 $P(1\|3)$	0.25 $P(1\|4)$
E_2 Better-engineered barcodes (0.25)*	0.30 $P(2\|1)$	1.0 $P(2\|2)$	0.20 $P(2\|3)$	0.26 $P(2\|4)$
E_3 Use of organic ink (0.55)*	0.28 $P(3\|1)$	0.40 $P(3\|2)$	1.0 $P(3\|3)$	0.60 $P(3\|4)$
E_4 Elimination of "black art" (0.30)*	0.55 $P(4\|1)$	0.20 $P(4\|2)$	0.25 $P(4\|3)$	1.0 $P(4\|4)$

*Initial marginal (ceteris paribus) probability.

You have completed two components of the CI matrix: The events critical to the forecast have been identified, and their initial (marginal) probabilities of occurrence have been estimated. The cells of the matrix will be used to record the *conditional probabilities*, that is, the probability that event i occurs given that event j occurs. These probabilities are the heart of CI. They portray the impact that the occurrence of any event has on the probability that any other event will occur.

The conditional probabilities must be estimated next. However, first note that the matrix diagonal entries all will be 1.0, since it is certain that event i will occur given that it has occurred. The first step is to compute the statistically acceptable range of conditional probability for each cell (pair of interactions) *above the diagonal.* These ranges will provide guidelines if you have no other basis from which to estimate the conditionals. This can be done using the marginal probabilities established previously for each event. To explain how to compute this statistical range requires the introduction of some statistical notation.

$P(i)$ = probability that event i will occur (the marginal probability of i)

$P(i|j)$ = probability that event i will occur given that event j has occurred (the conditional probability of i given j)

$P(\bar{i})$ = probability that event i does *not* occur

$P(i|\bar{j})$ = conditional probability that event i will occur given that event j does *not* occur

$P(i \cap j)$ = probability that both events i and j will occur (the intersection of events i and j)

$P(i \cup j)$ = probability that event i or j or both will occur (the union of events i and j)

By using the laws of conditional probability and the probability of compound events, Sage (1977) showed that limits exist to the range of statistically acceptable conditional probabilities. If the occurrence of event j enhances (increases) the probability that i will occur, then

$$P(i) \leq P(i|j) \leq [P(i)/P(j)] \tag{6.11}$$

On the other hand, if the occurrence of j inhibits (decreases) the probability that i will occur, then

$$1 + [P(i) - 1]/P(j) \leq P(i|j) < P(i) \tag{6.12}$$

Note that only the initial marginal probabilities $P(i)$ and $P(j)$ are necessary to compute these ranges, and they have already been estimated. Any computed value that is greater than unity is set to 1.00, that is, certainty.

Now you must estimate a conditional probability for each of the cells above the diagonal and compare them to the ranges computed from Equation 6.11 or 6.12. Estimates that violate the computed ranges should be retained if a solid rationale for them can be given. For example, in Table 6.5, the conditional probability $P(1|3)$ has been estimated as 0.44, which is within the statistically acceptable range, 0.35 to 0.64, computed from Equation 6.11. However, if you had estimated that it should be 0.82 and had evidence to support your estimate, you would enter 0.82 instead. Alternatively, you could elect to assign one of the extreme values of the range to such a probability. Thus, lacking strong evidence to support an estimate of 0.82, you might choose $P(1|3)$ to be 0.64 instead.

Now that conditional probabilities above the diagonal have been estimated (the $P(i\,|\,j)$ values), you can turn to those below the diagonal (the $P(j\,|\,i)$ values). Here, you can use Bayes' rule to help. If $P(i\,|\,j)$ was in the range established by Equation 6.11 or 6.12, Bayes' rule says that the corresponding probability below the diagonal should be

$$P(j\,|\,i) = [P(i\,|\,j)/P(i)]P(j) \tag{6.13}$$

If $P(i\,|\,j)$ was not in the range or if you do not agree with the value produced by Equation 6.13, you should subjectively estimate the value of $P(j\,|\,i)$. In other words, if the values computed using Bayes' rule are reasonable, keep them; if not, estimate values you believe to be more appropriate. For example, in Table 6.5, the conditional probability $P(3|4)$ was estimated as 0.60, within the range of 0.55 to 1.00 computed from Equation 6.11. Therefore, Bayes' rule can be applied to give a value of $P(4|3) = [P(3|4)/P(3)]P(4) = 0.33$. Table 6.5 indicates, however, that you apparently had a strong rationale to support a lower estimate of 0.25.

Just as the occurrence of an event can affect the probability that another event will occur, its nonoccurrence can have an impact as well. In the RFID example, for instance, if better-engineered barcodes fail to materialize, then the impetus for and probability of better technologies will decrease. So, you need to construct a nonoccurrence matrix, as shown in Table 6.7.

TABLE 6.7 Nonoccurrence Matrix

If This Event Does Not Occur				
The Probability of This Event Becomes:	E_1	E_2	E_3	E_4
E_1 Replacement technology (0.65.)*	0.00 $P(1\|\bar{1})$	0.33 $P(1\|\bar{2})$	0.24 $P(1\|\bar{3})$	0.42 $P(1\|\bar{4})$
E_2 Better engineered barcodes (0.75)*	0.22 $P(2\|\bar{1})$	0.00 $P(2\|\bar{2})$	0.31 $P(2\|\bar{3})$	0.25 $P(2\|\bar{4})$
E_3 Use of organic ink (0.45)*	0.70 $P(3\|\bar{1})$	0.60 $P(3\|\bar{2})$	0.00 $P(3\|\bar{3})$	0.53 $P(3\|\bar{4})$
E_4 Elimination of "black art" (0.70)*	0.17 $P(4\|\bar{1})$	0.70 $P(4\|\bar{2})$	0.36 $P(4\|\bar{3})$	0.00 $P(4\|\bar{4})$

*Initial marginal (ceteris paribus) probability of nonoccurrence: $P(\bar{i}) = 1 - P(i)$.

The last step is to estimate the entries for the nonoccurrence matrix. First, compute the entries statistically from the following equation:

$$P(i|\bar{j}) = [P(i) - P(j)P(i|j)]/[1 - P(j)] \tag{6.14}$$

Lacking evidence to the contrary, these values will be entered. However, if evidence supports a different estimate, that estimate will be entered instead.

Returning to the example:

$$P(2|\bar{1}) = [P(2) - P(1)P(2|1)]/[1 - P(1)] = 0.22$$

If you have no reason to estimate some other probability, then 0.22 should be entered into the nonoccurrence matrix.

Note that the diagonal entries in the nonoccurrence matrix will all be 0.00 since the probability of an event given that it has not occurred is 0. Negative probabilities predicted by Equation 6.14 should be set at 0, while predicted probabilities greater than 1 should be set at 1.

The next stage in CI analysis is to simulate the effects of these conditional relationships. You must determine whether the initial estimates of event marginal probabilities are mutually consistent given these perceptions of how events impact each other.

If all the entries in the two matrices agree with the results computed from Equations 6.11 through 6.14, then the initial marginal and conditional probabilities are mutually consistent. However, if one or more of the conditional probabilities differ from the computed results, you will have to "play" the matrices to determine a consistent set of marginal probabilities. A computer-based Monte Carlo simulation can be used to do this:

1. An event is selected randomly (say, Event 2 in Table 6.6).
2. A random number between 0 and 1 is generated and compared to the marginal probability of the event to determine if it occurs. Suppose the random number is 0.19, since $0.19 \leq 0.25$; Event 2 is assumed to occur. If the random number was greater than 0.25, it would be assumed that Event 2 did not occur.
3. The marginal probability of each remaining event is replaced by its conditional probability given that the event in Step 2 occurs or does not occur. That is, in the example, $P(i)$ is replaced by $P(i|2)$ if Event 2 occurs, or by $P(i|\bar{2})$ if it does not ($i \neq 2$). Thus, since Event 2 occurred in Step 2, the replacement values will be $P(1) = 0.40$, $P(3) = 0.40$, $P(4) = 0.20$.
4. A second event is selected randomly from those remaining (Events 1, 3, and 4), and Steps 1 through 3 are repeated. In this play, the probability used in Step 2 is the value produced in Step 3 of the previous play. Thus, if Event 2 occurred in the first play and Event 4 is selected in the second, the probability of Event 4 used in Step 2 of the second play is $P(4|2) = 0.20$.

5. The process described in Steps 1 through 4 is repeated until all four events have been selected. All marginal probabilities are then returned to their initial values and the game is replayed, typically 1000 or more times.
6. Each time the game is played, the events that occur are noted. The total number of occurrences divided by the number of games is taken as the final (marginal) probability for each event. The initial marginal probabilities are then replaced by the final marginal probabilities, which account for event interaction.

The conditional probability of an event occurring given two or more other events is required in Step 3 after the first two events have been determined. For instance, the examples woven into these six steps simulated that Event 2 occurred in the first round of play. If in the second round Event 4 is simulated to occur, then we need conditional probabilities such as $P(1|2 \cap 4)$ to proceed. The occurrence and nonoccurrence matrices only specify pairwise interactions—that is, joint probabilities of one event conditioned on one other event. Joint conditionals are called *second-order conditionals*. There also are third-order and higher-order conditionals—for example, $P(i|j \cap k \cap 1)$. These probabilities are difficult to determine. Instead, they usually are approximated by averaging second-order probabilities. For instance,

$$P(1|2 \cap 3) = [P(1|2) + P(1|3)]/2$$

A similar averaging procedure is used for higher-order nonoccurrence probabilities, such as

$$P(1|2 \cap \bar{3}) = [P(1|2) + P(1|\bar{3})]/2$$

These approximations are acceptable when the conditional probabilities being averaged are close in value.

Of course, the game is probabilistic, so results vary somewhat from game to game. Table 6.8 shows the marginal probabilities after various numbers of plays of the cross-impact matrix. Note that there have been changes in the marginal probabilities of all events, the largest being in Event 4. For example, $P(3)$ initially was estimated as 0.55 (Table 6.7), but $P(3)' = 0.49$ after 1000 plays (Table 6.8). The prime indicates that the probability is the final marginal value. As the number of plays increases, the marginal probabilities change very little. So, for this

TABLE 6.8 Marginal Probabilities

Marginal Probability	1000 Plays	10,000 Plays	100,000 Plays	1,000,000 Plays
$P(1)'$	0.32	0.32	0.32	0.31
$P(2)'$	0.29	0.29	0.28	0.28
$P(3)'$	0.49	0.50	0.50	0.50
$P(4)'$	0.41	0.41	0.42	0.41

example at least, 1000 plays seem to be about as good as 1,000,000 plays. If Equations 6.11 through 6.13 were used to generate all the conditional probabilities in the two matrices, you likely would have seen little difference between the initial and final marginal probabilities. This is because using those equations tacitly assumes that the conditional probabilities are consistent with the initial marginal probability estimates.

The CI game is attractive since it can be generated from relatively little data. It can be used to examine the interaction of events and to ensure, insofar as possible, that the probabilities to be used in, say, a scenario account for those interactions. The technology manager also could use the CI matrices to plan strategies to enhance or inhibit the probability of key events. Moreover, constructing the CI matrices can provide a useful format within which to frame discussions of interactions between events. Most importantly, merely constructing the matrices forces the manager or forecaster to consider interactions between events and thus can help define or refine the TDS. However, it is important to note that, even though the probabilities produced by CI are consistent with your perceptions of how they interact, they still may be incorrect.

Halverson, Swain, et al. (1989) point out that the effect of an event on the probabilities of other events is not necessarily determined by the size of its marginal probability. This stems from the somewhat trivial observation that an event with a low probability of occurrence has a high probability of nonoccurrence. Since the nonoccurrence of an event can have a major effect on other events, the effect of an unlikely event thus may be quite high (i.e., a black swan).

Halverson, Swain, et al. (1989) also observe that the conditional probability estimates may be more important than the marginal probability estimates. The initial marginal probability estimates likely will be changed by the conditional estimates during the game. Therefore, you should expend more effort on accurately estimating the impact of events on each other and less on estimating the marginal probabilities.

6.4.2 Qualitative Cross-Impact Analysis

This section is deliberately located after the quantitative CI section to encourage one to become familiar with the stochastic manipulations described there. But if you do not wish to pursue quantitative manipulations, there is still value to be obtained from a qualitative CI.

Here is a simple framework to explore interactions among factors pertinent to a technology forecast:

1. Convene a small group with diverse perspectives to think about the innovation system (e.g., RFID or alternative tagging futures).
2. Specify the possibly influential factors, distinguishing:
   a. Events from trends
   b. Externally from internally controlled factors

   Distinguish the *key* factors.

3. Set out a matrix of key factors and brainstorm each cell in terms of the interaction effects of factors (events or trends) on each other. This means a series of "if—then" brief discussions.
4. Reflect back on the innovation system. This could lead to changes in the key factor set. Most importantly, everyone should gain insight into important considerations and potentially pivotal factors (e.g., key influences, leverage points on system outcomes, and special sensitivities).

Such a qualitative cross-impact exploration could prompt a quantitative one. Or, if important trends seem sensitive to the occurrence of certain events, possibly a trend impact assessment could be used (Gordon 2009). Moreover, such thinking could lead to a system dynamics exercise (Section 6.6).

6.5 MONTE CARLO SIMULATION

The Monte Carlo method can be used to generate distributions for the outcomes of a probabilistic system by rerunning a model repeatedly and tabulating the results. It also can be used to sample a probability distribution for the occurrence of an event. The basis of the method is the approximation of a problem solution by sampling from a random process.

This section provides three examples. The first shows how random values are generated and displayed. The second shows how Monte Carlo simulation can be used to sample from multiple random variables, with a final result that is based on a composite of the individual samples. The third example is a Monte Carlo simulation that applies the methods of the first and second examples to a determination of the net present value of a method for implementing RFID in a hospital setting.

6.5.1 Generating and Displaying Random Values

This example is about generating 1000 uniformly distributed random values between 100 and 200, obtaining their frequency distribution in bins of size 20 (i.e., in intervals of 20), and charting the result.

In MS Excel, RAND( ) generates uniformly distributed random numbers between 0 and 1. In cell A1, enter the formula =100*(1 + RAND( )); then drag it to, say, A1000 so that 1000 values are produced. Notice that the resulting values are between 100 and 200. If you wanted *integer* values between 100 and 200, you could have used =RANDBETWEEN(100,200). Suppose that the first value of RAND( ) is 0.699067. Then the value in A1 will be 169.9067. If you press the F9 key, you will see a recalculation of all of the values in the column. Since values to many decimal places are used in this example, the values are nearly continuous, as is the uniform distribution.

What you have done is to simulate random draws from the uniform distribution with lower limit 100 and upper limit 200. This is known as a *Monte Carlo simulation*. The name comes from the use of randomness and the repetitive nature of the process, which is analogous to activities at a casino.

In the Monte Carlo simulation above, time is not a variable. In discrete-event simulation, however, the passage of time may be important. For example, when determining the number of tellers needed at a bank, customer waiting time is important. Thus, their arrival and departure times must be considered.

You can display your results as a frequency distribution, which makes it easier to see what has been produced. Enter the values 120, 140, 160, 180, and 200 in cells C1 through C5. The first cell will hold the frequency (number of occurrences) between 100 and 120; the second, the frequency between 120 and 140; and so on. It is highly unlikely that a value will fall on an endpoint since RAND( ) output is calculated to at least nine decimal places, although fewer may be displayed.

The next step requires that you have the Excel "Analysis ToolPak" from "Add-Ins" installed. If you haven't, go to the "Tools" tab and click on it. Then click on "Add-Ins" and put a check mark by "Analysis ToolPak." Next, click "OK." When you see "Data Analysis," click on it and select "Histogram." In "Input Range" you will see a tiny red square in an array to the right-hand side. Click on it and another "Histogram" dialog box opens. First, enter the Input Range. To do this, click on cell A1, then hold the Shift Key down and click on cell A1000. The result shown in the dialog box should be $A$1:$A$1000. Close that box and return to the "Histogram" dialog box.

Among the options in the "Histogram" dialog box, you need to set the "Bin Range." To do this, follow a procedure similar to the one in the previous paragraph. However, use the values in cells C1 through C5 as your inputs. When you click "OK," the frequencies in the various bins will appear on another sheet. As an example, the 1000 random values in the bins were distributed as 180, 208, 195, 211, and 206. (That is to say, 180 were in the first bin.)

Two points should be made here. When there are 1000 values, you expect that there would be about 200 values in each bin. However, in the example above, there are as few as 180 values in one bin and as many as 211 in another. With such a small sample size, differences like this can be expected. But if you generate many thousands of values, you could expect close to 20% of them to be in each bin.

One more step and you will be ready to display your results graphically. To do this, click on "Insert" and then click on "Chart." Next, click on "Column" and then click on the chart that appears at the left-hand side on the top row. Finally, click on "Finish." The result is shown in Figure 6.7.

6.5.2 Sampling Multiple Random Variables

In a classic problem, a news seller buys papers for $0.33 each and sells them for $0.50 each. Newspapers not sold at the end of the day are sold as scrap for $0.05 each. Newspapers can be purchased in bundles of 10. So, the news seller could purchase 50, 60, 70, and so on. The news seller must decide how

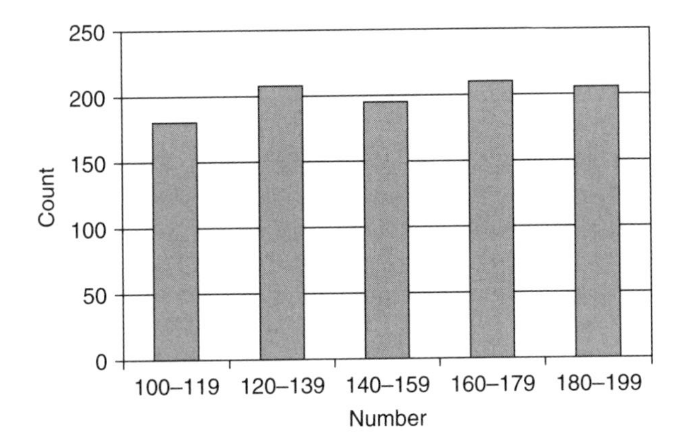

Figure 6.7. Chart of Random Uniform Values

TABLE 6.9 Newspaper Demand

Demand	Good Day	Fair Day	Poor Day
40	0.03	0.10	0.44
50	0.05	0.18	0.22
60	0.15	0.40	0.16
70	0.20	0.20	0.12
80	0.35	0.08	0.06
90	0.15	0.04	0.00
100	0.07	0.00	0.00

many to purchase each day. There are three types of news days: "good," "fair," and "poor," with probabilities 0.35, 0.45, and 0.20, respectively. However, the news seller cannot predict what kind of news day is coming. The demand for newspapers for various types of news day is shown in Table 6.9. Thus, if it is a good news day, the probability that the demand is for 80 papers is 0.35.

The policy to be determined is the number of newspapers to order each day. This problem could be solved analytically, but Monte Carlo simulation is used to provide insight into the technique. This presentation follows that of Banks et al. (2010) and uses a Microsoft Excel spreadsheet that can be downloaded at www.bcnn.net. Twenty days of profit (or loss) will be used as a trial, and numerous trials will be conducted to make a determination. Profit is determined by:

$$\begin{aligned}\text{Profit} = {} & \text{Sales revenue} - \text{Cost of newspapers} \\ & - \text{Lost profit from excess demand} + \text{Salvage scrap sales}\end{aligned}$$

The lost profit from excess demand is \$0.50 – \$0.33 = \$0.17 per newspaper. If the news seller has a policy of buying 50 newspapers and the demand is for 60, the lost profit from excess demand is \$1.70. Suppose that the news seller has

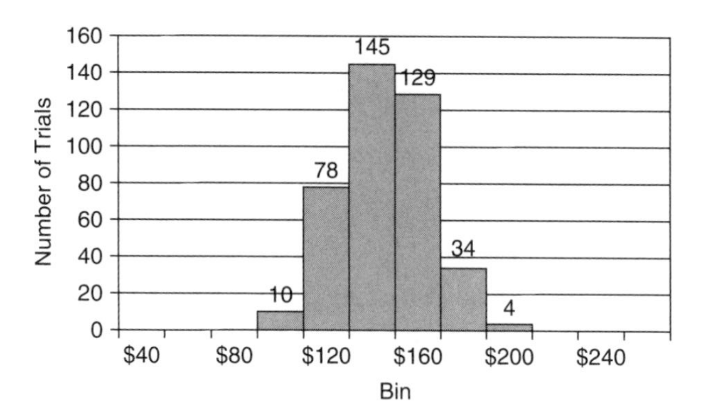

Figure 6.8. Histogram of 400 Trials of Policy

a policy of buying 70 newspapers each day, and on the first day, the demand is for 50. On that day, the profit is:

$$\text{Profit} = \$25.00 - \$23.10 - 0 + \$1.00 = \$2.90$$

After 19 additional days, the total profit is Trial 1. Suppose this simulation is performed for 400 trials and produces the histogram shown in Figure 6.8.

Different policies would be tried to determine the number of newspapers to buy each day. The one with the highest potential value would be chosen. Running the Monte Carlo simulation for one trial only is ill advised. Over the 400 trials, the minimum value was $86.60 and the maximum was $198.80. If an unusually low value or an unusually high value resulted, that might lead to the wrong decision. More reliable data are the average result of the 400 trials or perhaps the median value. Those results are $135.49 and $136.60, respectively.

6.5.3 RFID Application in a Hospital Decision

One of the areas of technology highlighted in this book is RFID. The book *RFID Applied* (Banks, Hanney, et al., 2007) gives an example based on a decision made by a large French hospital. The questions addressed by the hospital were whether to implement RFID and, if so, whether to implement and operate the system themselves or contract with an outside firm. There were 21 input variables used in the study, as shown in Table 6.10.

The example in the book has been extended by allowing many of the input values to be random variables that are uniformly or triangularly distributed. The time value of money is considered in the calculations (see Section 10.2.1). The major benefit from using RFID was that nurses and technicians would not have to search for missing MBEs (mobile biomedical equipment) as frequently, thus drastically reducing the rental cost of MBEs. For the 2000-bed hospital with 50,000 MBEs valued at some 200,000,000 € the net present value (NPV) of the savings from using RFID (the x-axis in Figure 6.9) was quite large.

TABLE 6.10 Input Variables

Project Life (years)
Number of beds
Number of nurses
Nurse's salary/year
Total number of MBEs*
Average cost of an MBE
Annual rent of an MBE
MBE's salvage value
Average MBE life (years)
Rental percentage
Inventory auditing cost/MBE
Maintenance cost
Number of biomedical technicians
Biomedical technician's salary
MBE average utilization
Shrinkage rate (stolen or lost MBEs)
Nurse's hours/year
Biomedical technician's hours/year
Nurse's MBE search time
Biomedical technician's MBE search time
Cost of capital (inflation free)
MBE recovery value

*MBE, mobile biomedical equipment.

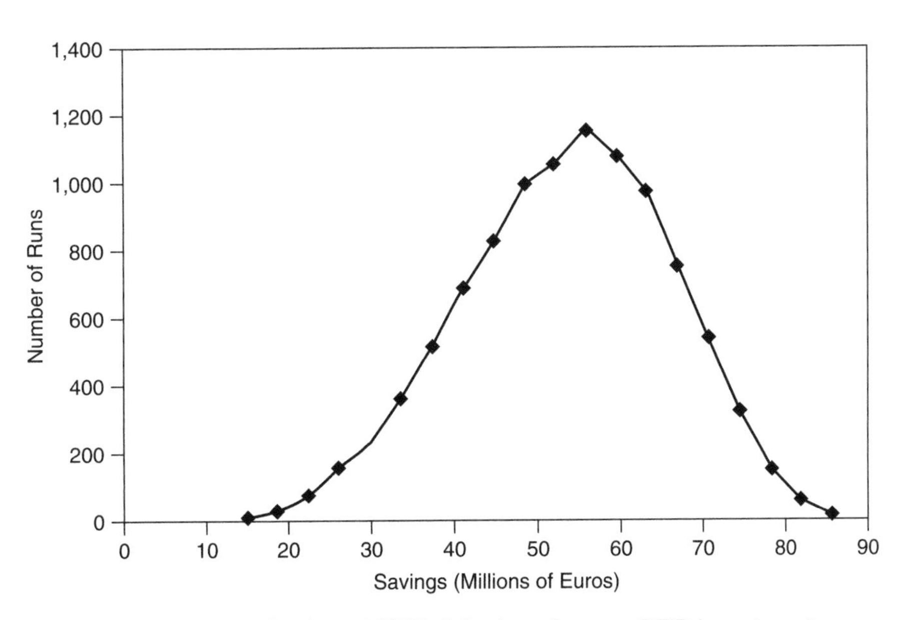

Figure 6.9. Distribution of NPV of Savings from an RFID Investment

Rather than a point estimate (e.g., "the NPV of the savings from the investment is 55,000,000 € over five years"), decision makers could be given a distribution of NPVs with this type of analysis. This shows them both the possible savings and the probability of achieving them. The distribution shows the probability of savings within a range of values. The bottom line is that managers have more information for decision making. The NPV of 1 € one year from today is 1/1.05 today, or 0.9524 € today. Thus, 0.9524 € invested at 5% interest will be 1 € in one year. The NPV of 1 € one year from today is 1/1.05 today, or 0.9524 € today. Thus, 0.9524 € invested at 5% interest will be 1 € in one year See Section 10.2.1.

6.6 SYSTEM DYNAMICS

System dynamics modeling is a method in which the world is viewed as a continuous, interconnected system of accumulations and associated fluxes. The accumulations are the stores of materials or information (pertaining to the materials) that make up the attributes of the system under study. These accumulations are termed *stocks* or *level variables*, while the fluxes are called *lows* or *rate variables*. In a business enterprise, for example, the inventory would be a level variable. The changes in the stock, perhaps through production or import on the one hand and sales on the other, are rate variables.

The evolution of such a system over time, that is, the system dynamics, occurs primarily by the interaction and feedback between the stocks and flows under the influence of implicit policies. These implicit policies are the usual ways that people seek to control and influence the behavior of the system. In this aspect, system dynamics differs significantly from other modeling approaches such as discrete event modeling or spreadsheet modeling. For a system dynamics study to be effective, it must include the interactions of people within the system (i.e., most actions determining the behavior of the system in the long term should be included) so that the full feedback structure of relevance is modeled. The only external factors, called *exogenous variables*, should be those over which people within the system have no control and that are not strongly influenced by interactions within the system. Clearly, this frequently brings a longer study time horizon and a view of a human society and its interaction with its biophysical or technical environment that is paradigmatic.

Forrester (2007a, 2007b), the founder of system dynamics, and Meadows (1976) have devoted attention to the explication of this paradigm. The paradigm needs to be taken into account when selecting system dynamics as a modeling method, as it places requirements on the system boundaries and the mix of human and technical factors to be included in the model. A further way in which the paradigm influences the choice of modeling method is its view that underlying gradual change is more relevant in determining the range of feasible long-term dynamic behavior than short-term variations. The validity of this assumption needs to be checked for the problem at hand. Further, information is needed from people about how they think and interact with the stocks in the system, and this information may not be readily available. When all of these considerations can be met, system dynamics is an appropriate and powerful modeling tool.

6.6.1 The System Dynamics Modeling Cycle

A system dynamics model is built in an iterative fashion. The first step is to conceptualize the sociotechnical system in terms of cause-and-effect relations, which are commonly visualized as causal diagrams. In a causal diagram the fluxes, accumulations, and other relevant factors (such as constants), are shown. If a flux (e.g., sales) depletes a stock (e.g., inventory), then sales is connected to the inventory with a directed arrow of negative sign. If a flux causes a stock to accrue (e.g., production rate or import rate), it is connected with a directed positive arrow. Similarly, if a stock influences a flux, the causal effect is depicted by a directed arrow of negative or positive sign, depending on whether the effect is to decrease or increase the flux. The resulting diagram represents the elements of the system to be included in the model and the causal relations between them. Thus, it helps to identify the feedback loops driving system behavior. A causal diagram of the adoption of cable-to-the-curb technology by households with an existing fiber-optic cable connection to their district is depicted in Figure 6.10.

The provision of a fiberoptic cable connection to a city district by the municipality means that district households then have the choice of accepting or rejecting the cable-to-the-curb technology. As the number of households with district connection (i.e., those that have yet to choose) increases, the number of requests for connection from those that accept the technology increases, as does the number of those awaiting connection. This causes the number of households with only district connection to decrease, completing the *acceptance feedback loop.*

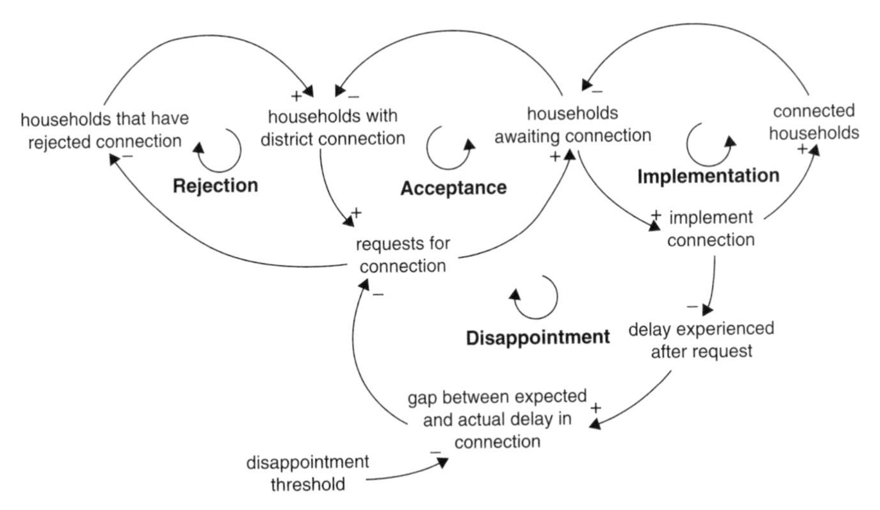

Figure 6.10. Causal Diagram of Technology Adoption

Similarly, as the number of requests for connection increases (decreases), the number of households that reject a connection decreases (increases) and the number with district connection decreases (increases). This represents the *rejection feedback loop*. The *implementation feedback loop* is formed when the rate of implementing connections rises in response to an increase in the number of households awaiting them. This causes the number of connected households to increase and the number of those awaiting connection to decrease. However, when the rate of implementing connections cannot keep pace with the increase in the households awaiting connection, the delay in connection increases and the gap between the expected and actual delays grows (and exceeds the disappointment threshold). This causes fewer people to adopt the technology and requests for connection to decline. The decline in connection requests leads to a decline in the number of households awaiting connection and a corresponding decline in the rate of implementing connections, completing the *disappointment feedback loop*.

The causal diagram depicts the elements to be included in the model and the level of aggregation. Implicitly, it represents the elements excluded and an appropriate choice of scale (time horizon, aggregation level, resolution, etc.). A good modeler revisits these choices, clarifying the problem and the desired system behavior and making assumptions as explicit as possible. This first iteration can lead to adjustments in the causal diagram and to a better and more explicit fit with the problem.

Specification of the model forms the second step in the modeling cycle. In this step, the causal diagrams are translated into stock-and-flow diagrams and the model equations are specified. Each accumulation in the causal diagram is depicted as a stock (or level) variable using a box. The fluxes are depicted as flow or rate variables using arrows with valves. Factors that are neither stocks nor flows can either be constants, termed *parameters*, and depicted using diamond shapes or auxiliary variables. The auxiliary variables usually have real-world counterparts and may partially determine the value of a rate, or they may be deduced from the value of a level variable. An example is the water level of a lake. The stock variable is the water volume, but if we know the bathymetry of the lake, we can derive the lake water level from the stock.

This specification of a model is commonly undertaken using readily available software packages such as STELLA®/iThink® (ISEE Systems 2010), Powersim Software (Powersim 2010), and Vensim® (Ventana Systems 2010), among others. The modeler uses icons to place a variable type on the diagramming page of the software and is then prompted to specify the relationships between variables using equations, graphs, and data. The stock-and-flow diagram thus depicts the underlying difference equations that form the system dynamics model. That is, a system dynamics model comprises a system of nonlinear, ordinary differential equations of the form

$$\frac{dx_i}{dt} = F_i(x, p, t) \quad i = 1, 2, \ldots, n, \tag{6.15}$$

where

$$x = (x_1, x_2, \ldots, x_n)^T, \; p = (p_1, p_2, \ldots, p_m)^T$$

are the state variables ($1 \times n$ column vector) and parameters ($1 \times m$ column vector), respectively. These equations are solved numerically using either the Euler or the fourth-order Runge-Kutta integration methods, depending on (1) the differentiability of the equations of the continuous model, (2) the accuracy requirements, and (3) the computational effort. Each stock is assigned an initial value (a constant) at the start of the simulation. These constants are part of the parameter set. Thereafter, the rates determine the changes in the levels of the stocks within each time step of the simulation, and a pattern of behavior emerges as the stocks vary over time.

So, beginning with a causal structure, translating it into an interconnected stock-and-flow system by specifying equations, and then assigning initial values and simulation settings, one can step through time and trace the system's evolution. By understanding how the strengths of the causal loops vary over time (owing to nonlinear effects), the influence of feedback on system behavior can be explored. This insight is helpful in designing new policies or new ways in which humans can interact with the system. But before new policies are designed, the degree to which the outcomes can be trusted must be established.

The third step in the modeling cycle is validation. In this step, the fitness of the model for the purpose for which it is designed is evaluated (Sterman 2000). A range of tests are undertaken. First, the consistency between the structure of the model, the causal diagrams, and the original problem is checked and the boundary adequacy is rechecked. Second, the dimensional consistency and the accuracy of the numerical simulation are verified and the parameter assessment is reviewed. Next, the ability of the model to behave plausibly when model inputs take on extreme values is assessed. This represents the beginning of the qualitative and quantitative validation of model behavior.

Here, the system dynamics paradigm of including the interactions of people with elements of the system and of exploring the system's behavior over time rather than predicting an exact value at a specific time plays a role. Accurate simulation of historical system behavior is not viewed as adequate for model validation. Instead, it is viewed as one of a range of tests needed to establish model validity. Emphasis is placed on generating hypotheses of model behavior and then testing whether the model generates these modes of behavior. Deeper insights into the system's behavior are sought by interpreting the model outputs in terms of the shifting influence of causal loops. Finally, the influence of uncertainties in parameter values on outcomes is tested using sensitivity analysis. Sensitivity analysis forms a bridge between investigating model trustworthiness, its validation, and its use in decision making.

Using a model is the final step in the cycle. During the validation, and particularly from a sensitivity analysis, the most influential parameters are identified. Indeed, parameter space can be searched for the parameter combinations that will deliver the "desired behavior" (Hearne 1987) leading to the design of parameter-based policies. More important, however, are the effects of changing the implicit policies captured in the causal structure, that is, investigating the effects of potential changes in the human interaction with system elements. These structural changes deliver the most benefit, according to Forrester (2007a, 2007b), and represent a strength of system dynamics modeling. Structural policy testing and communicating with those involved (client, problem owner, etc.) to increase their insights into the underlying dynamics of the sociotechnical system is the ultimate goal of system dynamics modeling.

6.6.2 A Technology Forecasting Example: The Cable-to-the-Curb Model

Consider an application of system dynamics in technology forecasting. The example involves using such a model in decision making on the provision of cable-to-the-curb technology to households in a small European city called Kennishoeve. The Kennishoeve city council wishes to ensure the success of the cable-to-the-curb initiative. Thus, the system dynamics modeling study focuses on implementation issues and addresses the following questions: Is the project feasible? Are the existing arrangements about staff, and about ordering and allocating cable between the municipality and the service providers, sufficient to ensure cable-to-the-curb technology adoption by more than 50% of the households within three years?

The causal diagram of the adoption of fiber-optic cable connection by households, shown in Figure 6.11, describes the cause-and-effect relations involved in accepting or rejecting the technology for households with a district connection. The municipality is responsible for providing district connection and the service providers for implementing the connection from district to household upon request. To implement the district connections and the household connections, fiber-optic cable material and staff are needed and could be in short supply. In

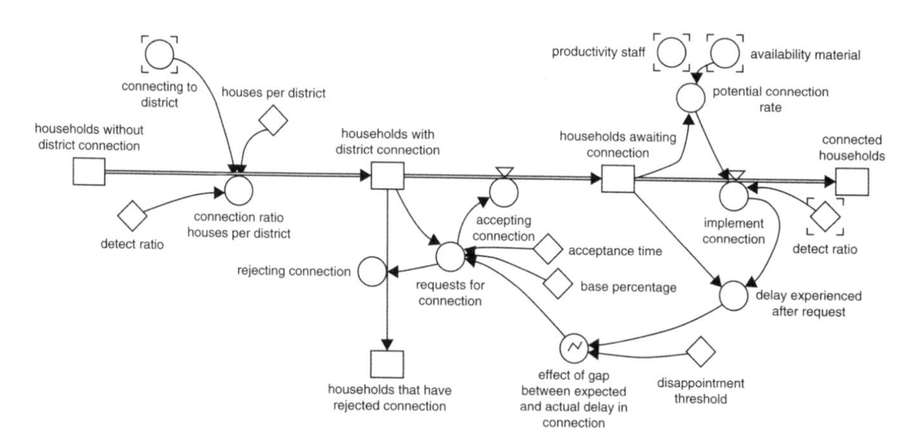

Figure 6.11. Stock-and-Flow Diagram

particular, when the supply of fiber-optic cable is insufficient to meet the demand, less could be supplied to Kennishoeve. The municipality has undertaken to order supplies for both itself and the service providers every two weeks and to store them. It has further undertaken to allocate cable between itself and the service providers in proportion to the demand for district and household connections. Additionally, the municipality is prepared to devote two teams of workers to laying the district connections, whereas the service providers intend to recruit and train additional staff in response to the requests for connections. The effects of these decisions are unclear at this stage. The influences of limitations in staff and material availability on the connection of districts and households are shown in the stock-and-flow diagram in Figure 6.11.

Each household begins in the stock of households without district connection and passes to the stock of households with district connection according to the rate at which the municipality can make these connections. The connection rate of houses per district depends on the number of houses (a constant), the percentage of defective cables (a constant), and an auxiliary variable "connecting to district" that is, in turn, influenced by the availability of material and the municipal teams. A household leaves the stock of households with district connection to move either to the stock awaiting connection or to the stock that have rejected connection. The rate of acceptance or rejection of connection is directly proportional to a base percentage (constant) and inversely proportional to the decision time (constant). The rate is modified by the effect of disappointment; that is, when the gap between expected and actual connection delay increases and exceeds the disappointment threshold, the rate at which connection is rejected increases. This effect is captured in a nonlinear graph function. The households awaiting connection move to the stock of connected houses at the implement connection rate. This is influenced by the availability of material, the productivity of staff, and the percentage of defective cables (a constant).

During the specification of the stock-and-flow diagram, the equations are specified per variable in the model. For instance, following specification of the rate equations accepting connection and implementing connection, the equation for the stock variable households awaiting connection is defined by the difference equation:

$$\frac{d}{dt}(\text{awaiting connection}) = \frac{d}{dt}(\text{accepting connection} - \text{implement connection})$$

The output of the cable-to-the-curb model (Figure 6.12) indicates that Kennishoeve would be unsuccessful in achieving the goal of connecting 50% of the households via fiber-optic cable within three years (connection shown by the thick solid line). This confirms the municipality's concerns about the project. The model explains the reasons for the failure: the delay in being connected is so long that many people decide to reject it. The disappointment loop is of overriding influence in the behavior of the system.

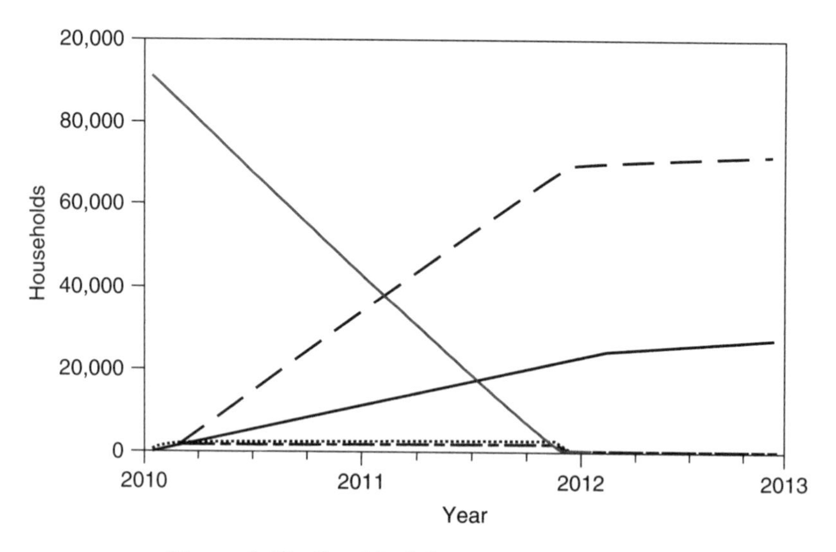

Figure 6.12. Graphical Output from the Model

In Figure 6.12, households without district connection = thin solid line; with district connection = shaded line; rejected connection = dashed line; awaiting connection = dot and dashed line; and connected = thick solid line.

Several alterations in operation and policy were considered to redress this result. Modeling demonstrated that the project is feasible provided that the constraining effects of material and staff availability are addressed, the expectations of customers are actively managed, and the municipality does not connect districts too quickly. The last result ran counter to the decision makers' intuition. They had planned to devote teams and material to the project to ensure its success. Instead, they needed to work steadily, but at a slower rate, and focus on ensuring a steady supply of cable to the service providers and to themselves throughout the project.

The value of the insights from the modeling study rest on the validity of the assumptions made in building the model. Interested readers are referred to the References for more extensive works, such as Forrester (1961), to the technology forecasting studies mentioned below, and to the system dynamics software websites for more examples of small models.

6.7 GAMING

The methods discussed in this section can be useful in exploring alternative futures. For example, decision trees are a recommended way of thinking about attaching revenues, costs, and probabilities to technology decisions. Bayesian estimation is a way of thinking about how given or observed information changes decisions. The value of perfect or imperfect information is discussed. Finally, real options analysis, from financial analysis, is introduced.

6.7.1 Decision Trees

Decision trees parade under various names; however, all of them are tracing techniques (Section 9.4.2) that depict relationships between various members of sets of entities. An example illustrates how decision trees can be used. Figure 6.13 depicts a tree for a decision about an identification scheme for a logistics system.

The two options for the system are RFID and Barcode. If RFID is chosen, there are two additional options. One is to use a new technology that has a reading range for passive RFID tags up to 100 meters. The net return for this system is $1,000,000, but the probability of success is only 0.3. The other option is to use the existing technology for which the reading range is up to 6 meters with a net return of $750,000. The probability of success is much higher, 0.7, even though some continue to call it "black magic."

The alternative to the RFID is to use a barcode system. Here too there are two options. A new generation of readers is one. These units can read tags up to 1 meter away and don't require that the reader be perpendicular to the tag. The net return is $600,000 because there are a great many misreads. The probability of success for this system is only 0.6. The other option is to use existing standard barcode readers. It takes a lot of labor to read tags with a standard barcode. The reader must be perpendicular to the tag, and it must be flush, or nearly flush, to it. So, the net return over the study period is only $350,000.

The expected returns for the four alternatives are $1,000,000(0.3) = $300,000, $525,000, $360,000, and $260,000, respectively. Clearly, using existing RFID technology is the best alternative based on expectation.

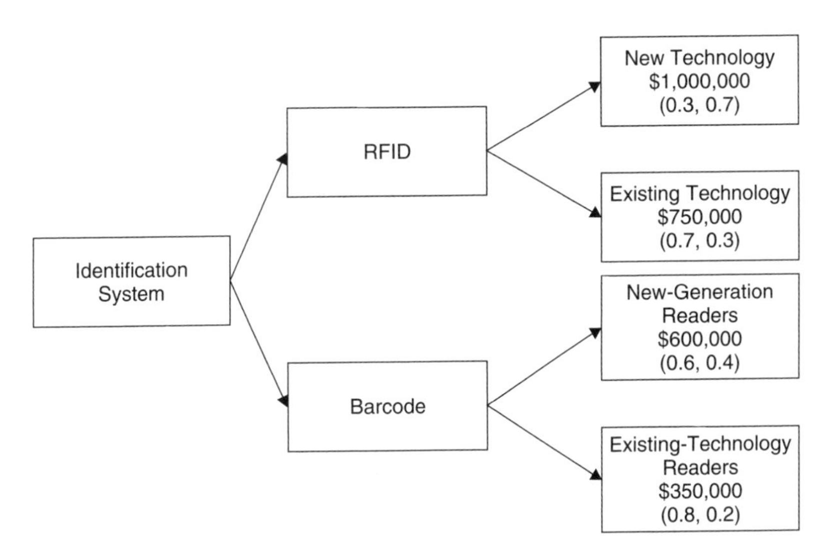

Figure 6.13. Decision Tree for an Identification System

This example does not consider the time value of money. What one usually would do in a situation like this is bring all of the expenditures and the returns over the study period back to the present by computing the NPV (Section 10.2.1) and make the decision based on which alternative has the highest NPV. Additionally, the probabilities involved in the decision in the example are either success or failure. There could be more than two outcomes, one of which always has a return of zero in the example, whereas the failure outcome might even produce a negative net return.

For instance, if the failures had net costs of \$400,000, \$300,000, \$200,000, and \$100,000, the expected returns would be \$1,000,000(0.3) – \$400,000(0.7) = \$20,000, \$435,000, \$280,000, and \$240,000, respectively. Based on expectation, using existing RFID technology is still the best alternative.

The purpose here is to demonstrate how a tree analysis can be used in making a decision about the adoption of technology. A real analysis should consider all of the possible outcomes, the numbers used should be as close to real as possible, and the time value of money should be taken into consideration.

6.7.2 Bayesian Estimation

Suppose you go to Sacramento, California, to begin your research comparing the heights of men visiting Sacramento versus the heights of men visiting La Paz, Bolivia. You know that the average height of men in the United States is about 177 centimeter. You expect about half of the men you see to be taller than that height and half to be less. Since height is a continuous measure, the probability that someone is exactly 177 centimeter is zero.

As you meander through the hotel lobby, you notice that far more men than expected are taller than 177 centimeter—a lot taller. Then you see a sign that says that "The Atlanta Hawks Basketball Team Meeting is in the Hawthorne Room." No wonder there are so many tall men. If you had said at the outset, "What is the probability that the next man you see in the hotel lobby is taller than 177 centimeter given that the Atlanta Hawks are staying in the hotel?" you would be making a Bayesian estimate. In symbols, $P(A|B)$, or the probability of A given B, which is called a *conditional probability*. The condition is that event B has occurred. The Bayesian estimate is given by the following (see Equation 6.13):

$$P(A|B) = [P(B|A)/P(B)]P(A) \tag{6.16}$$

Another example: suppose that the PSA (prostate-specific antigen) test for prostate cancer generates the following results:

- If a man 50 years of age or older is tested and *has prostate cancer*, the PSA test returns a positive result 90% of the time, or with probability 0.90.
- If a man 50 years of age or older *does not have prostate cancer*, the PSA test returns a positive result 30% of the time, or with probability 0.30.

At this point, you might think that only 30% of positive PSA test results on men 50 years of age or older are false, but let's go deeper into the analysis. Suppose that 5% of men 50 years of age or older have prostate cancer, so that a randomly selected man has a 0.05 prior probability of having the disease.

Let A represent the condition in which a man 50 years of age or older has prostate cancer and let B represent the evidence of a positive PSA test result. Then the probability that a man 50 years of age or older actually has prostate cancer given the positive test result is

$$P(A|B) = [P(B|A)P(A)]/[P(B|A)P(A) + P(B|\bar{A})P(\bar{A})] \quad (6.17)$$

Notice that $P(B)$ is a compound event requiring more computation than in the equation above.

$$P(A|B) = [(0.90)(0.05)]/[(0.90)(0.05) + (0.30)(0.95)]$$
$$P(A|B) = 0.136$$

These numbers are all made up. But if they were accurate, a man 50 years of age or older would have only a 13.6% chance of having prostate cancer if the test result was positive. It has long been the case that there are too many false positives with the PSA test. According to the National Cancer Institute (2010), only 25 to 30% of men who have a biopsy due to elevated PSA levels actually have prostate cancer.

Bayesian thinking is used quite often in technology forecasting. What will be the growth in sales on the Internet given no taxes on transactions? What will be the growth given that sales are taxed? What will be the adoption rate of RFID if organic ink successfully replaces copper in the tags, resulting in much cheaper tags? What will be the adoption rate if organic ink is not successful?

6.7.3 Value of Information

Suppose that you are going to South Florida to escape the cold weather and snow for a week. You have made plans to fly out on Saturday, and it's now Tuesday—four days to go before you and the family depart. You check the weather forecast on the Internet. It says that South Florida is going to have rain on Sunday and Monday, with a blast of cold air moving in on Tuesday, possibly lingering until Thursday.

But forecasts can be wrong. And the longer into the future you plan, the more error prone will be your forecast. Should you cancel the family vacation? Should you go and hope for the best? How much would you be willing to pay for a better forecast? Not better in the sense of more favorable but more accurate—say, 95% accuracy for the specific location where you will be staying rather than the 70% accuracy of the Internet forecast for a general region of the country. How much would you pay for perfect information—that is, 100% accuracy, perfect information. This is called *Value of Clairvoyance* (VoC).

As an example, suppose that you were going to make an investment in only one of the following: a common stock (CS), a mutual fund (MF), or a certificate of deposit (CD). Suppose that the market has a 50% chance of increasing, a 10% chance of staying even, and a 40% chance of decreasing. If the market increases, the stock investment will earn \$2000 and the mutual fund will earn \$1200. If the market stays even, the stock investment will earn \$400 and the mutual fund will earn \$300. If the market decreases, the stock investment will lose \$1600 and the mutual fund will lose \$1000. The certificate of deposit will earn \$350 independent of market fluctuations.

The expectations are computed as follows:

$$
\begin{aligned}
E(\mathrm{CS}) &= 0.5(\$2000) + 0.1(\$400) + 0.4(-\$1600) = \$400 \\
E(\mathrm{MF}) &= 0.5(\$1200) + 0.1(\$300) + 0.4(-\$1000) = \$230 \\
E(\mathrm{CD}) &= \$350
\end{aligned}
$$

The maximum of these expectations is CS. With no knowledge other than the probabilities indicated, invest in common stock with an expected value (EV) of \$400.

On the other hand, consider that you know ahead of time which way the market will turn. If you know the market is going up, just invest in the stock and make \$2000. (Actually, if you had such information, you could borrow money, sell assets, and raise money wherever possible, and make far more than \$2000.) If you knew, just before the beginning of the year, that the market was going to stay even or lose, just invest in the CD and make \$350.

The expected value given perfect information over a long time horizon is $EV|PI$, or

$$
EV|PI = 0.5(\$2000) + 0.1(\$350) + 0.4(\$350) = \$1175
$$

Consider that over a long time horizon, you only know the direction just before the beginning of the year. Then, with perfect information, you can expect to make an average of \$1175 per year. (Fifty percent of the time you will make \$2000, and the remainder of the time you will make \$350.) Having this knowledge each year is worth an expected \$1175 – \$400 = \$775. That is, \$775 is a reasonable amount to pay for perfect information before you know what the perfect information is.

There are two caveats. The first is that you don't know the direction of the stock market in advance. You just know about the general trend over many years (it's positive). If anybody tells you that they can predict the direction of the stock market in the short run, run away as fast as you can. There are too many interacting variables to succeed. Consider the effect on the stock market of the September 11, 2001, attack, as well as the dot.com bust of 2000 and the meltdown of the stock market in 2008.

The second caveat is that this example does not consider the time value of money (Section 10.2.1), that is, the investment in CS or MF could have been placed in a CD at the outset. The investment in CS or MF eliminates the opportunity to receive fixed income, say \$350 per year from a CD.

VoI is a useful concept in technology forecasting. For example, let's say that you are considering an investment in the RFID business. Specifically, you want to make tags using organic ink, greatly reducing the cost of the devices. You plan to invest $2,000,000 in the business. You can use the Internet to learn what is happening, and you have been to two RFID conferences in the last six months, paying close attention to what is said about organic ink. Another possibility is to use a company, like Gartner, Inc., to help you make a decision. You can see the range of services that they offer (Gartner 2010). These services aren't free, and the more directed the service, the higher its price. Thus, getting general information about RFID technology in a document prepared six months ago might cost $X, but getting specific industry advisory service directly pertaining to the investment that you are considering might cost $25X. The question that you need to answer is what the VoI is for each of your possible purchases from a company such as Gartner, Inc. In any case, the concepts learned in this section can be useful in making the decision.

6.7.4 Real Options Analysis

Real options analysis (ROA) has become a key management technique for estimating the value of investments. It can be utilized in situations where decision makers have flexibility in large capital budgeting in the presence of high uncertainty. A decision maker, for instance, could add a new wing to the building or decide not to. Alternatively, he or she could decide to add a new line of product or decide not to. In these cases, no money changes hands.

ROA is related to the concept of NPV. As we discussed at the end of Section 6.5.3, if the time value of money is 5%, would you rather have $1 today or one year from today? If you have $1 today, you can invest it and have $1.10 one year from today. Alternately, the NPV of $1 a year from today at 5% is $0.9524 today because, if $0.9524 is invested today at 5%, the interest after one year will be $0.0476 and you will have $1 in one year (see Section 10.2.1 for a more complete discussion of NPV).

With ROA, the uncertainty inherent in investment projects is usually accounted for by risk-adjusting probabilities (a technique known as the *equivalent Martingale approach*). Cash flows can then be discounted at the risk-free rate. In the NPV approach, this uncertainty is accounted for by adjusting the time value of money or the cash flows. NPV normally does not account for changes in risk over a project's life cycle and fails to appropriately adapt the risk adjustment.

The most widely used solution methods for ROA are closed-form equations, partial differential equations, and binomial lattices. These methods are rather complex from a computational standpoint. Fortunately, some new methods recently have been introduced that simplify the calculation of the real option value and thus make the numerical use of the methods easier for practitioners. These latest methods are the Datar-Mathews method (Matthews, Datar, et al. 2007) and the Fuzzy Pay-off method for real option valuation (Collan 2008). Additionally, software and Excel spreadsheet add-ins are available, as mentioned in the next section.

6.8 SOFTWARE SUGGESTIONS

6.8.1 Software for Regression

Microsoft Excel was used to build the regression models described in this chapter because it is so widely available. An Excel Trend macro has been offered (Technology Policy and Assessment Center 2010) to facilitate calculation of the models described.

Much more powerful statistical software is available. For example, Minitab is very easy to use and has a lot of capability. You can learn more about it at Minitab (2010). A trial download of the software is available at no cost. One of two gold standards for statistical software is SPSS. Find out more about it at SPSS (2010). The other gold standard is SAS. Find out more about it at SAS (2010).

Bayesian statistics can be computed within Minitab® 15 (Minitab 2010). A book by Albert uses the language R to compute Bayesian probabilities. R is a free software environment for statistical computing and graphics (R Project 2010). See *Bayesian Computation with R* (Albert 2009). It compiles and runs on a wide variety of UNIX platforms, Windows, and MacOS.

6.8.2 Simulation Analysis Software

For Monte Carlo simulation, two products are recommended: @Risk, which is an add-in to Microsoft Excel, and Oracle Crystal Ball. For system dynamics, three software packages are suggested: Powersim©, STELLA©, and Vensim®. Websites are shown in the references.

6.8.3 Software for Analysis of Decisions

Microsoft Office 2007 Visio (Microsoft 2010) or SmartDraw 2009 (SmartDraw 2010), among many others, could be used to draw a decision tree. Figure 6.13 was drawn initially using the Draw feature in Microsoft Office 2003.

6.8.4 Real Options Super Lattice Software

Software for ROA is available from Real Options Valuation, Inc. (Real Options Valuation 2010). The real options solved in this software include the following:

- American, Bermudan, Customized, and European Options
- Abandonment, Contraction, Expansion, and Chooser Options
- Changing Volatility Options
- Exotic Single and Double Barrier Options
- Financial Options, Real Options, and Employee Stock Options
- Multiple Underlying Asset and Multiple Phased Options
- Simultaneous and Multiple Phased Sequential Compound Options

Also, there are many Excel spreadsheets for solving real options problems. A partial listing of them is found at Real Options Software Webpage (2010).

6.8.5 Software Sites

For Monte Carlo analysis:

@Risk v. 5.5 (Palisade 2010)

Oracle Crystal Ball (Oracle 2010)

For system dynamics:

Powersim Software AS© 2009. Powersim Studio 8 (Powersim 2010)

STELLA®/iThink® v9.1.3. (ISEE Systems 2010)

Vensim® Version 5 2009. *User's Guide*, *Modeling Guide*, *Reference Manual* and *DSS Reference Supplement* (Ventana Systems 2010).

REFERENCES

Albert, J. (2009). *Bayesian Computation with R*, 2nd ed. Berlin, Springer.

Argote, L. and D. Epple. (1990). "Learning Curves in Manufacturing." *Science* **247** (4945): 920–924.

Banks, J., D. Hanny, et al. (2007). *RFID Applied*. Hoboken, NJ: John Wiley & Sons.

Banks, J., J. S. Carson, II, B. L. Nelson, D. M. Nicol (2010), *Discrete-Event System Simulation,* 5th ed. Upper Saddle River, NJ: Prentice Hall.

Bass's Basement Research Institute (2011), "Bass Model Functions in Excel Visual Basic," Retrieved 27 January 2011, from http://bassbasement.org/BassModel/VBBassModel.aspx.

Bass, F. (1969). "A New Product Growth Model for Consumer Durables." *Management Science* **15** (5): 215–227.

Christensen, C., M. Horn, et al. (2008). *Disrupting Class: How Disruptive Innovation Will Change the Way the World Learns*. New York, McGraw-Hill.

Collan, M. (2008). "A New Method for Real Option Valuation Using Fuzzy Numbers." *IAMSR Research Report.* Retrieved 13 January 2011 from http://iamsr.abo.fi/publications/openFile.php?pub_id=466.

Engineering Statistics Handbook. (2010). "What Are Variables Control Charts?" Retrieved 29 August 2010 from http://www.itl.nist.gov/div898/handbook/pmc/section3/pmc32.htm.

Fisher, J. C. and R. H. Pry. (1971). "A Simple Substitution Model of Technological Change." *Technological Forecasting and Social Change* **3**: 75–88.

Forrester, J. W. (1961). *Industrial Dynamics*. Waltham, MA: Pegasus Communications.

Forrester, J. W. (2007a). "System Dynamics—a Personal View of the First Fifty Years." *System Dynamics Review* **23** (2–3): 345–358.

Forrester, J. W. (2007b). "System Dynamics—the Next Fifty Years." *System Dynamics Review* **23** (2–3): 359–370.

Gartner (2010). "Gartner Research," Retrieved 19 May 2010, from http://www.gartner.com/technology/home.jsp.

Google. (2010a). "Google Patents." Retrieved 21 May 2010 from http://www.google.com/patents.

Google. (2010b). "Google Trends." Retrieved 19 May 2010 from http://www.google.com/trends.

Google. (2010c). "Insights for Search." Retrieved 21 May 2010 from http://www.google.com/insights/search/.

Gordon, T. J. (2009). "Trend Impact Analysis." Futures Research Methodology Version 3.0. Retrieved 13 January 2011 from http://www.millennium-project.org/millennium/FRM-V3.html.

Gordon, T. J. and H. Hayward (1968). "Initial Experiments with the Cross-Impact Matrix Method of Forecasting." *Futures* **1** (2): 100–116.

Halverson, T., J. J. Swain, et al. (1989). *Analysis of the Cross-Impact Model*. Technical Report. Atlanta, Georgia Institute of Technology, Technology Policy and Assessment Center.

Hearne, J. W. (1987). "An Approach to Resolving the Parameter Sensitivity Problem in System Dynamics Methodology." *Applied Mathematical Modeling* **11**: 315–318.

Hu, H. B. and X. F. Wang (2009), "Evolution of a Large Online Social Network," *Physics Letters A* 373 (12–13): 1105–1110.

ISEE Systems. (2010). "The World Leader in Systems Thinking Software." Retrieved 29 August 2010 from http://www.iseesystems.com/.

Lenz, R. (1985). *Rates of Adoption/Substitution in Technological Change*. Austin, TX, Technology Futures, Inc.

Levitt, S. and S. J. Dubner. (2005). *Freakonomics: A Rogue Economist Explores the Hidden Side of Everything*. New York, William Morrow/HarperCollins.

Linden Research. (2010). "What Is Second Life?" Retrieved 20 May 2010 from http://secondlife.com/whatis/.

Matthews, S., V. Datar, et al. (2007). "A Practical Method for Valuing Real Options: The Boeing Approach." *Journal of Applied Corporate Finance* **Spring**: 95–104.

Meadows, D. H. (1976). "The Unavoidable a Priori." In *Elements of the System Dynamics Method,* ed. J. Randers. Cambridge, MA: MIT Press: 23–57.

Microsoft. (2010). "Microsoft Visio." Retrieved 29 August 2010 from http://office.microsoft.com/en-us/visio/.

Minitab. (2010). "Software for Quality Improvement." Retrieved 29 August 2010 from http://www.minitab.com/en-US/default.aspx.

Miranda, L.C.M. and C.A.S. Lima. (2010). "A New Methodology for the Logistic Analysis of Evolutionary S-Shaped Processes: Application to Historical Time Series and Forecasting." *Technological Forecasting and Social Change* **77**: 175–192.

National Cancer Institute. (2010). "Prostate-Specific Antigen (PSA) Test." Retrieved 29 August 2010 from http://www.cancer.gov/cancertopics/factsheet/Detection/PSA.

Oracle. (2010). "Crystal Ball." Retrieved 29 August 2010 from http://www.oracle.com/us/products/middleware/bus-int/crystalball/index-066566.html.

Palisade. (2010). "@Risk: Risk Analysis Software Using Monte Carlo Simulation for Excel." Retrieved 29 August 2010 from http://www.palisade.com/risk/.

Powersim. (2010). "Powersim Software." Retrieved 29 August 2010 from http://www.powersim.com/.

R Project. (2010). "The R Project for Statistical Computing." Retrieved 29 August 2010 from http://www.r-project.org/.

Real Options Software Webpage. (2010). "Real Options Software Webpage," Retrieved 29 August 2010 From http://www.puc-rio.br/marco.ind/software.html.

Real Options Valuation. (2010). "Risk Simulator: Real Options Valuation." Retrieved 29 August 2010 from http://www.realoptionsvaluation.com/realoption.html.

Sage, A.P., (1977). Methodology for Large-scale Systems. New York: McGraw-Hill.

SAS. (2010). "Products and Solutions: Statistics," Retrieved 29 August 2010 from http://www.sas.com/technologies/analytics/statistics/index.html.

Semiconductor Industries Association. (2010). "Welcome to the Semiconductor Industry Association." Retrieved 29 August 2010 from http://www.sia-online.org.

SmartDraw. (2010). "SmartDraw: Communicate Visually," Retrieved 29 August 2010 from http://www.smartdraw.com/.

Smith, N. (2007). "Young Black Americans at Higher Risk for Colon Cancer." Retrieved 29 August 2010 from http://www.bet.com/Lifestyle/bodysoul/Colon_Cancer_Younger_Blacks_Life_BAS_article.htm?wbc_purpose=Basic&WBCMODE=PresentationUnpublished.

SPSS. (2010). "IBM SPSS Statistics." Retrieved 29 August 2010 from http://www.spss.com/software/statistics/.

Sterman, J. D. (2000). "Truth and Beauty: Validation and Model Testing." In *Business Dynamics: Systems Thinking and Modeling for a Complex World,* ed. J. D. Sterman. New York, McGraw-Hill: 845–891.

Srinivasan, V. S. and C. Mason. (1986). "Nonlinear Least Squares Estimation of New Product Diffusion Models." *Marketing Science* **5** (2): 169–178.

Technology Policy and Assessment Center. (2010). "Technology Forecasting and Assessment Tools." Retrieved 4 June 2010 from http://www.tpac.gatech.edu/tfat.php.

Trellian. (2010). "Keyword Discovery: Advanced Keyword Research Tool and Search Term Suggestion Tool." Retrieved 21 May 2010 from http://www.keyworddiscovery.com/.

U.S. Patent and Trademark Office. (2010). "USPTO FY 2008 Performance and Accountability Report." Retrieved 21 May 2010 from http://www.uspto.gov/about/stratplan/ar/2008/index.jsp.

Ventana Systems, I. (2010). "Vensim." Retrieved 29 August 2010 from http://www.powersim.com/.

7

FOCUSING PHASE: USING SCENARIO ANALYSIS

Chapter Summary: This chapter discusses a versatile technique for technology forecasting: scenario analysis. Scenario analysis is useful because it serves as a vehicle for integrating diverse sources of information and focusing the forecast toward meaningful futures. New ideas in scenario analysis are discussed. The chapter introduces the third phase of forecasting.

Recall once more the three-phase approach for any forecast advocated in Chapter 3. This chapter begins the third phase, focusing. Focusing zooms in on the few most significant directions of the technology; develops as deep an analysis as possible within forecast limits; makes definitive decisions about the direction of the technology; and provides a basis for implementing the forecast.

Guidance for selecting the most promising paths for the technology that is identified in earlier phases is given in Chapter 11, while a deeper analysis of their economic implication is provided in Chapters 8 and 10. Ways to identify and evaluate impacts on or by the technology on its environments are addressed in Chapter 9. This chapter concentrates on one method for focusing the study: the scenario. A scenario is a descriptive sketch intended to give a more or less holistic view of a possible future. The great strength of such "future narratives" is their richness—a richness that can set the context within which the meaning of individual factors can be assessed. Scenarios may be either largely verbal or composed of quantified forecasts that together convey a holistic perspective. While the former are more common, it is not unusual for the narrative to be "hung on hooks" provided by one or more quantified forecasts. Whatever form they take, scenarios blend insight with storytelling skill to provide a relatively complete picture illustrating possible outcomes. It is common to provide a set of three scenarios that span what the forecaster believes to be the range of possible futures: a surprise-free projection, the worst case projection, and, the best case projection.

As noted, this chapter introduces the focusing phase of forecasting activities. Scenario analyses play a critical role in the focusing activity. Scenarios provide a means for integrating a range of information—quantitative and qualitative, objective and normative—into a compelling forecast for decision making. In Section 7.2.2, more details about the scenario construction activity and its role in developing focused scenarios are provided.

For some, scenarios are inextricably tied to the management of uncertainty. Therefore, the chapter first characterizes the range of uncertainties present in a forecast and reviews the forecasting techniques that are most appropriate for managing the uncertainty. The chapter then discusses the scenario analysis technique by outlining both the steps of the approach and the various kinds of alternative analysis that are used by practitioners. In the final section, extensions to the scenario analysis approach are highlighted.

7.1 UNCERTAINTY

Technology forecasters need to answer questions about what is known and what is unknown: they must even ask if it is possible to know the future. Answers to these questions guide the selection of forecasting techniques and provide a paradigm for structuring the forecast. This section provides a framework for characterizing and managing uncertainty. The framework is then used to examine the position of scenario analyses among major families of technology forecasting techniques (Chapter 2). Finally, a word about the adaptive paradigm is provided.

7.1.1 Uncertainty Frameworks

Knowledge about new technological systems can be characterized quite simply; either the system and its behavior are known or they are not. Complete system knowledge means that system operation and the full consequences of its behavior are known, and the means to control or adapt it to achieve desired current and future outcomes are understood. Clearly, this is a tall order for any new technology. Thus, knowledge will be a matter of degree.

Table 7.1 compares what is known with what is knowable. This simple matrix, inspired by the Johari diagram (Luft and Ingham 1955), is sufficient to examine categories of technological knowledge.

TABLE 7.1 Categories of Knowledge

	System Is Knowable	System Is Unknowable
System Is Known	known knowns	known unknowns
System Is Unknown	unknown knowns	unknown unknowns

These are futures about which forecasters:

- Are quite certain—known knowns
- Recognize but have no information about—known unknowns
- Are sadly ignorant—unknown unknows
- Could gather facts if they knew they were needed—unknown knowns

This characterization of knowledge became popular after a briefing by U.S. Secretary of Defense Donald Rumsfeld in 2002. Rumsfeld's assertions were simultaneously mocked and championed. However, this categorization has a much older history in military planning. Ultimately, it may date back to the words of Confucius:

> To know that we know what we know, and that we do not know what we do not know, that is true knowledge.

Writers who write about uncertainty and decision making have acknowledged the continuum of knowledge from certainty to ignorance and have referred to it as the *level* of uncertainty (Walker, Harremoes, et al. 2003). Labels for the categories of uncertainty include *stochastic uncertainty* (known unknowns) and *scenario uncertainty* (unknown unknowns). Unfortunately, such labels confuse the source of uncertainty with the choices the technology forecaster makes to manage it. It is essential to separate the two for both practical and conceptual reasons.

Discussions of uncertainty often ignore the important category of unknown knowns. These include things that were once known but have been forgotten, as well as things one could know if one made the effort. This category includes tacit information that is often unacknowledged by organizations but may be perfectly apparent to individuals within them. These sources of uncertainty should not be neglected in developing a strategy for managing uncertainty.

7.1.2 Source and Nature of Uncertainty

There are limits to knowledge; therefore, there are uncertainties. Some may be merely programmatic, and thus depend on the time and resources applied to the forecasting project. This class of uncertainties can be characterized as *epistemic*—having to do with the current state of knowledge. Some limits result because the world is complex and inherently unpredictable; thus, there are uncertainties related to *variability*. Finally, there is uncertainty that results from errors including measurement error, as well as errors in judgment and execution. This source of uncertainty is called *prejudicial uncertainty* by some authors.

Together, these three sources are known as the *nature* of uncertainty (Walker, Harremoes, et al. 2003). Walker and coauthors also introduce an uncertainty called *location,* which involves uncertainties in using models to guide decision making. Other authors have considered whether the source of uncertainty is primarily technological, economic, legislative, or social in character (Linstone 1988; van der Sluijs, Risbey, et al. 2003).

7.1.3 Uncertainty and the Adaptive Paradigm

The existence of unknown unknowns should cause technology managers to examine their approaches. While the *plan and act* paradigm is appropriate for many kinds of uncertainty, a new approach is required to deal with deep levels of uncertainty. This new approach—known by many as the *adaptive paradigm*—involves making small, robust decisions and then adapting strategies that remain successful even when new information emerges (Dewar 2002). Traditional techniques for economic valuation (discussed in Section 10.3) are often quite brittle when underlying assumptions are invalidated. Scenario analysis techniques can assist in such a situation by affording a broad baseline of possible future outcomes on which to base economic assumptions.

Collingwood (1981) discusses a fundamental dilemma for those who would anticipate the future impacts of new technology. Often the impacts of new technology are poorly understood before the technology is widely adopted. Yet, after the technology has been adopted, it is fully entrenched in networks of power and authority, making changes in control very difficult. For instance, the true impact of the automobile was poorly understood when it was first introduced. Now its impact—on family life, urban development, noise, congestion, and air quality—is all too clear. Nonetheless, many regions are inextricably dependent upon the use of the automobile. Collingswood's recommendations are that society should maintain a wide range of technological options and show flexibility in making incremental choices across these technologies. We will be discussing the impacts of new technology further in Chapter 9.

The forecasters do not necessarily need new techniques when supporting decision making within the adaptive paradigm. Rather, they need to constantly question assumptions and to test whether existing assumptions still hold as the future grows clearer. This is not a bad idea whatever form of decision making is supported. The next section offers some recommendations for selecting specific families of forecasting techniques in light of the kinds of uncertainty faced by decision makers.

7.1.4 Techniques for Addressing Uncertainty

Five families of forecasting techniques were introduced in Section 2.2.1. Considering which techniques among them are most effective in dealing with various types of uncertainty can help the forecaster manage that uncertainty.

The known knowns are managed by descriptive approaches and matrix techniques. Unknown knows are best handled by monitoring and expert opinion. The best techniques for dealing with known unknowns vary according to the source of uncertainty. If the key issue is variability, statistics and trend analyses may be best. If it is knowledge, then modeling and simulation may be the most useful approaches. If human error is the key issue, then tracing models such as roadmapping or decision-aiding techniques may be the best bets. The unknown unknowns are arguably best handled in one of two ways—using creativity techniques

(Section 4.3) or the scenario techniques described in this chapter. Either can be used to address a wide variety of uncertainties.

The versatility of scenario analysis can be confusing because there are so many different approaches to it. The rest of this chapter discusses the methodology of scenario analysis and explores the dilemma of adapting this versatile technique to achieve the specific needs of a technology forecast.

7.2 SCENARIOS

A usage editor notes that "the over-use of the word 'scenarios' in various loose senses has attracted frequent hostile comment" (Oxford English Dictionary 2011). Indeed, in technology forecasting, the term often is used too loosely. However, *scenarios* is a technical term that refers to a specific family of techniques. It is therefore more useful, and certainly more consistent with current usage, to attempt a clear and unambiguous definition of how scenarios are created and used.

The Oxford English Dictionary provides three definitions of uncertainty. The second is the one with which the remainder of the chapter will be concerned (OED Online 2011):

> A sketch, outline, or description of an imagined situation or sequence of events; esp. . . . (b) an outline of an intended course of action

Scenarios are a natural outgrowth of efforts to help organizations adapt to their rapidly changing environments. In assisting organizations in making decisions, forecasters face several problems. Qualitative techniques, such as checklists, often fail in the important tasks of engaging decision makers and facilitating organizational learning. On the other hand, solely quantitative approaches often are not appropriate. Moreover, narrow assumptions about the future can result in overly rigid prescriptions for future actions. Recognition of these issues has led to development of several distinct scenario methodologies.

Strength, Weakness, Opportunities, Threats (SWOT) analyses were some of the precursors to modern scenario analysis approaches (Spies 1994). Newer approaches include the Shell scenario approach (Wack 1985a, 1985b; Schwartz 1991; Van der Heijden 2005), as well as the RAND planning approach (Dewar 2002; Lempert, Popper, et al. 2003). In the following sections, a methodology that integrates both of these perspectives is presented.

7.2.1 Steps in Creating Scenarios

The narratives, in many ways, are the most characteristic feature of scenario analysis. However, a lot of preliminary work must precede them. There are five steps in constructing a scenario:

1. Identifying variables
2. Developing levels of measurement

3. Characterizing significant variables
4. Creating scenarios
5. Writing the narrative

Each is discussed in turn below.

Step 1 entails identifying meaningful external variables for the technological system in question. The technology delivery system constructed for the forecast will help, but the forecaster should think broadly and comprehensively, as if participating in a brainstorming exercise (see Section 4.3.7). For instance, he or she might search through the space of possible social, political, economic, institutional, legal, environmental, or technological variables that might affect the system. The variables that are identified are called *scenario variables*. Subsequent steps involve presenting, structuring, and refining them.

Step 2 involves developing levels of measurement. Recall that measurement scales were discussed in Section 5.4 (see also Section 11.4.1). The goal in this step is to combine scales and variables to pursue one or more questions that are central to the forecast. Figure 7.1 shows six scenario variables with associated levels of measurement. The figure is part of a hypothetical business scenario planning exercise.

7		Global Objectives				
6		Major European Dominance		Dramatic Reduction in Total Contract Risk	World Industry Benchmark	
5	Very Strong (>£6)	Discrete but Significant Advantages	Best Practice in Industry	Significant Improvement	UK Industry Benchmark	Extensive Major Opportunities
4	Strong (£5.50)	Incremental Growth	Better than Average	Today's Standard	Preferred Choice of Customers	Strong but Patchy
3	Neutral (£5)	Maintenance of Position	Industry Average	Significacly Worse Than Today	Minor Difficulties Only	Some Problems but Essentially Sound
2	Weak (£4)	Localization	Below Requirements	One Major Contract Failure	Some Major Shortcomings	Fragile Markets
1	Very Weak (<£4)	Survival	Poor, Requiring Major Restraining	Multiple Major Failures	Customers Reluctant to Place Work	Major Problems in Essential Markets
Scenario Variables	Share Price	Extent of Objectives	Internal View of Abilities	Risk on Contracts	External Credibility	Procurement Environment
Short Key	S	O	A	R	C	E

Figure 7.1. Identifying and Measuring Scenario Variables
Source: Adapted from Powell and Coyle (1997)

The scenario developer should ask questions. First, which variables have the greatest impact on the system? Some developers go further to ask whether these impacts are desirable or undesirable. Second, which variables are the most uncertain, or alternatively, which are most controllable by the decision maker? Impact assessment, which is discussed in Section 9.5, also involves selecting high-impact variables. Risk assessment (Section 10.3) addresses the quantitative or qualitative likelihood of certain events coming to pass. These techniques can be productively integrated within a scenario framework.

There are two classes of variables in scenario analysis: *scenario variables,* which are inside the scenario itself, and *conditioning variables,* which are used to help determine the impact of scenario variables. It is important to be clear about the underlying assumptions guiding the choice of high-impact scenario variables. Conditioning variables typically are not described in the scenario. They are further discussed in the next section concerning the types of scenario analysis.

The choice of questions and the resultant measurement across scenario variables enable the developer to characterize the significant variables as Step 3 of the process. Often a matrix is used to structure the ratings. For instance, a simple two-point ordinal scaling, for which all variables are placed in a matrix structure, results in Table 7.2.

The high-uncertainty, high-impact variables are included for further analysis. The low-uncertainty, high-impact variables are used as *color variables* and are discussed further below. The low-impact variables are excluded from further analysis.

Step 4 of scenario analysis involves creating the scenarios themselves. For each of the variables discovered in Step 3 that are to be included, the builder must consider a set of discrete outcomes. These outcomes should capture the variety of high-impact outcomes associated with the variable. For instance, just two variables with two outcomes result in the scenario matrix shown in Table 7.3.

Since scenario creation involves examining variables and their permutations, the number of scenarios to be considered can increase very rapidly. The scenario developer must therefore critically examine the choice of scenario variables and

TABLE 7.2 Taxonomy of Variables for Scenario Analysis

	Variable Is Low Impact	Variable Is High Impact
Variable Is High Uncertainty	Excluded variables	Included variables
Variable Is Low Uncertainty	Excluded variables	Color variables

TABLE 7.3 Simple 2x2 Dichotomy for Developing Scenarios

	Variable 2, Outcome A	Variable 2, Outcome B
Variable 1, Outcome A	Scenario 1	Scenario 2
Variable 2, Outcome B	Scenario 3	Scenario 4

the relationships between them. In particular, the potential scenarios must be evaluated for consistency and plausibility. Some combinations of variables may be very unlikely or infeasible and can be eliminated.

This process of selecting combinations of scenario variables for consistency and plausibility is at the very heart of the scenario creation process. It is also at the crux of the focusing stage of technology forecasting. By detailed consideration of plausible scenario logics, the forecaster helps to integrate diverse sources of information and opinion into a single and actionable forecast. There are a number of different techniques that can aid the forecaster in the focusing task.

Combinations of scenario variables can be used in such creativity exercises as lateral thinking and checklists (Sections 4.3.2 and 4.3.3). The most compelling scenarios can be adopted for further refinement and exploration. Some forecasters use a form of morphological analysis (discussed in Section 4.3.5) to select the logically consistent combinations of variables. Expert opinion is useful here (Section 5.1). Simulation and modeling techniques can provide a principled approach to selecting coherent scenario variable combinations. Linear regression, particularly when used with confidence intervals (Section 6.2), can provide boundaries on plausible scenarios. Simulation techniques are also useful for grounding scenarios within a coherent scenario logic. Cross-impact techniques (Section 6.4), Monte Carlo simulation (Section 6.5), and system dynamics (Section 6.6) can all be used productively in this context.

Strategic scenarios (a type of scenario discussed in Section 7.2.2) require an integrated perspective on economic, strategic, and institutional factors. These concerns are further addressed in chapters that follow. Economic modeling is discussed in Section 8.2. Strategic and institutional analyses are discussed in Section 8.4. Normative scenarios (also discussed in Section 7.2.2) require that the forecaster work with decision makers to attain a firm idea of which kinds of futures are most valued and why these futures are most valued. Hedonic techniques (Section 8.3.) can help here. Likewise, more formal techniques for multicriteria decision making (as discussed in Section 11.4) can guide the process for the consistent valuation of future outcomes.

An example of this process of examining scenario variable combinations is offered by Powell and Coyle in Figure 7.2. As can be seen, each scenario variable (S, O, A, R, C, E) has one of up to seven different levels. The figure shows the permissible combinations of the row variables with the column variables. Since the matrix is symmetric, the lower diagonal is not presented. Along the diagonal, each level of each variable is presented. Certain combinations of variables are logically infeasible. For instance, a value of 1 on scenario variable S (share price) implies that the possibility of the company's achieving its objectives (scenario variable O) must be limited at best. This is scored with combination S1 and O1, O2, O3 in the figure, as seen at the intersection of variables S and O, first entry (S1).

Step 5, the final step of scenario analysis, is to develop narratives in which the scenarios are richly and plausibly presented. The color variables—the high-impact, high-certainty variables—are introduced at this stage to provide additional wealth of detail.

	S	O	A	R	C	E
S	1	123----	1234-	1234--	123---	1234-
	2	-234---	1234-	-234--	-234--	1234-
	3	--345--	-234-	--34--	--34--	-234-
	4	---45--	-234-	---45-	--345-	--345
	5	---4567	--345	---456	--3456	--345
O		1	12---	123---	12----	12---
		2	12---	123---	12----	12---
		3	-234-	-234--	-23---	123--
		4	-234-	-234--	--34--	-23--
		5	--34-	---45-	--345-	-234-
		6	---45	---456	---45-	-2345
		7	---45	----56	----56	-2345
A			1	12----	12----	12345
			2	-23---	-23---	12345
			3	---4--	--3---	12345
			4	---45-	--34--	12345
			5	----56	----56	12345
R				1	1-----	12---
				2	12----	123--
				3	123---	1234-
				4	-23---	-2345
				5	--345-	--345
				6	--3456	--345
C					1	12---
					2	12---
					3	1234-
					4	12345
					5	12345
					6	12345
E						1
						2
						3
						4
						5

Figure 7.2. Matrix of Acceptable Combinations of Scenario Variables
Source: Adapted from Powell and Coyle (1997)

7.2.2 Types of Scenarios

The following section describes a variety of ways in which scenarios can be applied. There are alternative ways of conceptualizing the embedding of technology in society; one of them, the technology delivery system (TDS), provides a system analytic perspective. It can be used to clarify how scenario analysis assists in isolating and communicating key variables in the forecast.

The TDS was described in detail in Section 2.2.1. To review, there are three components to the TDS: the technology *system* involves the operational parts of delivering the technology; the *arena*, or societal context, contains the political and decision-making entities that make decisions concerning the funding, design, and governance of the technology; and the *external environment* contains external forces that can affect or perhaps disrupt the operation of the technology system. The arena generally involves multiple stakeholders with diverse perspectives, including institutions that monitor and report whether the technology is delivering desirable outcomes for society. Each of these subsystems communicates with the other subsystems. See Section 2.2.1 for schematic representations of the TDS.

The TDS delivers outcomes that are evaluated by key stakeholders in the arena. Outcomes are quantities that potentially can be objectively measured. Values are communicated from the evaluation elements of the arena to the decision-making institutions. Values, unlike outcomes, are rarely measurable by objective units of analysis. Further, the communication of values is socially, strategically, and politically challenged by the need to aggregate societal preferences into a set of priorities or objectives. The decision-making elements of the arena implement policies affecting the system in light of perceived limitations or shortcomings in the way it is operating.

The TDS allows description of three distinct types of scenarios. Each isolates system inputs and outputs for further analysis, communication, and reflection. The first, called the *external forces scenario,* isolates external forces in light of the outcomes delivered by the system. Decision makers encounter the challenge of unknown unknowns from this part of the system. Note that the impact of external forces scenarios is conditioned on a judgment of the key outcomes of interest. Thus, outcomes of interest become key *conditional variables*. External forces scenarios are useful for portraying the current and future state of the system in light of fundamental uncertainties. Such scenarios also are useful for developing and testing alternative simulation models of system performance.

Another class of scenarios is called *strategic scenarios*. These involve a concrete set of policies, strategies, or tactics enacted to affect the outcomes produced by the technology system. The key uncertainties here involve the ability to achieve different outcomes of interest since system performance is often a source of persistent uncertainty. Considering the roles and responsibilities of stakeholders as the system achieves, or fails to achieve, desired outcomes is an important aspect of this kind of scenario building. Thus, the outcomes of interest are the key scenario variables. Strategic scenarios are conditioned on a set of assumptions about the status quo and the strategies, policies, or tactics that will be deployed in any given circumstance.

The third type of scenario involves isolating the outcomes of interest as scenario variables and considering how these outcomes impact stakeholder values. These are known as *normative scenarios*. In this type, values are the key scenario variables and are conditioned on the outcomes of interest. Values may be challenged when new information becomes available, when new outcomes become apparent, or when baseline delivery of essential outcomes is challenged. Thus, normative scenarios are best used to ask difficult questions among stakeholders concerning the valuation of new technology. The examples in the next section demonstrate how normative scenarios were used in South Africa to envision new coalition governments in the postapartheid era.

A typical technology development scenario most likely will require elements of all three types of scenario. The forecaster should use their TDS to consciously design their scenarios according to the needs of their forecast. This involves recognizing the sources of uncertainty that may affect technological outcomes, selecting the correct variables for the scenario, and then interpreting the variables appropriately, given the system assumptions.

To summarize, there are three types of scenario analysis—external forces, strategic, and normative scenario analysis. Since the 1990s, it has become clearer that each type serves a distinct purpose. As Robinson (2003) notes, more recent trends in scenario analysis have involved the purposeful combination of these three types given the needs of the study.

7.3 EXAMPLES AND APPLICATIONS

The following subsections provide three examples of scenario analysis. The goals are to demonstrate the diversity of approaches used in scenario analysis and to highlight the wide range of social and technical systems to which it has been applied. The first example involves the use of scenarios in renewable energy planning (de Vries, van Vuuren, et al. 2007). It entails external forces scenarios. The second example involves the creation of strategic scenarios of technology change in the field of pervasive computing (Satyanaran 2001). The final example presents normative scenarios that were used to facilitate social change in South Africa (Mont Fleur Scenarios 1992).

7.3.1 Scenarios for Renewable Energy Planning

External forces scenarios involve testing key assumptions and their effects; this type of scenario is often highly quantitative and relies upon using models or simulation techniques. In this case, the associated models are the emission scenarios produced by the International Panel for Climate Change (IMAGE-team 2001).

De Vries and coauthors (2007) used the models to create robust projections of renewable energy production in the first half of the twenty-first century. The three renewable energy sources considered were wind, biomass, and solar energy. As the authors note, previous emissions studies were based on critical assumptions about land use and technological viability. Therefore, to critically test assumptions and to better understand a range of archetypal futures, they varied the underlying assumptions in established numerical models.

The key scenario variables are agricultural (meat consumption, crop intensity, fertilizer use), technological (technology development, agricultural management), economic (GDP and food trade), and population (total world population). Correlations between variables resulted in two major axes—material/economic concerns opposed by environmental/social concerns—and a global orientation opposed by a regional orientation. A two-by-two dichotomy on each of these axes results in four major scenarios. The authors adopted the material-economic-global scenario) as their default scenario and used it to contrast the other three.

These scenarios are used to anchor the authors' conclusions about the regional prevalence of different renewable energy sources. The authors conclude that across the world as a whole, renewable energy will be sufficient to meet energy demands. Specific regions, however, will be challenged. Japan and Southeast Asia will find less than 10% of their energy needs met by renewable sources. Further, the intermittent supply caused by variability in solar or wind power may further jeopardize OECD nations, Eastern Europe, and South Asia.

The authors find that solar energy has the highest potential for affordable energy production worldwide. Economic potentials are particularly high in the desert regions of North Africa and the Middle East. Nonetheless, this solar energy future also is critically dependent upon the specific world future. Should the world adopt a more regionalist perspective, the spread of cost-reducing innovations in solar power will slow, and renewable-energy regions will be forced to use their land for food production. These events could completely negate the capacity to produce low-cost solar energy in 2050. Regardless, biofuels (in tropical regions) and wind energy (in vast temperate plains regions) will still operate at a lower economic potential.

7.3.2 Pervasive Computing Scenarios

Pervasive computing, also known as *ubiquitous computing* (Weiser 1991), involves computing power that is so widespread, and so carefully integrated with the background of daily life, that it becomes ever-present. Driving technologies for pervasive computing involve distributed systems as well as mobile computing. Pervasive computing progressively builds upon these older technologies.

Distributed computing entails building systems that enable remote communication and provide remote information access, are fault tolerant, afford good security, and provide high access availability. Mobile computing adds the additional technical challenges of mobile networking and information access. Given the power requirements, computer systems also must acknowledge their energy needs and adapt accordingly. Furthermore, mobile computing applications must account for their own local computing environment, as well as the social context of their use. Pervasive computing extends these problems to include building smart spaces where computing is embedded directly in architecture. Pervasive computers should be invisible as well as locally scalable. The penetration of pervasive computing into our homes, businesses, and cities may be a lengthy and uneven process. Pervasive computing applications must therefore cope with the problem of uneven conditioning (Satyanaran 2001).

In a 2001 article in *IEEE Personal Computing*, Satyanaran provided two scenarios describing "killer applications" of pervasive computing. The first involves a business traveler at an airport terminal. She is faced with many large files that need to be sent, as well as a pending long flight. Her laptop advises her to move a short distance down the concourse. Here she is able to find a clear, uncongested signal, allowing her to send her communications before boarding the plane.

The second scenario describes a professor in transit across a university campus. He is pressed for time, and his palmtop computer assists him. While he moves across the campus, the palmtop sends his presentation ahead. In addition, it sets up the lighting and projection equipment for him ahead of time. As a final touch, the room's facial recognition software notifies the palmtop that there are new persons in the room. Accordingly, the palmtop suggests to the professor that he had better drop the slide with the sensitive budgetary information. The professor agrees.

The two scenarios seem like science fiction—not because the technical capabilities are out of reach, but because of two key features of the technical environment: *proactivity* (the laptop anticipates problems given limited time and limited connectivity) and *self-tuning behavior* (the palmtop recognizes the professor's habitual behaviors and his usual audiences and adapts the environment accordingly). The scenario process in this context was successful because it describes new modes of software and system operation while isolating the key decisions needed to facilitate new pervasive computing applications. The status quo in this situation involves computing capabilities already present in 2001, which need to be utilized in new ways to meet users' needs.

7.3.3 Scenarios for Social Change

Scenarios played a pervasive role in helping the South African people and government facilitate constructive social change. Scenarios stimulated a vision of a healthy transition from apartheid and guided the development of transformational projects. Part of the value of scenarios in this setting was in anticipating emerging issues of governance. But much of the value also involved the ability of scenarios to affect the norms and worldview of the South African leadership. A number of different South African scenarios exist, including the 1987 Anglo-America scenarios, the 1991 Nedcore/Old Mutual scenarios, and the 1992 Mont Fleur scenarios (Spies 1994). The last are briefly reviewed below, with an eye toward using normative scenarios in technological forecasting.

The Mont Fleur scenarios were developed by researchers at the University of the Western Cape, and were financed by the Swiss Development Agency and the Friedrich Ebert Foundation, a German political foundation promoting political pluralism and societal education. The variables considered in the scenarios include government social policy, government consensus or disconsensus, citizen radicalism, business confidence, and economic growth. Government consensus corresponds to an appropriate social policy as well as a correspondingly moderate response from society as a whole. Business confidence is restored, and economic growth is sustainable. The full set of variables collapse to two primary dimensions, resulting in a two-by-two dichotomy and four principal dimensions for scenario exploration. The researchers formulated four scenarios of the South African future called "lame duck," "ostrich," "Icarus," and "flight of the flamingos."

The lame duck scenario presents a slow democratic transition away from apartheid, leading to indecisive policies and correspondingly slow economic investment and growth. The ostrich scenario involves a hardening of the South African old guard biases on the perception of selective international support for apartheid. As a result, the country grows fragmented and violent, and the old guard leadership eventually returns to the bargaining table. The Icarus scenario involves a quick fix by government, which embarks on an extensive spending program for social and economic endeavors. Although initial results are spectacular, government outlays are not maintained, inflation rises, and a return to authoritarian government ensues. The final scenario, flight of the flamingo, presents a case where a political settlement is made with clear and consistent policies. Clear, consistent, and well-targeted social investments are made, with a corresponding increase in business confidence.

The apparent choices were clear, and the consequences were believable for the South African leadership in the 1990s. The scenarios were plausible for the South African leadership because of their conditioning on the outcomes of interest for the South African government—economic growth and a peaceful and stable society.

7.4 SCENARIOS: EXTENSIONS AND ADVANCED TECHNIQUES

This section builds upon the previous discussion of scenario analysis by discussing extensions and advanced techniques. The first subsection discusses how to use scenarios as part of a larger technology forecasting study. Scenario analysis fits well with simulation, modeling, and expert opinion approaches. Next, two related techniques in the broader family of scenario-based forecasting tools are considered. The first, backcasting, involves an iterative approach of discussing futures and strategies with decision makers. The second creates more richly structured scenarios by examining possible paths or transitions between future scenarios.

7.4.1 Scenarios in Multimethodology Forecasts

Scenarios analysis can be applied alone or in combination with other technology forecasting approaches. Much of the value of scenarios is realized when they are employed with other techniques. This subsection examines how scenarios can be used to enhance other techniques. The first case explores the benefits of scenarios with simulation and modeling. The second considers the use of scenario analysis in concert with expert opinion techniques.

Multimethodologies Involving Simulation and Modeling. Scenarios, simulation, and modeling make a fruitful combination for a number of reasons. Many simulation and modeling techniques require precise estimates of parameters and exact specifications of system structure. Such precision may not be consistent with the fundamental uncertainty surrounding many aspects of technological change. Simulation and modeling studies provide quantitative results based on a reproducible and objective methodology to which scenarios can add a robust examination of multiple futures. Combining approaches can achieve the benefits of both and produce robust quantitative projections of the future.

A related virtue of scenarios is their ability to synthesize and integrate various pieces of information. As discussed earlier, scenarios may be used across the entire TDS and are therefore useful for portraying technical inputs and outputs, as well as softer features of the environment such as norms and decisions. Further, qualitative information can be readily wrapped in a scenario narrative and may therefore be considered alongside more quantitative simulation or data analytic results.

This also suggests a major value that scenarios can add—effectively communicating results to decision makers. As noted many times before, this communication is critical. Even the best modeling and simulation project is of no value if no one hears or heeds it. Communicating a model's structure, assumptions, and numerical results always presents a challenge, especially if decision makers are unfamiliar with the format. Scenarios provide a vehicle to present key results via a readily comprehensible set of narratives.

The renewable energy study by de Vries, van Vuuren, et al. (2007) showcases many of the virtues of a multimethodology study that combines simulation, modeling, and scenarios. While anticipating huge growth in solar power, the study shows that this future is sensitive to features of the political and economic environment. The potential futures are compellingly presented in four specific scenarios that provide vehicles for useful discussions with decision makers. The scenarios integrate some of the more qualitative and evaluative features of these futures (e.g., ecological considerations and local autonomy) with quantitative elements of energy production.

Multimethodologies Involving Expert Opinion and Public Participation. Scenarios enable researchers to present short, pointed, and tangible accounts of future technologies and communicate them effectively to diverse publics. Questionnaires allow stakeholders to react to these possible futures. The combination of clearly communicating possible futures and eliciting reactions to them offers clear benefits.

Siegrist, Keller, et al (2007) demonstrate the use of scenarios in tandem with surveys and expert opinion techniques in their study of the perceived risks of new nanotechnology products. In order to make nanotechnology tangible and more comprehensible to laypeople, they present short narrative scenarios of 24 nanotechnology applications ranging from the mundane (sunscreens) to high tech (nanocapsule treatments for cancer). Survey and expert opinion activities then questioned respondents about the perceived likelihood and magnitude of health risks in each case. The authors also probed the knowledge level, control, worry, and trust features of future nanotechnology environments.

Study results revealed systematic patterns of risk perception related to dread and distrust. Dread is related to the perceived high probability of involuntary risks seen as disproportionate to benefits and ethically unjustifiable. The distrust dimension taps laypeople's perception of national regulatory agencies, as well as the ability of individuals to control their exposure to future risks. Experts showed a different pattern of risk perception and generally were better able to distinguish the unique features of the technology and separate them from their personal perceptions of risk.

The study illustrates many of the benefits of combining short scenarios with expert opinion methods and questionnaires. A sometimes poorly understood technology was communicated in short, clear narratives to a mixed population of experts and laypeople. These scenarios were used to help respondents imagine new applications of nanotechnology, some of which have yet to be marketed. The scenario perspective afforded a more concrete examination of risks and values than might have been otherwise possible.

Robinson (2003) combined scenario analyses with explicit processes of community engagement and direct public participation in futures valuation. The goal of the study was to evaluate a range of possible social and sustainability futures for the Georgia Basin, an area that includes Puget Sound and the cities of Vancouver, British Columbia, and Seattle, Washington. The study also sought to identify potential policies for achieving desirable futures.

7.4.2 Extensions of Scenario Analysis

Scenario analysis techniques have been extended in a number of ways. One extension involves more facilitated discussion with decision makers concerning their desired futures. Once desirable (or disastrous) futures are identified, a richer discussion about strategies and tactics becomes possible. Note that normative and strategic scenarios already include some of these evaluative and planning elements. These kinds of extensions include backcasting techniques and network-based futures analysis, which are discussed below.

Backcasting. Backcasting involves working backward to determine how desirable futures can be achieved. In some contexts, the term refers to verifying simulation models using historical data. However, this form of backcasting is not addressed here.

The technique was developed for use in studying energy efficiency and alternative energy sources. However, backcasting now is widely applied to a range of problems (Robinson 2003). It may be an appropriate technique to address problems that are fundamentally uncertain – where there are "unknown unknowns." Backcasting also may be suitable for situations in which the capacity for intentional decision making is especially high.

Backcasting contains elements of both strategic and normative scenario analysis. Since studies can consider the role of specific policies in driving desirable futures, they can be strategic in character. However, since backcasts also enable users to make explicit their valuations of potential futures, they can be normative in character as well (Robinson 1988, 2003).

The backcasting process involves seven basic steps. The first five are those of the basic scenario analysis process previously described. The sixth step is interacting with decision makers or stakeholders to evaluate the desirability of the futures portrayed by the scenarios produced in the first five steps. The final step is to identify the policies that will lead toward the desirable futures. This might involve consultations with decision makers or experts, or it might be done through simulation models. Both good and bad decisions may be identified at this stage.

If decision makers succeed in implementing their policies, a range of different possible futures may emerge. In such cases, it may be useful to revise earlier steps of the scenario analysis process in light of experience and explicit policy choices. Thus, backcasting can be iterative.

Network-Based Futures Analysis. In this technique, state transitions are added to scenario analysis. Powell and Coyle (1997) describe a seven-step process for performing network-based futures analysis:

1. Identifying variables
2. Developing levels of measurement
3. Characterizing significant variables
4. Creating states
5. Determining possible transitions between states
6. Appling business judgment to simplify the topology
7. Writing the narrative

Most of these steps are common to scenario analysis. However, futures produced by the technique are referred to as *states* rather than *scenarios* to emphasize that the preferences of key players can produce transitions between them. After the first four steps of the scenario analysis process have been completed, there often are more scenarios or states left in the analysis than usual.

Steps 5 and 6 are unique to the network-based futures approach. In Step 5, the analyst must describe the states that can be reached from one another. Decision makers may make choices that open up new possibilities or foreclose others, while external events may shape the range of possibilities. This step entails recognizing that some pairs of futures may be mutually exclusive. The discussion of interpretive structural modeling (Section 11.4.2) details a useful method for achieving this step.

Figure 7.3 presents an example transition matrix. In that matrix, 10 possible futures are arrayed along the top and the left-hand side. If a "1" appears in a matrix cell, it means that the column scenario can be reached from the row scenario. A "0" means that it cannot. For instance, scenario A can be reached from scenario B, while scenario J cannot be reached from B. In the example, the transitions have been conceptualized as symmetrical; if scenario A can be reached from scenario B, then B also can be reached from A. Thus, the matrix is also symmetric. Of course, diagonal cells are all 1 since any future can be

	A	B	C	D	E	F	G	H	I	J
A	1	1	1	1	0	0	0	0	0	0
B	1	1	1	1	0	0	0	0	0	0
C	1	1	1	0	1	1	1	0	0	0
D	1	1	0	1	1	1	0	0	0	0
E	0	0	1	1	1	1	1	1	0	0
F	0	0	1	1	1	1	1	1	0	0
G	0	0	1	0	1	1	1	0	1	1
H	0	0	0	0	1	1	0	1	0	0
I	0	0	0	0	0	0	1	0	1	1
J	0	0	0	0	0	0	1	0	1	1

Figure 7.3. Graphical Representation of a Transition Matrix
Source: Adapted from Powell and Coyle (1997)

reached from itself. If transitivity can be assumed (i.e., if A can reach B and B can reach C, then A can reach C as well), the number of responses needed to fill the matrix cells can be reduced.

In Step 6, the analyst simplifies the topology by combining functionally equivalent states. Figure 7.4 displays the example scenarios and their transitions arrayed vertically from least to most desirable. Some scenarios can be grouped together according to business purposes and logic. For instance, scenarios A and B involve exactly the same transitions. After consultation with business experts, the analyst could choose to group these scenarios to simplify the network. The procedure, like scenario analysis, is completed with a narrative describing each state and its possible transitions.

7.5 CONCLUSIONS

The scenario analysis technique is versatile and rich. Of all the forecasting techniques, it may be best positioned to deal with fundamental uncertainty. Scenarios can clarify the range of uncertainties about the future and help forecasters, decision makers, and stakeholders acknowledge the unknown unknowns of the world. Scenario analysis can be effectively combined with other techniques—either by providing a narrative context for simulations or by providing an integrated framework for the norms and values of decision makers and key stakeholders. Since scenario analysis often is quite loosely applied, this chapter has sought to add structure by identifying three different types of scenario analyses that can be mixed or matched according to the needs of the forecast.

One of the principal virtues of the scenario analysis technique is that it can be used to integrate strategy, judgment, and value and effectively communicate the result. Scenario analysis is most effective when it engages decision makers and stakeholders and stimulates learning about the future. In this regard, the aims and outcomes of scenarios may be very different from those of the more quantitatively oriented technology forecasting.

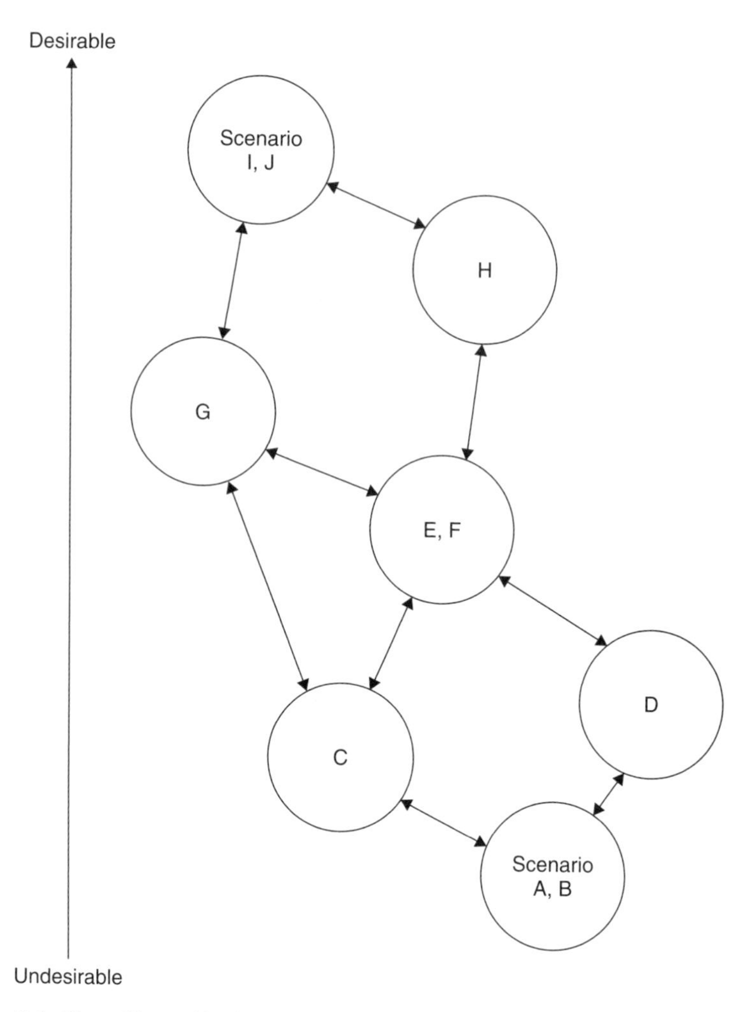

Figure 7.4. Transitions, Ranked and Simplified
Source: Adapted from Powell and Coyle (1997)

The goal of the analyst is not necessarily to pursue precision when precision may not be obtainable. Instead, the analyst should be involved in helping decision makers pursue robust and adaptive decision making. Scenario analysis is an effective technique with which to accomplish this end.

REFERENCES

Collingridge, D. (1981). *The Social Control of Technology*. Basingstoke: Palgrave Macmillan.

de Vries, B.J.M., D. P. van Vuuren, et al. (2007). "Renewable Energy Sources: Their Global Potential for the First-Half of the 21st Century at a Global Level: An Integrated Approach." *Energy Policy* **35:** 2590–2610.

Dewar, J. A. (2002). *Assumption-Based Planning: A Tool for Reducing Avoidable Surprises*. Cambridge, Cambridge University Press.

IMAGE-team (2001). *The IMAGE 2.2 Implementation of the SRES Scenarios: A Comprehensive Analysis of Emissions, Climate Change and Impacts in the 21st Century*. Biltohoven, The Netherlands, National Institute for Public Health and the Environment (RIVM).

Lempert, R. J., S. W. Popper, et al. (2003). Shaping the Next One Hundred Years: New Methods for Long-Term Quantitative Policy Analysis. *RAND Monograph MR-1626-RPC*. Santa Monica, CA, RAND.

Linstone, H. A. (1988). "Multiple Perspectives: Concept, Applications and User Guidelines." *Systems Practice* **2**(3): 307–331.

Luft, J. and H. Ingham. (1955). "The Johari Window, a Graphic Model of Interpersonal Awareness." *Proceedings of the Western Training Laboratory in Group Development*. Los Angeles, UCLA.

Mont Fleur Scenarios (1992). Bellville, Institute for Social Development, University of the Western Cape.

Oxford English Dictionary (2011), "Scenarios: Usage comment by R. W. B.," Retrieved 13 January 2011, from http://www.oed.com/view/Entry/172215?rskey=DivzHf&result=2&isAdvanced=false#eid24174925.

Powell, J. H. and R. G. Coyle (1997). "A Network-Based Futures Method for Strategic Business Planning." *Journal of the Operational Research Society* **48**(8): 793–803.

Robinson, J. B. (1988). "Unlearning and Backcasting: Rethinking Some of the Questions We Ask about the Future." *Technological Forecasting and Social Change* **33**: 325–338.

Robinson, J. B. (2003). "Future Subjunctive: Backcasting as Social Learning." *Futures* **35**: 839–856.

Satyanaran, M. (2001). "Pervasive Computing: Vision and Challenges." *IEEE Personal Computing* **8**: 10–17.

Schwartz, P. (1991). *The Art of the Long View*. New York, Doubleday.

Siegrist, M., C. Keller, et al. (2007). "Laypeople's and Experts' Perception of Nanotechnology Hazards." *Risk Analysis* **27**(1): 59–69.

Spies, P. (1994). "Experience with Futures Research in South Africa." *Futures* **26**(9): 964–979.

van der Heijden, K. (2005). *Scenarios: The Art of Strategic Conversation*. Hoboken, NJ, John Wiley & Sons.

van der Sluijs, J. P., J. S. Risbey, et al. (2003). *RIVM/MNP Guidance for Uncertainty Assessment and Communication: Detailed Guidance*, vol. 3, Utrecht, Utrecht University.

Wack, P. (1985a). "Scenarios: Shooting the Rapids." *Harvard Business Review* **63**(6): 139–150.

Wack, P. (1985b). "Scenarios: Uncharted Waters Ahead." *Harvard Business Review* **63**(5): 73–89

Walker, W. E., P. Harremoes, et al. (2003). "Defining Uncertainty: A Conceptual Basis for Uncertainty Management in Model-Based Decision Support." *Integrated Assessment* **4**(1): 5–17.

Weiser, M. (1991). "The Computer for the 21st Century." *Scientific American*, **265**(3): 66–75.

8
ECONOMIC AND MARKET ANALYSIS

Chapter Summary: Changes in technology are probably the most important sources of change in the structure and performance of the world's economies. Economic factors, in turn, are crucial to the introduction and acceptance of new technologies within the global marketplace. This chapter describes the interrelationship between economics, markets, and technology and shows how economic and market forecasting can aid technology forecasters and managers. Economic analysis and other tools for anticipating market responses are described, with emphasis on their use by managers.

The chapter begins by describing how technological progress has enhanced economic well-being, albeit sometimes with considerable collateral damage. From the broad view of national and global economies, the discussion shifts to how economic forces play out in the markets for specific technologies. Next, the role of social institutions in the growth of technology applications is discussed and the issues of forecasting within an institutional context are raised. Finally, specific models that are useful in forecasting are described as aids in determining both the likely environment for adoption and the more specific opportunities for particular technologies.

8.1 THE CONTEXT

World resources are scarce relative to human needs and desires; therefore, choices are necessary. Economics really is a science of choices—how they are made and how they ought to be made. Because economics must deal with the complexities of human behavior, accurate prediction is difficult. The tools of the economist and the technology manager are almost never as precise as those used by the physical scientist. However, economists and others have developed models that can make extrapolative predictions about the path of a technology's development and have provided simulations to illuminate the interrelationships that will determine its success or failure. While the results of these models cannot replace managerial judgment, they *can* inform that judgment.

Demand is the willingness to buy plus the ability to pay. The process of technological innovation occurs only when consumers, businesses, and governments demand the benefits of a new technology and pay for its cost. Technologies launched when profits are high and unemployment is low would seem to have a greater chance of success than those offered when consumers are insecure and businesses are struggling to survive. However, new technologies can bring societies out of a recession or depression as the push for new production stimulates investment, employment, and consumption. Technology managers must understand these relationships and how to integrate the implications of economic analysis into management decisions.

In the mid-twentieth century, Nobel laureate economist Robert Solow (1957) concluded that most of the increase in the standard of living in the United States was due to technological progress. He also concluded that 87.5% of the increased output per capita from 1909 to 1949 was due to technical change; the remainder was due to capital investment. While subsequent analysts have used far more sophisticated methods, most economists would still agree with Edwin Mansfield (1989, p. 700) that "the rate of technological change is perhaps the most important single determinant of a nation's rate of economic growth." Clearly, technology has created new products to enhance the quality of life. It also has dramatically improved the productivity of labor and capital resources.

People can produce more in less time if they know how to use resources more effectively. In the early twentieth century, for instance, Henry Ford could raise the wages of automobile assemblers to $5 a day because of production innovations. While many now view the manufacturing sector as a declining component of developing economies, the reality is that phenomenal productivity gains were still being made in the closing decade of the twentieth century. Manufacturing labor productivity in the United States grew at a rate of 3% per year between 1990 and 1994, while other nonfarm businesses managed only a 1.2% annual increase in output per worker (Duesterberg and Preeg 2003, pp. 18–19). Moreover, from 1995 to 2000, manufacturing productivity surged at a rate of 4.3% per year, while the remainder of nonfarm businesses achieved only a 1.7% average annual rate. Declines in manufacturing employment during this period largely happened because the application of technology reduced the need for human resources.

New technology can overcome the inherent limitations imposed by the law of diminishing marginal returns. This law says that if the input of a resource is increased in a production process for which supplies of some other resources are fixed, at some point the additional output that results from one more unit of input will decline. With technological improvement, however, both capital and labor resources can be used more efficiently. Thus, increased inputs together with improved technology can mean greater productivity and higher worker income. What's more, recent advances in technology have offered products that exhibit increasing, not decreasing, returns from the expansion of sales. For example, it is very expensive to develop the first unit of software but very inexpensive to reproduce additional units. Similar results can be seen in electronics, pharmaceuticals, and other products.

The effects just described can lead to dramatically falling prices or even free products like search engines and social networking sites that have rapidly grown to become enormous businesses based on advertising revenue alone. They also can lead to more consolidation of economic power and increased economic volatility as new technologies ignite booms that are inevitably followed by downturns. Phenomena related to such increasing returns have serious implications for economic policy (Arthur 1994). For more discussion, see Arthur (2010). Recently, Arthur noted that technologies that produce nonequilibrium behavior in economic systems resemble evolutionary models like those of biological phenomena. The resulting marketplace complexity makes it critical that technical managers grasp the range of possible future scenarios if they are to make sound business decisions.

In the first half of the twentieth century, Schumpeter noted the disequilibrium effects of technology and discussed the consequences in *Capitalism, Socialism and Democracy* (Schumpeter 1942). He coined the term *creative destruction* to describe how new technologies not only open up new opportunities, but also shatter old ways and destroy old products and their associated organizations. The decline of traditional giants in the American automobile, steel, and newspaper industries provides dramatic evidence that this phenomenon is powerful and ever-present. Walmart currently rides the crest of a wave of growth made possible, to a substantial degree, by its superior uses of technology in logistics and retail operation management. But it is likely that new technologies will spawn new business models that will topple Walmart, just as it eclipsed its competitors in earlier eras. Technology basically changes the nature of economic systems from inherently stable environments where careful and conservative planning can assure business longevity to ones where chaotic change can rapidly topple even the strongest.

Nolan and Croson (1995) developed a six-stage strategy that heavily depends on information technology to dramatically increase organizational flexibility in order to be the source of change rather than its victim. These six stages, with some illustrative comments, are:

1. Downsize (competition requires that costs constantly be reduced).
2. Seek dynamic balance (e.g., supply chains need to change rapidly with product change).
3. Develop a market access strategy (organizational change needs market direction).
4. Become customer driven (customers want solutions, not technologies).

5. Develop a market foreclosure strategy (competitors must be restrained from seizing advantages).
6. Pursue global growth (surrendering global markets invites more rapidly growing competition).

A quick review of this list shows that many of the survivors in today's rapidly changing markets have been implementing the strategy that Nolan and Croson described.

8.1.1 Markets and Innovation

Awareness of the effects of technological change on the global economy is important. However, most decisions that technical managers make involve specific markets. Indeed, a prerequisite for a discovery to become widely accepted is that there be adequate demand for the benefits that the new technology produces. Demand is necessary to justify the investments in development and implementation that a new technology will require.

Markets are generally immune to the excitement created by new technology; instead, they respond to the value it creates. Fuel cells, alcohol from cellulose, and many applications of nanotechnology often seem to be at hand, but their growth has been hindered by the lack of a compelling connection between their perceived benefits and the cost to the consumer. On the other hand, the explosive growth of Google, Facebook, Twitter, and many video games provides ample evidence of the possibility for very rapid growth when that connection is favorable. Forecasting and managing technologies require an understanding of how the market will pull them to success. Achieving this understanding can be at least as hard as predicting the breakthroughs that will enable new technologies. One problem is that people resist change. The slowness with which education has embraced the potential of the Internet is an example. Much of the current online education tries to emulate classroom instruction, largely because students and teachers expect it. Thus, preference for the familiar has slowed what ultimately will be a huge impact on learning enabled by the ocean of information on the Web and the various techniques that might be developed to convey it.

This chapter will survey some of the ways that managers can look into the future of markets to assess their receptiveness to technology. This is not a well-defined discipline. Rather, it is an evolving set of qualitative and quantitative approaches that can help reduce uncertainty when tempered by sound judgment. Each market and each technology has its own characteristics, and the analyst will have to identify which ones will be important. This is an uncertain task. The experiences of technology entrepreneurs who have built successful businesses based on new technologies suggest that while managers must predict what will be embraced, they also must retain the flexibility to shift to a different application or even a different market as opportunities emerge. Most successful products were not those with which the founders launched their businesses.

Clayton Christensen and coauthors have published, *Seeing What's Next,* which includes theory that, they argue, can be used to make predictions with respect to disruptive technologies—those that completely change industries (Christensen, Roth, et al. 2004). They use an analytical process that includes three different types of buyers who could provide initial demand: nonconsumers, undershot consumers, and overshot consumers. These groups potentially can provide a niche for a disruptive technology to capture and use as a base for growth. Henry Ford's development of the Model T for middle-class buyers at a time when automobiles were toys for the rich is an example of introducing technology to nonconsumers. The undershot consumer is one who is dissatisfied with important aspects of existing products and is ready to pay a premium to eliminate that dissatisfaction. Some might argue that the entry of the I-phone into the cell phone fray is an example of capturing undershot consumers and that the appearance of organic foods in supermarkets is another. Then there are the overshot consumers, whom the authors regard as particularly attractive. These are consumers for whom the existing products are too complicated or provide extra features that they do not need, often at a cost they cannot afford. For example, health care in the United States is delivered by highly qualified specialists at very high cost. In many cases, the expensive diagnostics and treatments have overshot the ultimate consumers' understanding, in some cases their needs, and certainly their financial means. Christensen, Grossman, et al. (2009) have applied the theories of disruptive technology to this problem in *The Innovator's Prescription*. In *The Atlantic Monthly,* Christensen and Hwang (2009) point out that insulin-dependent diabetics already have used advances in science and technology to cost-effectively improve their management of their own disease. They suggest other ways in which disruptive change could reform health care by making solutions *directly* accessible to patients.

It's clear that finding consumers who will value the benefits of new products, processes, or services enabled by a new technology will be critical to supporting that technology's growth. While there has to be a market for the growth of an innovation, there also are market attributes that will affect the rate of that growth. Michael Porter's famous five-forces model is a good inventory of the potential competitive issues for a new technology. The five forces are *barriers to entry, supplier power, buyer power, substitutes,* and *rivalry* (Porter 1985).

Barriers to entry for an innovation can be difficult for new technologies, just as they are for firms trying to enter a market. If there are significant economies of scale both for the new technology and for the one it is replacing, the incumbent technology has the advantage of a large existing user base as well as an established infrastructure. Similar barriers to growth exist if the cost of switching to the new technology is very high. There also may be patents that protect the existing technology by surrounding it with proprietary supporting technologies. Institutional limits such as tariffs and outright prohibitions to entry sometimes may exist as well. A good technology forecast must consider all of these potential barriers to market penetration and growth.

Buyer power and *supplier power* also must be evaluated. Most technologies require supporting technologies that may be supplied by others. Powerful suppliers must see the growth of the innovation to be in their interest for them to support it. Likewise, if the innovation is only suited to a few powerful buyers, their decisions can dramatically affect the technology's prospects. For example, many military technologies critically depend upon decisions by the U.S. Department of Defense. Further, decisions may be the result of political forces rather than the attributes of the technology. History has shown that truly disruptive technologies do best when they are targeted at nonconsumers, that is, those who presently do without the benefits of the innovation. The lesson is that the way to avoid buyer power is to sell to buyers who have little of it.

Substitutes need not present any of the specific barriers mentioned above to inhibit a particular innovation. Rather, they simply may offer the same benefits. For instance, consider the wide range of proposed solutions to the problem of reducing dependence on fossil fuels. Forecasts for a new technology must consider its chances for market presence in relation to what is being offered by substitutes for it.

Rivalry refers to the nature of the interactions among the competitors in the market. It might seem that a total lack of competitors is good because the way is clear. But it usually means that there is no market, and one must be created. This is both time-consuming and expensive. Generally, however, there are competitors. Then the new technology must overcome the notion that if current practice works, there is no need to change. Even if customers see benefits from adopting new approaches, they may decide to do it themselves. Industrial innovations must frequently overcome the desire to build rather than buy. These factors can slow the growth of a new technology, particularly when there are economies from scale of production, scope of product offerings, or learning that could accrue to the innovation's creator.

Some markets are characterized by the consolidation of control into a few producers who often are slow to change and do not aggressively seek to capture new market shares through innovation. Innovation in these markets is viewed as risky. If the innovation doesn't work, money will be lost, and if it does work, it will be quickly copied. The automobile market in the United States exhibited this competitive posture throughout much of the twentieth century, as did the steel industry. Participants in other markets may use innovation as a market weapon. For instance, businesses producing computer chips and computer games seem to be constantly trying to out-innovate their competitors.

Innovations are unlikely to be pushed into markets by their elegance, originality, or inherent performance. They are accepted and grow because there is a real need for their capabilities and a competitive environment that is receptive to their acceptance and growth. Understanding the future paths of innovation requires understanding the markets and the forces within them that will affect those paths.

8.1.2 Technology and Institutions

The relationship of technology and institutions is a two-way street. Institutions constrain, control, and promote technologies, but technologies also have profound impacts on institutions. That is why the technology forecaster needs to consider the institutional context. It also is why an economic analysis of innovations must include institutional considerations.

Much has been written about the interactions of technology and social institutions; a complete discussion goes well beyond the scope of equipping managers to make better decisions about technology. The preceding section discussed a major institution that affects innovation—the market—and the need for market receptiveness for a new technology to prosper. It also is worth noting that technology sometimes creates new markets or dramatically impacts established ones. For example, many consumers now buy online, and auctions for vendor inputs now are used in supply chains. These two important markets were unimagined 25 years ago. Moreover, even though most cars still are purchased from traditional dealers, the volume of automotive information available on the Internet clearly has affected their marketing and sales.

Understanding the mutual impacts of technologies and the market is an important institutional concern. However, there are two other institutions that interact with technology that merit special attention—government and education. The U.S. Food and Drug Administration (FDA) is a good example of the impact of government institutions. Responding to consumers' concerns, government created this powerful agency with authority to inspect and approve that can modify, impede, or even stop innovations. Medical device technology, for instance, is strongly affected by how designs might be viewed by FDA regulators, as are the development and manufacturing processes for drugs. Regulatory constraints affect wide ranges of technology. Even financial innovations are likely to be regulated after their role in the financial crises of 2008–2009. Other examples of the alphabet soup of agencies and the technologies that might be affected are:

- Federal Communications Commission (FCC)—broadcasting ranging from cell phones to electronic devices
- Federal Energy Regulatory Commission (FERC)—electric power transmission
- Environmental Protection Agency (USEPA)—technologies that impact air and/or water quality
- Occupational Safety and Health Administration (OSHA)—anything impacting workers
- Consumer Product Safety Commission (CPSC)—consumer technologies, including toys, clothing, furniture, and appliances (U.S. Consumer Product Safety Commission, 2011).
- National Transportation Safety Board (NTSB)—all types of transport (highway, air, water, rail)
- Department of Energy (USDOE)—nuclear technologies are a particular focus

A quick way to identify institutional regulations and actions is to go to the website of the Code of Federal Regulations (GPOAccess 2010). This site includes a search engine to find regulatory language relevant a particular innovation.

The United States has simultaneously subsidized development of ethanol fuels using corn from domestic farms while restricting ethanol imports from countries like Brazil that produced it from a variety of nonfood sources. The United States also has restricted technology exports when the technologies have potential military applications.

The United States is not the only country with extensive regulatory institutions that impact new technologies. Virtually every nation has them to some degree. The types of regulation and the stringency with respect to particular innovations vary from country to country. Good examples of problems that can arise are Time Warner and Google disagreements with Chinese policies on censorship (Mufson and Whoriskey 2010). Countries may not explicitly restrict trade with formal laws, but they can slow it through their customs procedures, which can inhibit the growth of a technology. Since markets tend to be global, the proper management of technology requires at least some familiarity with the range of regulation across countries.

Government institutions do not always constrain innovation. They also are major promoters. Citizens around the world have become aware that innovation is a major source of economic growth, and politicians are anxious to appear to be advocates. For instance, the Internet, one of the most important innovations of the last few decades, was virtually created by the U.S. Department of Defense. More recently, government agencies have made hundreds of millions of dollars in economic stimulus funds available to companies to develop such things as electric hybrid cars and other alternative energy technologies (U.S. Department of Energy 2010). In spite of severe budget constraints, state governments are continuing to subsidize the commercialization of new technologies in hopes of becoming the centers of new high-growth industries. See, for example, the State of Indiana's 21st Century Fund (2010).

No institution may be more important to innovation than education. Most people think of great research institutions as critical to the discovery of new knowledge that can fuel innovation, and there is a lot of evidence to support this view. However, discovery is not innovation, and much must be done to actually implement the benefits of new knowledge discovered in university laboratories. The structure and incentives of universities typically do not encourage scholars to participate in the development processes needed to create marketable products and services. Scholarly reputations are built by publication and generating grants to take new knowledge to the next level. While the practical problems of getting a product ready for market may be just as challenging as discovery, they do not carry the same scholarly prestige. Technology transfer offices work hard to get technologies licensed for commercialization, but they are not designed to alleviate the product development difficulties that will arise. Opportunities exist for universities, firms, states, and nations to find ways to exploit higher education knowledge resources more effectively. In the meantime, technology managers must be aware of both the tremendous potential and the practical obstacles of taking innovation from university research laboratories to marketable products.

The policies and practices of educational institutions at all levels also impact technologies through their effects on workforce quality and consumer characteristics. Education must prepare students to succeed in the knowledge-based global economy in which they will compete. Opportunities abound for innovators to find ways to teach people what they need to know. Nor is there a shortage of problems, such as the inability of many workers to learn the basic science and mathematical skills needed to produce and use new technological approaches. During much of the past two decades, workplace competencies have been major concerns of employers, especially in industries experiencing rapid technological change, like manufacturing. Technology managers need to take account of the capabilities of workers and consumers as new products and services are created. The growth of new technologies could well be affected by the availability of people with the skills to take advantage of them.

Government policies that encourage or discourage immigration also affect the workforce. For example, in the past, the United States was a magnet for the best and brightest young persons in the world. In his book *Flight Capital,* David Heenan (2005, pp. 1–2) noted that Chinese and Indian immigrants were in charge of over 25% of Silicon Valley's high-tech firms and that half of the American winners of Nobel Prizes in the late 1990s and early 2000s were foreign born. Moreover, more than half of the Ph.D.s and 45% of the physicists, computer scientists, and mathematicians were from other countries. Heenan goes on to note that the United States has been less welcoming since September 11, 2001, and that other countries are increasingly encouraging their citizens to pursue careers at home. Immigration has advantages for the diversity and the creativity it fosters, but its effect on the availability of sheer brain power also is a major issue for those assessing the future of technologies. The impacts of immigration restrictions and the increased attractiveness of foreign economies mean that the technology delivery systems will need to include the effects of immigration on the likely supply of talent to support innovation.

Another critical part of the institutional context is access to intellectual property. Patents generally are the barriers to access that first come to mind, but copyrights, trademarks, and trade secrets also can be very important. In the United States, a utility patent provides a monopoly for 20 years after a patent application has been filed. This application must describe how the new idea or ideas will work and be useful for the purposes outlined in the claims. This means that the invention and what it does will no longer be secret, which is one reason that the institution of patenting exists. This is why some innovators prefer not to disclose discoveries and to rapidly exploit their technologies before competitors can match them. The technology manager should realize that patent laws in other parts of the world differ from those of the United States. Those seeking global markets for their innovations will need to take this into account.

Existing patents can be used to inhibit the commercialization of new technologies, and the costs and delays they can create are serious management problems. Michael Heller, in his book *The Gridlock Economy: How Too Much Ownership Wrecks Markets, Stops Innovation and Costs Lives,* describes how patents and other property rights can thwart valuable advances (Heller 2008). However, the value of patents can be overestimated. They are only as good as the breadth of their claims and the ability of the patent holder to defend them. This ability can be problematic if the violations occur in other parts of the world where the patent may not provide any assurance of protection. Moreover, patent disputes can be very expensive, and organizations with the most resources have a distinct advantage. Finally, technology sometimes changes so quickly that much of the 20 years' protection provided by the patent is meaningless. Thus, an understanding of the patent landscape is probably needed to forecast the future prospects of a new technology.

Other forms of intellectual property protection need to be considered as well. Copyrights protect creative works for 70 years beyond the death of the creator and have been used to protect software innovations. Trademarks also can be used to ensure that the advantages of a technology accrue to the brand associated with its creator. The "Intel Inside" stickers on computers, for example, provide important protection for Intel since the buyer cannot easily distinguish the maker of the chip inside a computer. Having access to all of these forms of intellectual property protection can provide some assurance to innovators that investments in developing new technology will get adequate returns.

Intellectual property and its protection are increasingly key components of another important institutional consideration—international agreements. Governments frequently discuss trade agreements that affect the size of global markets available to innovations. Some technologies related to national defense, for example, have been restricted to domestic markets. Increasingly, agreements are being made to allow the flow of technologies throughout the world, and protection of the underlying intellectual property is improving. Future areas of liberalization might include technical standards and regulatory constraints. Such agreements could provide for more rapid globalization of innovations.

8.2 FORECASTING THE MARKET

While many technologies directly affect the choices available to the ultimate consumer, many others are further removed and reside in the supply chain that develops, delivers, and services products. The primary market for some technologies is business; for others it may be governments, and for still others hospitals or educational institutions. To forecast the size and growth of a market for a new technology, it is important to *know who the customer is going to be*. For example, if the innovation is a medical device used to improve a surgical procedure, is the customer the patient, the surgeon, the hospital, or some other party? Actually, the real customer in such cases may be the insurance companies that decide whether or not it will be paid for. Forecasting the market means identifying the decision-making customers, determining how much they will pay, understanding how many there will be, and knowing how much they will buy over time.

8.2.1 The Consumer/Customer Marketplace

Consumers are virtually overwhelmed with choices. This is true both of the individuals who are generally thought of as the targets of new products and services, and of the decision makers who act as customers within the businesses that produce those products and services. Actually, new technologies are more often targeted at producers rather than at ultimate consumers. However, customers' behavior has similarities whether they are buying a new technology for themselves or acting as agents for a business. Customers all have limited funds to spend and a vast array of products available to them. Traditional economic analysis suggests that they will rationally consider prices, their resources, their tastes, and the characteristics of existing products and services relative to those offered by a new technology. In reality, the process is much more complicated. New technologies must go through a diffusion process while their advantages become apparent. First, there must be *awareness* of the new technology and its supposed advantages. Then *persuasion* must occur so that people will part with their scarce resources to try it. Then time is needed to make and implement *decisions* to purchase and use the innovation. Commercial success follows only after a *confirmation* period characterized by repeat buying by satisfied customers and/or their recommendations to others.

This diffusion process tends to produce an S-shaped pattern of technology growth. A relatively long period of constant demand can pass until enough people embrace the product for growth to occur. Then, as repeat buying and the broadcasting of benefits occur, demand can grow at an increasing rate for a period. Eventually, an inflection point is reached and the market continues to grow, but at a decreasing rate as the innovation and its marketplace mature. Finally, the market peaks and demand may decline as other innovations take over the market. While this all sounds smooth and predictable, the reality of technology adoption can be discontinuous and very uncertain. For example, Geoffrey Moore (1999) described the process of moving from "early adopters" who just like to buy the latest technology to an established position with mainstream buyers. He points out that there can be a chasm in the growth curve as early buyers are reached and the mainstream is not yet ready to buy. Then, when mainstream buying begins, network and bandwagon effects can create a tornado of extremely rapid growth that is difficult to manage. Early introducers can be highly vulnerable to later market entrants as a result of this volatility of market development.

Another aspect of innovation marketplaces is that they are increasingly global. While an early focus on a market niche may be strategically and practically advisable, it is important to remember that most markets are global. Many innovators don't possess the resources to offer their innovation worldwide. However, if they don't move quickly to capture global markets, competitors will. The increased sales and profits from foreign sales can be used to speed the development of subsequent releases of the technology, as well as to fund aggressive marketing campaigns to capture domestic markets from early movers who do not reach far enough for dominance. Thus, forecasts of a technology's growth must include developing the resource base to capitalize on early successes.

Once the marketplace characteristics are determined, the next job is to identify the marketing and pricing strategies that fit them. As mentioned, a niche strategy often is advisable for a new technology. Ideally, the niche is one that can be quickly dominated and effectively protected. The niche may be at the upper end of the market, where affluent customers will pay a premium for new technology and the entrant can have some "first mover" advantages. This can help establish a brand with prestige that may carry over as the marketing strategy for larger groups of lower-income consumers. This high-value niche market entry strategy is often accompanied by premium pricing to "skim the cream." As production costs come down, sales volume increases, lower income groups become more attractive, and prices can be lowered with cost reductions. This approach discourages competitors. An example of this approach is offered by Bose™, which began selling stereo speaker innovations to the top of the market but now tries to reach much broader audiences for its products. The niche sought could also be a specialty market. For example, "green" products now appeal to large groups of often affluent consumers and image-conscious businesses that not only prefer environmental alternatives but will pay a premium for them. Similar niches can be found among those seeking more organic products and/or local origination.

The other end of the marketplace, where customers look for the lowest cost, also can support an effective strategy for growing successful innovations. A century ago, Henry Ford made his automobile prices low enough for people of modest means by developing innovations like integrated supply chains and standard parts. Sam Walton dramatically improved retail logistics technologies to get lower-priced goods first to rural consumers and later to almost everyone. Amazon used the Internet to lower overhead and other costs so as to offer lower prices for books and then for many other retail items. This low-price strategy also has worked for Japanese, Korean, and now perhaps Chinese auto producers; for producers of steel based upon increased recycling, like Nucor; and for direct sellers of goods, like Dell. In many cases, these innovators, like Ford, also advance production, distribution, and other technologies.

Other strategies that have been followed by innovators include creative imitation. For examples, Wendy's, Burger King, and Subway introduced their own differentiation to the business model and associated technologies of McDonald's fast food innovation, while Target arguably is an upscale variation on Walmart. Historically, the rise of General Motors in the early twentieth century was an imitation of Ford with a new emphasis on variety and style. In the twenty-first century, LinkedIn has creatively imitated Facebook by aiming at more mature users who want to stay in touch for professional reasons. These imitators often do more than clever marketing. They also may enhance the technology of products and the processes to make and deliver the product or service. The twenty-first century may open many more such strategic opportunities as technologies enable more and more flexibility to configure products for particular customer wants without sacrificing economies of mass production.

8.2.2 Qualitative Techniques for Appraising Market Potential

This section began by stressing the importance of knowing who the customer for the innovation will be. Once you know this, how do you answer questions about what they will do so that strategies can be designed and implemented? The first rule of forecasting the market is to talk to potential customers. Later paragraphs discuss market research tools that are indirect, objective, and seemingly sophisticated ways to infer what customers will do. However, they are no match for talking to those who will actually make the decisions to buy. Ask them about what they are currently doing, their needs, their likes and dislikes, and what the features provided by the new technology would be worth to them. These questions can be asked in person, by phone, or even by e-mail. Although many may decline to talk, others will respond and their input is priceless.

Section 5.1.2 detailed several techniques that are useful in gathering information from potential customers and experts. However, they are neither easy nor inexpensive. After the issues have been identified in small-group settings, surveys can be constructed for a much larger customer sample. Wording the survey questions is extremely important, and professional expertise is valuable in ensuring the reliability of the information gathered. See the list of things to avoid in posing survey questions given in Section 5.1.2. The surveys can be conducted through the mail, e-mail or websites (e.g., SurveyMonkey 2010). There are firms that will conduct interviews, and their expertise may be needed to get a significant sample. For more information on such techniques, one can consult any number of books, like *The Market Research Toolbox: A Concise Guide for Beginners* by McQuarrie (1996).

While primary sources are critical for determining customers' wishes, secondary sources will be used to estimate market size. For example, once the target customers are identified, census information (U.S. Census Bureau 2010a) can be used to approximate their number in the United States. More specific data, such as the number of female heads of households in certain income classes, can be inferred from the way population information is arrayed. Similarly, the number of manufacturers in a particular industry can be found from the Census of Manufacturers (U.S. Census Bureau 2010b). Similar global data can be gathered from organizations like the United Nations (United Nations Population Division 2010) or from the Organization for Economic Cooperation and Development (OECD; Organization for Economic Cooperation and Development 2010). Trade associations and their literature also are very good sources for statements about the size of markets. If competitors are operating in the market, items about them in the press can provide evidence of both their sales and the size of the market. There also are lots of financial analysts' reports on companies and entire industries, many of which can be found online through the usual search engines.

A few words of caution are appropriate about the temptation to promote a very large market for a new technology. The best approach is to find a well-defined niche where an innovation can gain a dominant position that can be used as a base from which to expand to other related markets. Product features are much easier to define for a market niche, and major competitors may be willing to overlook an entrant that seems confined to a small portion of the market. Some new-technology proponents like to tout the market for their product as being so large that even 1% of it will provide a bonanza. However, the success of small firms in large markets is rare unless there is a particular 1% that offers them a special niche. Otherwise, the costs of dealing with the entire market while capturing only a small fraction of it eventually lead to failure.

The growth rate of markets is also extremely relevant. Mature markets exhibit slow growth and generally also are slow to change. In new markets, however, technologies can grow explosively, as demonstrated by the market for search engines in the first decade of this century. But getting a first mover or early position in the market is no guarantee that a particular innovation will win a dominant position as the market grows. There are many cases (e.g., spreadsheets, personal computers) in which early technology producers stumbled or were overtaken by later, more successful entrants. Nevertheless, it is still attractive to be an early player if the technology also can be defended as the market grows. Quantitative techniques for estimating the path of this growth, such as the Fischer-Pry and Gompertz models, were introduced in Section 6.3 and are further discussed in the next section.

Qualitative approaches can involve the assessments of early buyers as well as the speed of early growth. The forecaster might ask if the base of users attracted by the technology is broad and representative of many different groups or if it is just a particular niche, perhaps merely those who want the latest technology. Such early qualitative evidence will be different for different types of technology. For example, a clue that a social networking Web technology may take off is that 10,000 users were attracted within a few months after its launch. The early clues to the success of a medical technology may be that particular physicians who are pacesetters in their areas of practice are embracing the innovation and insurance companies and/or the government health reimbursement programs are willing to pay for it.

8.2.3 A Quantitative Approach—Adoption and Substitution: S-Curve Models

Talking to customers, gathering qualitative information, and using secondary sources are critical. However, there are ways to get more precise pictures of the paths of technology growth. Approaches that have proven to be useful in forecasting technologies include both extrapolative and simulation models. Two that can be particularly useful are the Fisher-Pry and Gompertz models (see Section 6.3.1). With limited data, one can use them to extrapolate a forecast of the growth of technology that has the S-shape, which was described earlier as the pattern of technology adoption and diffusion.

First, it is important to note to which stage of technology growth these models apply. Bright (1978) and later Martino (1983) described the stages of technology growth shown in Exhibit 8.1. The diffusion models are useful for Stages 4 through 7.

Exhibit 8.1 Stages of Technology Growth

Stage 1: Scientific findings; determination of opportunity or need
Stage 2: Demonstration of laboratory feasibility
Stage 3: Operating a full-scale prototype or field trial
Stage 4: Commercial introduction or operational use
Stage 5: Widespread adoption
Stage 6: Proliferation and diffusion to other uses
Stage 7: Effect on societal behavior and/or significant involvement in the economy

Source : Based on Martino (1983)

Extrapolation requires statistical estimation of model parameters, so enough purchases of the technology are required to provide a reasonable sample. As noted in Section 6.3.1, some suggest that the sample should be 10% of the potential market. However since this implies waiting well into the period of market penetration, the models usually need to be applied earlier, with less data. This obviously can be done, but the results are likely to be less accurate. However, there is nothing to keep the forecaster from using the models repeatedly as more data become available, and that is a good idea in any case.

The Fisher-Pry model (Fisher and Pry 1971) is used to forecast the rate at which one technology will replace another. The model graphically depicts an S-shaped variation over time, with a slow beginning, a rapid ascent, and a leveling off at the finish. The model's output is typically converted to a graph showing percentage capture from 0 to 100 as a function of time measured in calendar years from the starting point of the innovation. The Gompertz model is most appropriate when existing products are replaced because they are worn out rather than because of advantages of the new technology. Therefore, it is sometimes referred to as a *mortality model*. The Gompertz model also produces an S-shaped curve.

Although they are both S-shaped, the Fisher-Pry and Gompertz models focus on different processes. Fisher-Pry curves accurately depict the common diffusion situation in which the accumulation of sales facilitates an increasing growth rate of purchases as customers become familiar with the product. The Gompertz model is better for situations in which the technology's advantages are established early and penetration is characterized by robust early growth followed by a period in which the rest of the market waits for the existing technology to wear out before purchasing the new.

8.3 FORECASTING THE ECONOMIC CONTEXT

The preceding section described extrapolative models that can be used to increase one's ability to forecast how technologies will grow based upon how they have already grown. The models are *naive* in that they do not suppose any understanding of the causal relationships determining the pattern. Instead, they use historical patterns of growth for earlier technologies and predict that the growth of the new technology will follow the same pattern. This has some obvious limitations. For example, the growth rate of adoption of a new technology no doubt will be affected by the availability of equity capital and/or credit, which may be different than in the past. Periods of high economic growth, for example, are characterized by rising stock markets and a lot of new initial public offerings (IPOs), most of which provide the funds to rapidly grow new technology businesses. Moreover, consumers and businesses have access to money to invest in new products and processes. Recessions, on the other hand, are characterized by a tempering of ambitious plans, a focus on making do with what one has, and a lack of credit and capital to try new things. Technology growth also is related to the production context in which it will be applied. Section 8.1.1 described how Porter's five-forces framework can affect the success of a new technology. Moreover, the technology delivery system (TDS) that is a focus of this book includes both the general economic environment as well as the conditions of specific competing and collateral technologies and the industries that are built upon them. This section will address the types of models available to analyze and predict the future economic context of an emerging technology.

8.3.1 Macroeconomic Forecasting

The first question the forecaster may want to consider is the outlook for overall conditions. Economists have developed large, sophisticated models of the economy, and consulting firms monitor general economic conditions and can relate them to a specific industry or firm. Most of these models produce results that are short-term relative to most technology cycles with which managers must deal. However, some perspective on the economic outlook for the next one to two years will be important for the investment and operating decisions surrounding the launch of new technologies.

The basic relationship for the macroeconomic outlook for a country is the familiar equation for gross domestic product (GDP), the total production of goods and services for a calendar year:

$$GDP = C + I + G + N \tag{8.1}$$

where:

C = aggregate consumption, including durable goods, nondurables, and services

I = investment, including new residential and commercial construction, machinery, equipment, and inventories

G = government expenditures, including federal, state, and local

N = net exports, or exports less imports

Macroeconomic forecasting involves estimating the changes that are likely to occur in these variables. However, this simple equation is only the tip of an apex formed of hundreds of equations and thousands of pieces of data. Consumption forecasts, for example, use the results of consumer surveys done at the University of Michigan Survey Research Center as well as past relationships among consumption, income, and wealth. Other variables in the equation are still very difficult to predict. For instance, interest rates and other financial market attributes are critical to overall demand for output but are notoriously hard to project, even though there are central banking authorities that dramatically affect or even determine them. Further, models for demand in one country are increasingly dependent on what is happening in the rest of the world, for which analogous models are used to predict output.

The choices for technical managers are wide. But building one's own forecasting models is unlikely to be justified. Even the purchase of forecasts from consultants can be very expensive in relation to the benefits for most firms engaged in launching new technologies. Moreover, there is a lot of evidence that following the consensus of forecasts in readily available business publications, government sources, or websites is likely to produce most of the useful information on general economic trends. While some sources, such as the Congressional Budget Office (see below), project as far out as 10 years, it is good to remember that uncertainty rises significantly for long-range projections. Furthermore, political as well as economic considerations sometimes play a role. There are a range of sources for economic forecasting online and in print (Wiley Online Library 2009; Conference Board 2010; International Monetary Fund 2010; Kiplinger 2010; U.S. Congressional Budget Office 2010).

8.3.2 Input-Output Analysis

In input-output analysis, the economy is modeled as a static structure that represents the flows of production from one industry to another and finally to the demand components in Equation 8.1. The model breaks down the interacting transactions (flows) necessary to make a product so as to show which components of the economy must supply what types of production. The transactions must share a common unit of measurement, and monetary units are the most convenient. (Try to imagine, for example, how the tons of steel and glass and the kilowatt-hours of electricity necessary to make a car might be combined. Clearly, it is much easier to add so many dollars' worth of steel, glass, and electricity to the costs of other components to arrive at the cost of producing a car.)

As an example, consider an economy that is composed of only two industries, *X*1 and *X*2. Table 8.1 shows what happens to the output of these two industries. The table is similar to the interindustry transactions tables published by the U.S. Department of Commerce in the Survey of Current Business (Bureau of Economic Analysis 2010a). While this example is limited to two industries for simplicity, the same principles apply to the government's data. Wassily Leontief received the 1973 Nobel Prize in economic sciences largely because of his work in developing input-output tables. These tables are used to better understand the industrial sectors of economies like that of the United States. They also are frequently used to analyze the impacts of changes in final demands for output on specific industries or regions. More information about the current input-output tables of the U.S. government can be found at Bureau of Economic Analysis (2010b), and a numerical example of the approach is available at Jensen (2001).

The first row of Table 8.1, for example, shows how much of the total output of *X*1 ($1000) is sold from one firm to another; within industry *X*1, ($100), to industry *X*2 ($200), and to each of the four final demand sectors in Equation 8.1. The second row gives a similar breakdown for industry *X*2 ($2000 total output). The "Wages" and "Other" entries in the table are incomes that represent the distribution of the value added by each industry. Value added is determined by sales minus the cost of the intermediate goods purchased from other industries. Note that the total of each industry's column equals the total of that industry's row. The direct requirements of one industry can be calculated for the output of another from the information in the interindustry transaction table.

Table 8.2 presents these requirements for our hypothetical two-industry economy as computed from the information in Table 8.1. Entries show the share of each dollar of output of an industry (column) represented by input to the other (row). For example, a dollar of output from *X*1 requires $0.10 (100/1000) of its own output as an intermediate input plus $0.50 of industry *X*2 output plus $0.30 of labor and $0.10 of other income, such as rents and profits. (The direct requirements table for the United States is also published in the Survey of Current Business.) However, while this direct requirements table is useful, it does not tell the whole story.

The limitations of the direct requirements table can be seen if one asks: "How much would industry *X*1 have to produce to sell an additional $100 of output

TABLE 8.1 Interindustry Transactions in $

	*X*1	*X*2	*C*	*I*	*G*	*N*	Total Output
*X*1	100	200	400	100	100	100	1000
*X*2	500	600	200	300	200	200	2000
Wages	300	700					
Other	100	500					
Total	1000	2000					

TABLE 8.2 Direct Requirements Table

	Output in $ of Industry	
Input To	*X*1	*X*2
*X*1	0.10	0.10
*X*2	0.50	0.30
Wages	0.30	0.35
Other	0.10	0.25
Total	1.00	1.00

to consumers?" Since $0.10 of each dollar of $X1$ output goes for intermediate goods produced within the industry, its total output must be at least $110 to produce an additional $100 of output for consumers. However, there is more to consider. Table 8.2 shows that each dollar of output by $X1$ uses an input of $0.50 from industry $X2$ and that $X2$ requires $0.10 \times 0.50 = \$0.05$ of output from industry $X1$ to produce it. Thus, the total output of X must be at least $115, but that requires additional input from $X2$—and so forth. A convergent pattern of increasing output levels clearly has been established. This simple example helps to explain why the direct requirements table is a building block for the more useful total requirements table shown as Table 8.3.

This problem can be better understood by casting it in terms of two equations in two unknowns. First, assume that the final demands for the outputs of industries $X1$ and $X2$ are as portrayed by the interindustry transactions table (Table 8.1), except that the level of consumer purchases of $X1$ output is $100 higher (that is, C = $500 instead of $400). This results in the following equations:

$$X1 = \frac{100}{1000}X1 + \frac{200}{2000}X2 + 500 + 100 + 100 + 100$$

or

$$X1 = 0.10X1 + 0.10X2 + 800 \tag{8.2}$$

and

$$X2 = \frac{500}{1000}X1 + \frac{600}{2000}X2 + 200 + 300 + 200 + 200$$

Likewise,

$$X2 = 0.50X1 + 0.30X2 + 900 \tag{8.3}$$

These equations are constructed using direct requirements data from Table 8.2 and the increased level of consumer purchases that was desired. They define actual industry output levels for a $100 increase in consumer demand for output of $X1$. Solving Equations 8.2 and 8.3 simultaneously gives values for $X1$ and $X2$ of $1120.69 and $2086.21, respectively. Comparing these values with those in Table 8.2, the result of a $100 higher level of consumer sales of $X1$ would

TABLE 8.3 Total Requirements Table: Direct and Indirect Effects per Dollar of Final Demand

An Increase in Demand for Output of this Industry Is Produced	By an Increase of $1.00 of Demand for Output of This Industry	
	$X1$	$X2$
$X1$	1.2069	0.1724
$X2$	0.8612	1.5517

be to increase the output required of $X1$ by \$120.69 and that of $X2$ by \$86.21. Thus, every dollar of increased final demand for output from $X1$ would increase the output for that industry by \$1.2069 and would increase the output of $X2$ by \$0.8621.

A similar exercise, assuming that final demand for $X2$ increases, reveals that the direct and indirect effects on outputs total \$0.1724 of additional output for $X1$ and \$1.5517 of additional output for $X2$.

The solutions to these equations can be used to construct the total requirements table (Table 8.3). This table shows the direct and indirect effects of a \$1 increase in the final demand for each industry's output. The total effects of a \$1 increase in final demand for the output of an industry can be found by reading down the column for that industry in Table 8.3.

Matrix algebra can be used to apply these principles to the more realistic case in which the number of industries is quite large. In this approach, Equations 8.2 and 8.3 can be written as

$$X = AX + Y \tag{8.4}$$

where X is a 2×1 matrix, the elements of which are $X1$ and $X2$ in the example; A is a 2×2 matrix, the elements of which are the entries of the first four cells in the direct requirements table (Table 8.2) expressed as fractions of industry total output; and Y is a 2×1 vector made up of the sums of the final demands for each industry's output from the components C, I, G, and N.

Rearranging and solving the equations for the *total* outputs of $X1$ and $X2$ as functions of the final demand in Y:

$$X = (I - A)^{-1}Y \tag{8.5}$$

The elements of the $(I - A)^{-1}$ matrix are the entries in the total requirements table. They show the direct and indirect effects of an increase in demand for the industry heading each column. For our simple example, the four values in Table 8.3 are the components of this matrix.

Computers are well suited to this kind of task even when the amount of data is very large. The U.S. Department of Commerce considers about 100 industry groups rather than the two industries in our example. Using the approach outlined in the example above, their computations produce the total requirements table that appears in the *Survey of Current Business*. That table, for example, might reveal that a \$1 increase in final demand for cardboard containers and boxes leads to direct and indirect output increases of \$1.05608 for the industry itself, a \$0.51828 increase in paper and allied products output, and a total increase of \$0.06008 in the output of the chemical products industry. Effects on all the other industry groups could be examined as well.

The implications of this total requirements table for those doing market planning should be clear. Each element in the row for the industry reveals the relationship of that industry to the final demands for output from each industry listed in the table. For example, suppose a new box-making technology is expected to double the output of boxes in the next decade. Since 6% ($0.06008/$1) of box-making costs are due to the chemicals industry, this change will be important for at least some firms that produce chemicals. Total requirements tables have been used to analyze the effects of economic changes ranging from decreases in defense spending to expanded investment in environmental protection technology. Many years ago, Isard (1972), for example, showed how input-output analysis could be used to analyze the environmental effects of economic development. This analysis involved an extension of the model to explicitly include ecological inputs in a manner analogous to industrial and factor inputs. Input-output tables also can be used to examine the effects of new technologies. For example, if one considered the impact of using ethanol from corn as a major source of fuel, one could see that such a technology not only raises the demand for corn but also affects many other industries, including those involved in petroleum refining and distribution for tractors, fertilizing, and transport.

Input-output information can be helpful in technology forecasting and management in a number of ways. First, the example above shows how it might identify bottlenecks or unanticipated input price challenges in implementing a new technology. However, it can also identify opportunities for alleviating such shortages with innovative approaches. At the same time, the likely decline of a particular industry might lead to declines in its suppliers. These declines might release resources that can be applied in other places, including new technology-based businesses. Therefore, input-output tables are useful both for the impact assessment discussed in the next chapter and for analyzing the opportunities and the potential for growth examined in this chapter.

However, there are some severe limitations to the use of input-output tables by technology managers. Perhaps most important is that *the table assumes that the proportions of inputs to outputs remain constant and that relative producer prices do not change*. These limitations are somewhat alleviated by frequent updates of the tabulated data, but as the technology manager is well aware, change can be rapid. This is not to say that the information provided is not helpful, but rather that the conclusions from input-output analysis should be treated as approximations for the short-term future.

8.3.3 General Equilibrium Models

The limitations of traditional input-output models led to the development of more realistic, and necessarily more complex, general equilibrium models. Most of the applications of these models have been to the study of macroeconomics and fiscal and monetary policy (see, for example, Del Negro and Schorfheide 2003). These general equilibrium models can be used as bases for exercises that go beyond the input-output approach by introducing mathematical functions to account for substitution. While these models are still static in character, they provide more realistic estimates of the effects of changes on an economy. More sophisticated, and therefore even more complex, models introduce both a dynamic aspect to study processes of adjustment over time and a stochastic aspect to simulate unforeseen random shocks like wars, changes in energy supplies, or the introduction of new technologies. These are called *dynamic stochastic general equilibrium models.* They have contributed to the understanding of how technology might impact economies, but they do not yet seem to have much usefulness for technology managers. While they strive to introduce microeconomics of individual markets into the analysis of national and world economies, their use to examine a specific technology does not appear practical. The overwhelming data collection they require and their sensitivity to assumptions about exogenous variables make their cost-benefit balance doubtful for technical managers. For the near future, they will probably be used to improve the performance of macroeconomic forecasting models and consulting that is based on their outputs. In time, like input-output models, they may become more amenable to application to particular technologies.

8.3.4 Hedonic Technometrics

Coccia (2005) notes that the concept of *technometrics* has been developing over the past several decades. For example, the journal *Technometrics* focuses on the application of statistics to the physical, chemical, and engineering sciences. However, for the purposes of this book, the term *technometrics* refers to a systems concept of technology that implies that innovations evolve from existing technologies (Sahal 1985). Sahal stressed the need to measure the value of technologies to inform decisions about R&D and government technology policy.

The roots of hedonic technometrics can be found in economics. Very early in the use of price indices, economists wrestled with the fact that quality changes made conclusions about price changes unrealistic. For example, the prices of electronic products like stereos and TVs may be similar to what they were 50 years ago, but the changes in capabilities make the prices incompatible for comparisons. The hedonic approach basically says that the value of a product or an underlying technology is based upon the attributes that lead to satisfaction for the ultimate users. Automotive technology advances should perhaps be measured by such things as fuel efficiency, safety, comfort, frequency of repair, and so forth. Griliches (1957, 1971) used such an approach to analyze progress in the adoption and diffusion of hybrid corn seeds. Grilliches also investigated quality and price indices. Lancaster (1966) actually proposed a theory of consumer choice based upon optimization of a linear combination of product characteristics. Such work has led to applications of the concept over the past 25 years in market research on such things as product differentiation and features in new product development. See, for example, the paper by Kristensen (1984).

While there have been practical applications of the hedonic approach to make decisions about product features at the firm level, a lot of the technometric literature is focused on getting measures of innovativeness per se at firm, sector, or national levels. Rand Corporation scholars developed techniques to combine parameters or characteristics of technology and relate them to the passage of calendar time. Others, like Martino (1983), extended the work on technology as a composite of surfaces related to attributes. That is, they tried to get metrics of various features of technologies and then mathematically aggregate those metrics as a way to describe the product and how it is different from previous ones. Martino also was able to show that indicators of technological change, like patents, papers, and R&D expenditures, can provide one to three years of advanced warning of a market shift to managers. Coccia's (2005) approach is to examine the impacts made by technologies after the fact to measure their importance in a way analogous to the seismic measurement of earthquakes.

Hedonic technometrics is making contributions to understanding technology as a multidimensional function. In time, it will provide more substantive tools for management that reflect Brian Arthur's insights into the origins of technological development (Arthur 2009). For now, it appears that more research is needed to find practical, less expensive approaches to draw useful quantitative conclusions for decisions about specific technologies.

8.4 FORECASTING IN AN INSTITUTIONAL CONTEXT

Section 8.1.2 described the relationship of institutions and technology. Certainly the institutional constraints, impacts, and reactions to and from innovation are critical components of the TDS. This is a very broad topic, and different technologies will encounter very different institutional contexts. For instance, the development of medical innovations from stem cells research has encountered major resistance from religious institutions and their political allies, while there is relatively little religious sensitivity to alternative sources of energy. In some situations, the institutional context can be very positive. Singh and Allen (2006) discuss the importance of universities in Pittsburgh's postindustrial economy, and many other regions with twentieth-century industrial success are looking to their universities as important institutional assets in growing technology-based businesses. Much also has been written about the ways that government can either enable or constrain innovation. India, for example, is increasingly a force in information technology (IT), but the booming IT economic sector did not develop until the government liberalized its restrictive policies toward business. Christensen (Christensen and Overdorf, 2000) and others have talked about the importance of a supportive environment to the nurturing of disruptive technologies in the United States. The constraints and encouragement can come in cultural and other unofficial ways, as well as through specific government action.

8.4.1 Institutional Arrangements and the Market

The purpose of raising the issue of institutions and technology again is to focus on how one might forecast how the institutional context will affect the prospects for innovations. Much of the answer may be qualitative, and the use of scenarios (Chapter 7) will be helpful. However, there also may be some metrics and models that will aid decisions about institutional obstacles and assistance that will affect the growth of an innovation throughout the value chain of development, implementation, and consumption of products and services provided by new technologies.

Perhaps the first question to ask about innovations and a particular institutional context is whether the innovation is being launched in a good location. Michael Porter (2001) noted that innovation and industries grow in the context of agglomeration economies. *Agglomeration economies* are the lower costs that result when producers of the same product locate in proximity to one another. This enables such things as more specialization in the supply chain and higher potential for knowledge sharing that reduces resource requirements and bottlenecks. Launching new technologies requires specialized labor and other resources that are not evenly distributed across the landscape. For example, orthopedic devices like replacement joints are most likely to have originated in the little town of Warsaw, Indiana, because it had the skilled machinists to work with the exotic metals required for these products. Thus, measuring agglomeration economy variables like quantities of specialized labor and number of firms in a particular industry segment cited by Porter could be important to a technology forecast. The experience and characteristics of the workforce also are important. Shane (2008) and others pointed out that most successful businesses depend on the actions of people with experience in the industry. So, measuring the availability of experienced professionals in the launch area also seems worthwhile. Moreover, Richard Florida (2002) developed data-driven creativity analyses of a region that show that creativity translates into innovation and economic development. Since the initial innovation usually will require continuing creativity to change the product or process or its content, a measure of the area's institutional commitment to fostering creativity might be important.

Increasingly, the bases of value and wealth in the world economy are knowledge and its applications. This means that the most significant source of wealth creation through technology is the brainpower of the workforce. A shortage of people with the appropriate education and experience can be an important obstacle to the growth of innovations. A wide range of technical and nontechnical abilities is important. For example, a technology push may only require great technical minds, but commercially successful innovation requires operational and business expertise that goes well beyond the genius needed to make the break through discoveries. Furthermore, innovation requires workforces that can handle rapid change, as well as highly complex and often ill-defined work assignments and environments. Such workers have different distributions of talents than a traditional workforce analysis might suggest. For example, in a twentieth-century assembly operation, the most creative professionals may have been 10 or 100 times better at adding value than their least talented colleagues. Understanding the potential of the workforce to grow a new technology means focusing on the best, and that implies more than just looking at census data. The work of Florida (2002) showed that the best members of the creative workforce gravitate to areas with cultural advantages for them. So, the analysis of institutional context should use an assessment of cultural attributes to supplement data on workforce availability. Such cultural data are also likely to be useful in identifying geographic market niches where innovations are more likely to be embraced.

Predicting the outcomes of political and social opposition to or encouragement of innovations is even more complex than trying to anticipate market demand. There is extensive literature about social choice and the alternatives. Elinor Ostrom was a cowinner of the 2009 Nobel Prize in economics for her scholarship and experimentation on the ways that groups arrive at solutions to difficult resource allocation problems. Her work shows that there are alternatives beyond the usual dichotomy of markets or government action. What's more, the same technologies can inspire different collective actions in different parts of the world. (For example, Europeans have opted for increased nuclear power generation despite the inherent risks, while that technology has been stalled for decades in the United States.) Much still needs to be learned about societies and their institutional actions in relation to technologies, but predictions must be made, however tentative their implications might seem. Research on game theory and agent-based models is providing insight into things like the way rivals, regulators, interest groups, and others might react to a new technology. A brief overview of these tools should help to convey the state of the art in projecting the institutional context of innovation.

8.4.2 Game Theory

Game theory (see also Section 6.7) is a set of mathematical tools that have been used by economists and other social scientists to model the processes and outcomes that occur when there is interdependence among the entities involved. One of the best-known and simplest illustrations is the prisoners' dilemma game. As in other games, there are players, rules, and payoffs. In this case, the players are prisoners accused of a crime and separately interrogated. The potential outcomes are no consequences if no one talks; severe consequences if someone else talks and you don't; and less severe consequences if you confess and implicate someone else. Even though no one talking produces the best payoff, rational, self-interested prisoners will each try to be the first to confess and give evidence. Organized crime has prevented this result by making death a consequence of cooperating—thereby changing the rules of the game.

Thanks to mathematics and computers, games can be much more complex, probabilistic, and dynamic than the simple example described here. Thus, they can be used to model actions and reactions in the institutional environment. There have been useful applications of these tools in corporate strategy, international relations, and other areas. While it is not clear that game theory and available software can provide accurate predictions of institutional responses to innovation, their construction and manipulation can provide useful insights into the relationships and the range of outcomes that are possible. Game playing also may suggest some rule modifications that can produce more favorable results in the real world. Game theory has been used, for instance, as a basis for providing guidelines for business strategy (Brandenburger and Nalebuff 1996). Others have used game theory models to anticipate and predict institutional and political change (Bueno de Mesquita 2009).

8.4.3 Agent-Based Models

Similar conclusions about current applicability probably can be drawn about agent-based models or multiagent simulation. These are computable models of social systems with multiple agents that use game theory, complexity theory, and other tools to simulate social interactions of multiple agents within the society (Axelrod and Tesfatsion 2010). Their basic purpose is to help the forecaster understand the behavior and interactions of social organizations as systems. Assumptions are made about the participants in the system, and simulations permit analysts to understand the implications of those assumptions. As in game theory, these insights might lead to system improvements. In the case of technology forecasting and management, one might think about constructing the TDS as a system that could lend itself to computation. Monte Carlo methods (Section 6.5) could be used to add probabilities to the simulations. Examination of the various solutions generated and their sensitivity to different assumptions could lead to useful changes in the variables of the TDS that are controlled by the technology manager.

8.5 CONCLUSION

New technologies, even great ones, seldom if ever generate their own growth from inherent technical advantages. Therefore, economic and market analyses are essential in forecasting and managing the future of technologies and the businesses that are built on them. There are both qualitative and quantitative tools to assist in envisioning the technology's future, but they are neither precise in their predictions nor cost free. This chapter has discussed many of these tools, which can be used to help assess the receptiveness of the environment to the unfolding commercialization of a technology.

The most important conclusion of the chapter is that innovation depends on customers, and efforts to project the future of innovations must include understanding those customers and what they will value. While there are qualitative techniques for gathering this information, direct contact with potential customers generally should be part of the investigation. Quantitative techniques also exist and can be useful in predicting what will happen to the technology and its environment. For example, there are S-shaped curve functions that can be used to extrapolate the technology's growth from existing information. Quantitative techniques can also be helpful in forecasting the economic environment. These include standard macroeconomic forecasting and information on direct and indirect effects available from input-output analysis. Although presently they may not be of great utility to the forecaster, progress is even being made in projecting the institutional environment with such tools as game theory and agent-based models.

Not all of these approaches will be applicable to every technology delivery system. Furthermore, none of them is without cost. However, it is useful to know what types of tools are available and to understand that even the simplest, easiest, and least expensive methods (like talking to potential customers) can make a real difference in the quality of management decisions.

REFERENCES

21st Century Fund. (2010). "The Indiana 21st Century Research and Technology Fund." Retrieved 1 September 2010 from http://www.21fund.org/.

Arthur, W. B. (1994). *Increasing Returns and Path Dependence in the Economy.* Ann Arbor, University of Michigan Press.

Arthur, W. B. (2009). *The Nature of Technology: What It Is and How It Evolves*. London, Allen Lane.

Arthur, W. B. (2010). "External Professor, Sante Fe Institute." Retrieved 1 September 2010 from http://tuvalu.santafe.edu/~wbarthur/.

Axelrod, R. and L. Tesfatsion. (2010). "On-Line Guide for Newcomers to Agent-Based Modeling in the Social Sciences." Retrieved 1 September 2010 from http://www.econ.iastate.edu/tesfatsi/abmread.htm.

Brandenburger, A. M. and B. Nalebuff. (1996). *Co-opetition: A Revolution Mindset That Combines Competition and Cooperation: The Game Theory Strategy That's Changing the Game of Business*. New York: Currency Doubleday.

Bright, J. R. (1978). *Practical Technology Forecasting: Concepts and Exercises.* Austin, TX, The Industrial Management Center.

Bueno de Mesquita, B. (2009). *The Predictioneer's Game: Using the Logic of Brazen Self-Interest to See and Shape the Future.* New York, Random House.

Bureau of Economic Analysis. (2010a). "Industry Economic Accounts." Retrieved 1 September 2010 from http://www.bea.gov/industry/.

Bureau of Economic Analysis. (2010b). "Industry Economic Accounts Information Guide." Retrieved 1 September 2010 from http://www.bea.gov/industry/iedguide.htm.

Christensen, C., J. Grossman, et al. (2009). *The Innovator's Prescription*. New York, McGraw-Hill.

Christensen, C. M., Roth, E. A., and S. D. Anthony. (2004). *Seeing What's Next: Using Theories of Innovation to Predict Industry Change,* Cambridge, MA, Harvard University Press.

Christensen, C. and J. Hwang (2009). "Power to the Patients." *The Atlantic,* 5 October 2009, Retrieved 14 January 2011, from http://www.theatlantic.com/magazine/archive/2009/10/power-to-the-patients/7727/

Christensen, C. and M. Overdorf. (2000). "Meeting the Challenge of Disruptive Change." *Harvard Business Review* **78**(2): 66–76.

Coccia, M. (2005). "Technometrics: Origins, Historical Evolution and New Directions." *Technological Forecasting and Social Change* **72**: 944–979.

Conference Board. (2010). "Data and Analysis." Retrieved 1 September 2010 from http://www.conference-board.org/data/.

Del Negro, M., F. Schorfheide (2003). "Take Your Model Bowling: Forecasting with General Equilibrium Models," *Economic Review*, Federal Reserve Bank of Atlanta, issue 4: 35–50.

Duesterberg, T. J. and E. Preeg. (2003). *U.S. Manufacturing: The Engine for Growth in a Global Economy*. Westport, CT, Praeger.

Fisher, J. C. and R. H. Pry. (1971). "A Simple Substitution Model of Technological Change." *Technological Forecasting and Social Change* **3**: 75–88.

Florida, R. (2002). *The Rise of the Creative Class: And How It Is Transforming Work, Leisure, Community and Everyday Life*. New York, Basic Books.

GPOAccess. (2010). "Code of Federal Regulations: Main Page." Retrieved 1 September 2010 from http://www.gpoaccess.gov/cfr/.

Griliches, Z. (1957). "Hybrid Corn: An Exploration in the Economics of Technological Change." *Econometrica* **XXV**: 501–522.

Griliches, Z. (1971). *Price Indices and Quality Change*. Cambridge, MA, Harvard University Press.

Heenan, D. (2005). *Flight Capital*. Mountain View, CA, Davies Black Publishing.

Heller, M. (2008). *The Gridlock Economy: How Too Much Ownership Wrecks Markets, Stops Innovation and Costs Lives.* New York, Basic Books.

International Monetary Fund. (2010). "World Economic Outlook Reports." Retrieved 1 September 2010 from http://www.imf.org/external/ns/cs.aspx?id=29.

Isard, W. (1972). *Ecological-Economic Analysis for Regional Development.* New York, Free Press.

Jensen, I. (2001). "The Leontief Open Production Model or Input-Output Analysis." Retrieved 1 September 2010 from http://online.redwoods.cc.ca.us/instruct/darnold/laproj/fall2001/iris/lapaper.pdf.

Kiplinger. (2010). "Economic Outlook, Indicators, Forecasts." Retrieved 1 September 2010 from http://www.kiplinger.com/businessresource/economic_outlook/.

Kristensen, K. (1984). "Hedonic Theory, Marketing Research, and the Analysis of Complex Goods." *International Journal of Research in Marketing* **1**(1): 17–36.

Lancaster, K. (1966). "A New Approach to Consumer Theory." *Journal of Political Economy* **74**: 132–157.

Mansfield, E. (1989). *Economics: Principles, Problems and Decisions*. New York, W.W. Norton.

Martino, J. P. (1983). *Technological Forecasting for Decision Making*. New York, McGraw-Hill.

Moore, G. (1999), *Crossing the Chasm: Marketing and Selling High-tech Products to Mainstream Customers,* rev. ed. New York: HarperCollins.

McQuarrie, E. (1996). *The Market Research Toolbox: A Concise Guide for Beginners*. Newbury Park, CA, Sage.

Mufson, S. and P. Whoriskey. (2010). "Drawbacks to Doing Business in China." *The Seattle Times.* 15 January 2010. Retrieved 14 January 2011 from http://seattletimes.nwsource.com/html/businesstechnology/2010797922_chinaunease15.html.

Nolan, R. and D. Croson. (1995). *Creative Destruction: A Six-Stage Process for Transforming the Organization*. Cambridge, MA, Harvard Business School Press.

Organisation for Economic Co-Operation and Development. (2010). "Welcome to OECD Stat Extracts." Retrieved 1 September 2010 from http://stats.oecd.org/Index.aspx.

Porter, M. (1985). *Competitive Advantage: Creating and Sustaining Superior Performance.* New York, The Free Press.

Porter, M. (2001). *Clusters of Innovation: Regional Foundations of U.S. Competitiveness.* Washington, DC, Council on Competitiveness.

Sahal, D. (1985). "Technology Guide-Posts and Innovation Avenues." *Research Policy* **14**: 61–82.

Schumpeter, J. A. (1942). *Capitalism, Socialism and Democracy.* New York, Harper.

Shane, S. (2008). *The Illusions of Entrepreneurship: The Costly Myths That Entrepreneurs, Investors, and Policy Makers Live By.* New Haven, CT, Yale University Press.

Singh, V. and T. J. Allen. (2006). "Institutional Contexts for Scientific Innovation and Economic Transformation." *European Planning Studies* **14**(5): 665–679.

Solow, R. M. (1957). "Technical Change and the Aggregate Production Function." *Review of Economics and Statistics* **39**: 312–320.

SurveyMonkey. (2010). "Free Online survey Software and Questionnaire Tool." Retrieved 1 September 2010 from http://www.surveymonkey.com/.

United Nations Population Division. (2010). "United Nations Population Information Network." Retrieved 1 September 2010 from http://www.un.org/popin/.

U.S. Census Bureau. (2010a)."U.S. Census Bureau," Retrieved 1 September 2010 from http://www.census.gov/.

U.S. Census Bureau. (2010b). "Manufacturing, Mining, and Construction Statistics." Retrieved 1 September 2010 from http://www.census.gov/mcd/.

U.S. Congressional Budget Office. (2010). "The Budget and Economic Outlook: Fiscal Years 2009 to 2019." Retrieved 1 September 2010 from http://www.cbo.gov/ftpdocs/99xx/doc9957/01-07-Outlook.pdf.

U. S. Consumer Product Safety Commission. (2011), "CPSC Home Page." Retrieved 13 January 2011, from http://www.cpsc.gov/

U.S. Department of Energy. (2010). "Secretary Chu Announces $256 Million Investment to Improve the Energy Efficiency of the American Economy." Retrieved 1 September 2010 from http://www.energy.gov/news2009/7434.htm.

Wiley Online Library. (2009). "World Economic Prospects." Retrieved 1 September 2010 from http://onlinelibrary.wiley.com/doi/10.1111/j.1468-0319.2009.00761.x/abstract.

9

IMPACT ASSESSMENT

Chapter Summary: This chapter presents an overview of impact assessment as designed for use in technology forecasting as a decision-making tool. It lays out the distinction between impacts on technology and impacts of technology. Then it presents the structure of a comprehensive approach to impact assessment. Finally, it addresses three steps in impact assessment: impact identification, analysis, and evaluation.

Impact assessment is a systematic examination of the effects on or of new developments such as technologies, processes, policies, organizations, and so on. Impact assessments are classified as policy studies, since they can affect the policies of the organizations that conduct them as well as those of other stakeholders. In most cases, impact assessments should result in actions. Assessments may be freestanding or part of another study such as a technology forecast.

Impact assessment is a necessary component of technology forecasting. It identifies areas in which significant impacts may occur, their likelihood, and their significance. The forecaster must evaluate these impacts, consider measures to enhance or inhibit them, and factor them into the planning process for developing the technology.

9.1 IMPACT ASSESSMENT IN TECHNOLOGY FORECASTING

A valid forecast must include the impacts on and of a technology's introduction and adoption because external factors in the technology delivery system (TDS), such as societal values and institutional regulation, may impinge on its development. The technology forecaster also must consider product acceptance, impacts on broad segments of society, and changes arising from the adoption of the technology. Thus, impact assessment is part of each of the three phases of technology forecasting (see Section 3.2) and relies heavily on the TDS. In the *exploring phase*, IA is "quick and dirty," and identifies likely impacts on the development of the technology by external factors and likely impacts of the technology on users and, more broadly, on society as a whole. Only impacts considered to be significant from the perspective of the forecast are retained for further analysis. A preliminary determination of the magnitude of the impacts may be in order. In the *analyzing phase,* impacts of concern are analyzed in greater depth and the results are factored into the forecast. In the final *focusing* phase, important impacts are further analyzed and may be evaluated. Results of this phase are typically factored into plans for the development and deployment of the technology.

Since impact identification has general utility in technology forecasting, this chapter strives to give the forecaster a good start by presenting useful techniques. Impact analysis and evaluation in a technology forecast depend on the goals of the forecast and the significance of the impacts identified. Thus, sections covering the principal impact areas are written at a general level. Other sources will be required to do impact analysis in depth. However, the material presented here still should prove adequate for quick and dirty work.

9.2 IMPACTS ON TECHNOLOGY AND IMPACTS OF TECHNOLOGY

The distinction between impacts on technology and impacts of technology is both useful and important. Impacts on a technology are societal constraints, facilitators, and direction changers from outside the developing organization that may mold the form of the technology and alter its implementation. For example, impacts on U.S. stem cell technology development included government restrictions due to the moral objections of a significant segment of the population. Impacts of a technology are the effects of the developed technology on its users and the broader societal environment. For example, impacts of stem cell technologies may include decreases in mortality rates from some diseases and the reduced use of other therapies.

Impacts on a technology may affect it at any stage of its life cycle, even in its maturity. For instance, the automobile ended horse and buggy technology (except among the Amish). Impacts of a technology result from its broad utilization after development. In a technology forecast, the study of impacts on it is helpful in understanding the shape the technology implementation may take. Likewise, understanding the impacts of technology helps the forecaster understand how users and society as a whole are impacted by the technology and thus how they will react to it. A feedback loop often exists since impacts of the technology may lead to impacts on it, such as modifications to mitigate its negative impacts or enhance its positive ones.

9.3 A COMPREHENSIVE APPROACH TO IMPACT ASSESSMENT

Porter, Rossini, et al. (1980, pp. 54–55) proposed a comprehensive 10-step approach to impact assessment. The focus of this approach is on freestanding impact assessments, but it applies equally well to assessments within a forecast. While the work of these authors emphasized impact assessments of technologies, the approach can be extended to anything that may have societal impacts, including policies. The steps are listed in Table 9.1.

Several observations are in order concerning the implementation of these steps. First, they should not be considered a linear progression. Given the complexities, steps often must be redone based on knowledge gained in subsequent steps. For instance, impact evaluation may suggest mitigation efforts to alter the technology, changing its description and requiring that the study be redone.

Second, the emphases given to each of the 10 steps may vary greatly from assessment to assessment. Indeed, sometimes it may be appropriate to truncate some steps or to skip them entirely. In addition, many other study strategies have been proposed for impact assessments (Jones 1971; Coates 1976; Armstrong and Harman 1977). Third, technology forecasting is part of impact assessment and vice versa. The two are linked in a feedback relationship that can be thought of as a spiral. Forecasting the direction of the technology is helpful in understanding its possible impacts, while studying the impacts is helpful in understanding potential directions of the technology.

In a technology forecast, many of the 10 steps of an impact assessment will be performed as a matter of course. Any forecast requires a problem definition, as well as technology and societal context descriptions to set up the TDS. Moreover, communicating results is a critical component of utilizing the forecast. Impact identification, analysis, and evaluation are covered in this chapter. These activities are the heart of an impact assessment. Their role in a forecast will depend on the goals of the forecast. At a minimum, impact identification will prove useful, and analysis of major impacts and their evaluation usually are helpful as well. Useful

TABLE 9.1 Ten Steps of Impact Assessment

1. Problem definition
2. Societal context description
3. Societal context forecast
4. Technology description
5. Technology forecast
6. Impact identification
7. Impact analysis
8. Impact evaluation
9. Policy analysis
10. Communication of results

material for many topics in impact assessment may be found at the website of the International Association for Impact Assessment (2010).

9.4 IMPACT IDENTIFICATION

What may impact the development of the technology? What will be the outcomes? What outcomes does the development of the technology lead to? These are the questions to be answered in impact identification. There are two major approaches to impact identification, *scanning* and *tracing*. They are complementary, and both typically will be useful in a forecast. Both begin from the TDS.

Scanning is a broad-brush approach to quickly identify all possible impacts flowing from or to the TDS. Tracing considers causal relationships between development actions and their impacts—their effects. This process creates a causal trail from developments to outcomes that may go through multiple steps or stages.

9.4.1 Scanning Techniques

Scanning methods search the impact field to minimize the probability that significant impacts will be overlooked. The simplest approach is to make a checklist of all candidate impact areas (Section 4.3.3). This list may be brief and at a high level of abstraction or highly detailed and concrete. Another possibility is to list all the parties affected by the development. Either list can be a starting point for identifying impacts. Combining these approaches produces a two-dimensional matrix with impact areas on one axis and stakeholders on the other. Some cells of the matrix may contain no impacts, while others may contain multiple impacts. It may prove useful and convenient to separate the identification of impacts on technology and impacts of technology by constructing separate lists of impact areas and stakeholders.

It is important to consider that the perspectives of the stakeholders may vary so widely that there is no privileged reference frame from which to view the consequences. Thus, a significant consequence for one stakeholder might not concern another. Consideration of the multiple perspectives of the stakeholders (Linstone 1988) becomes important in all phases of impact assessment, especially in evaluation.

A list of impact areas may include such general categories as technological, economic, environmental, social, political, behavioral, institutional, legal, regulatory, culture/values, and health, which can be subdivided as needed. The list of impact areas depends on the intent of the forecast. The lists for impacts on technology and impacts of technology typically will be different. These lists can be generated from the TDS and supplemented by additional monitoring or by contact with experts (see Section 5.1 for techniques). Creativity techniques such as brainstorming (Section 4.3.7) offer convenient and unstructured ways to generate lists. The parties at interest identified when constructing the TDS make up the other axis of the matrix.

As an example of this approach, Edelson and Olsen (1983) identified the impacts of different energy conservation measures by interviewing more than 50 stakeholders. The interviews elicited 20 potential impact areas for further study. In another example, Lough and White (1988) identified impacts from decommissioning nuclear power plants. They began with a literature search and refined it using the Delphi approach (Section 5.1.2). Seventeen participants represented different parties at interest (consulting firms, electric utilities, public utility commissions, and the federal government). The process yielded 19 potential impacts of varying significance. The assessors felt that using this many impacts to rank alternative decommissioning strategies would be confusing. Therefore, they reduced the set to four critical impacts: (1) cost, (2) occupational radiation exposure, (3) institutional (personnel requirement, regulatory and liability obligations), and (4) public attitudes.

9.4.2 Tracing Techniques

Tracing methods emphasize causal relationships in which impacts may be expressed as a chain of causes and effects. Tree techniques (Section 6.7.1) have proven very helpful in tracing. Relevance trees graphically depict the linkages between various members of sets of elements, moving from level to level via some relationship such as "leads to impacts on." At the top of the tree is the technology/development activity being assessed. On the next level are the direct impacts on the areas under consideration or on the parties at interest. The third level exhibits the second-order impacts, that is, the impacts of the direct impacts, and so on. Completing such a tree requires answering a series of "what if" questions for each node at each level. The answers to these questions are derived from the information developed during the exploration phase and used to develop the TDS. The branches originating at the node of an impact tree are not necessarily complete or mutually exclusive.

Forecasters should avoid being seduced by the "treeing game" lest they end up, as in one impact assessment, with a 30-page impact tree of no practical value. To prevent this, they must restrict themselves to the most important branches at each level but avoid a premature closure that misses important impacts.

A tree can be further developed by assigning subjective probabilities to the occurrence of each impact (see Section 6.7.1). Depending on the study needs, the timing and/or severity of each impact also may be estimated. These efforts can help determine which impacts are most significant so as to narrow the list for detailed study.

One advantage of the tree representation is that it simultaneously displays the linkages between impacts and the scope of the impact field. Further, the structured framework of the method tends to make the search for impacts self-propagating. On the down side, early omissions may cause significant impact areas to be ignored, and the process requires knowledge of the impact field structure that may require time and resources to develop. Figure 9.1 illustrates a hypothetical impact tree representing the impacts associated with surface coal mining.

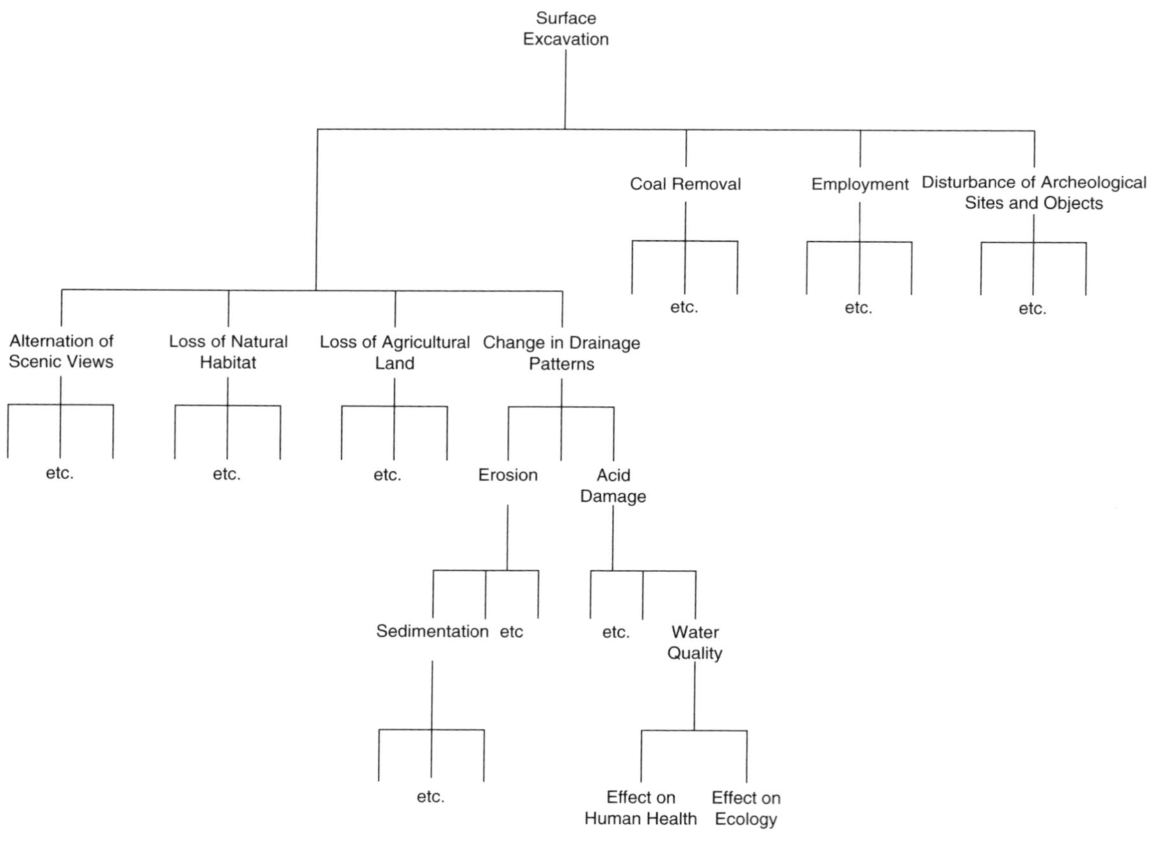

Figure 9.1. Impact Tree for Coal Mining

9.4.3 Narrowing the Impact Set and Estimating Effects

The primary roles of impact identification are in the first two forecasting phases. Scanning is most useful in the *exploring phase* to create a broad list of impacts for which subjective estimates of significance and likelihood can be made. As a first step, significance and likelihood can be limited to high and low levels. This simple dichotomy significantly streamlines the effort and may even be adequate for the simplest analyses. High-significance/high-likelihood impacts rate further consideration, while low-significance/low-likelihood impacts can be ignored. Low-significance/high-likelihood impacts can be set aside for future consideration as the TDS changes. The most challenging impacts are those with high-significance and low-likelihood. Unanticipated impacts of this type caused the breakdown of the U.S. housing market and the huge impacts on international financial markets that began in 2007. The possibility of such impacts must be factored into decisions going forward. It is often useful to quickly trace the most important impacts at this point of the forecast to see where they may lead.

In the *analyzing phase* of the forecast, the impact identification work can be briefly reviewed for completeness and cogency. Then a serious, but not resource-consuming, effort can be made to further trace highly likely/highly significant ones and any "black swans" that may emerge. Limiting the impacts under consideration by using the judgment of experts is very helpful here. Serious impact analysis begins in this phase

In the third forecasting phase, *focusing,* there is little need for additional impact identification unless there is significant new information. However, the impacts identified in earlier phases provide many of the scenario variables used in constructing the final narratives (Chapter 7). The forecaster's emphasis tightens to the most important impacts from the most likely courses of action that have been identified. These are subjects for more in-depth impact analysis and for impact evaluation.

9.4.4 A Final Word

Sometimes asking the right questions is as important as finding the right answers. This is certainly true in impact identification. The creativity techniques introduced in Section 4.3 can be helpful in this regard. Habits developed while systematically and reflectively asking questions about the consequences, especially the unintended ones, of a technological innovation will serve a technology manager well as the technology is implemented and used. Potentially important outcomes of new technologies, such as using the Internet for social networking, may appear in the process of impact identification. Having a grasp of higher-order possibilities allows the technology manager to make timely and cogent responses to the emergence of new opportunities.

9.5 IMPACT ANALYSIS

Impact analysis starts where impact identification leaves off. After learning what impacts are likely to occur, the next step is to analyze them to learn what will happen, how significant it will be, and how likely it is to occur. Answering these questions is the purpose of impact analysis. Once again, the TDS provides the major source of information.

When treating impact analysis, it is convenient to separate impacts by area since analytical methods usually are area specific. In this section, nine impact areas are considered: technological; economic; environmental; social; institutional; political; legal and regulatory; behavioral, cultural, and values; and health. This list is not exhaustive, and in a typical forecast not all of these areas will prove worth considering. The impacts could be further sorted by stakeholder group and/or according to whether they are impacts on or of the technology. This section begins with a brief discussion of analyzing impacts on and impacts of the technology. Then each of the nine impact areas is individually treated.

9.5.1 Analyzing Impacts on and Impacts of the Technology

Impacts on the development of a technology often arise both from within the family of technologies being developed and from complementary and competing technologies. Technological barriers and facilitators from within the family can alter the type and rate of technological development. The cost and availability of complementary technologies needed to build them can have similar effects. For example, the cost and availability of flash memory can determine how personal computer storage is configured and thus computer price points.

Laws, regulations, and incentives also have major impacts on technology developments. Environmental and safety regulations for automobiles, for instance, are driving forces in automotive pollution control and safety technologies. Likewise, laws and regulations coupled with societal values may substantially impede or alter the direction of a technology development, as they have in the medical use of stem cells. Publicly sponsored premarket research and subsidies can be significant drivers in some technological areas as well. The development of space vehicles in the United States before and after Sputnik is a clear example of how national goals and the resulting investments can drive technological development. Another example is detailed in Exhibit 9.1.

Impacts of a technological development have a different dynamic. The widespread implementation of an important technological system may have significant impacts on almost all sectors of society and may create strong feedback loops that drive or inhibit further technological development. Nowhere is this more obvious than in the development of the computer and the systems that incorporate it. The concept of a logic machine using electronic circuit elements was first implemented using vacuum tubes. Subsequent development of computers depended on the technology of miniaturization, first via transistors and then by microprocessors. Miniaturization, in turn, was itself fostered by the needs of space technology. Economic and institutional demands for increased

Exhibit 9.1 Alternate Energy Development in the United States

Alternative energy development in the United States provides an interesting example of how politics, ideology, and economic policies can impact technological development. To minimize dependence on imported fossil fuel, the Carter administration subsidized the development of alternative energy technologies, and substantial work was undertaken during the 1977–1981 period. This support and the subsidies were stopped by the Reagan administration, which championed free market mechanisms. Then, faced with significantly rising energy prices in the 2000s, the G. W. Bush administration chose to subsidize nuclear fission energy—despite major unsolved technological problems in the storage of nuclear wastes—and to encourage increased oil exploration. Only in 2009, with the new Obama administration, did the federal government show renewed interest in alternative energy technologies such as solar energy. This "on again, off again" performance by successive administrations meant that only alternative energy technologies with near-term market potential, such as wind energy, were seriously pursued over the long term.

computing power produced investments that pushed the silicon microprocessor to unanticipated levels of feature size. Enormous amounts of computing power were unleashed. One generation's supercomputer became a basic tool for secretaries and schoolchildren in the next. This technological feedback loop was possible because of the eager adoption and widespread implementation of microprocessor technology. These impacts are huge and continue to mount. Exhibit 9.2 traces some of these higher-order impacts.

To glimpse the difficulties in identifying and analyzing impacts, even first-order ones, imagine trying to anticipate the impacts traced in Exhibit 9.2 when computers were room-sized behemoths powered by vacuum tubes, which consumed significant power and produced little useful output.

Exhibit 9.2 Some Higher-Order Impacts of Microprocessors

The Internet is one outcome of the technological and institutional impacts of microprocessor use in computing and communications. And the transformation that these technologies has undergone shows no sign of slowing. They have created huge industries and have made major economic impacts in almost every sector of society around the world. As usual, the impacts have been mixed. Not all impacts of microprocessors have been positive. For instance, they and the systems in which they are embedded have created significant new sources of environmental waste. Yet they also have contributed to the solution of many environmental problems, such as automotive energy use and emissions control.

9.5.2 Analyzing Technological Impacts

As touched on above, changes in one technology beget changes in others. Two generic types of such impacts are vertical and horizontal ones. When combined, they may be termed *integrated impacts*.

Vertical impacts flow from the natural development and succession within a family of technologies. A fine example of this is the progress from the vacuum tube to the transistor to the integrated circuit and, subsequently, to larger-scale integration.

Horizontal impacts result when advances in one technology affect other families of technologies. The nature of these interactions can greatly vary. For instance, impacts may encourage some technologies by creating demands for new capabilities while inhibiting or even destroying others. Advances in automotive technology led to the widespread use of automobiles, which helped to solve the environmental problems caused by the accumulation of horse waste in cities. However, the automobile proved equally adept at creating pollution and, in time, this led to a demand for pollution abatement technologies.

Integrated impacts are those produced by the coupling of vertical and horizontal impacts. Technological opportunities and bottlenecks are areas with high potential for integrated impacts. A technological opportunity arises when demand is present or when supporting technologies outside the technological family are available to enhance or transform a particular development. For instance, microprocessors were incorporated in automotive technology to improve pollution control and fuel use. A technological bottleneck occurs when a missing innovation, typically from outside the family, prevents the advance of technology on a broad front. For example, battery technology is a current bottleneck in electric car technology.

Mapping the set of technologies within and related to a given technology is a good first step in analyzing all three types of impacts. Then the analyst must probe the relationships in depth to understand the development processes and the interactions among the technologies. Patent records are an excellent place to begin this analysis.

Patent analysis requires access to a comprehensive computerized database (see Section 5.2.1 for suggestions). In turn, it is limited by that database. A U.S. patent database, for instance, will not fully reflect worldwide developments. Industrial practices also vary. In some technological areas patenting may be aggressively pursued, while in others it may be neglected in favor of other strategies, such as keeping trade secrets.

Spinoff technologies are an important horizontal technological impact. Luchsinger and Van Blois (1988) presented some spinoffs from military-developed technology:

- *Energy*: Commercial nuclear power deriving from submarine reactor technology
- *Nutrition*: Radar magnetrons being recast as microwave ovens

Exhibit 9.3 Patent Analysis

Patent records provide information about many aspects of the impacts of technology (see Section 5.2.1). For instance, they can be used to provide measures of:

- *Activity*: For a selected technological area (among the hundreds specified by the U.S. Patent Office), general measures (such as the total patenting rate or the number of firms engaging in patenting) can be tracked to assess how rapidly a technology is changing. Alternatively, the patent activities of key competitors, suppliers, or customers can be tracked. Are others getting ahead? Is a key supplier decreasing activity in an area important to your firm? Is a customer patenting in an area in which it purchases products from your firm?
- *Immediacy*: If most patents cited in recent patent applications are themselves recent, rapid technological advancement can be anticipated.
- *Dominance*: Whose patents are being cited? If one firm's patents are often cited, it may be the forerunner in the area. When considering an acquisition, this indicator can be used to verify that the target firm's R&D is influential or that it can be used to complement those of your acquiring firm.
- *Technological Linkage*: What other technological areas are cited in or cited by patents in the target area? For example, at some point, fuel injection patents began to cite computer technologies. At one time, fuel injection patents were referenced by aircraft developments, not automotive ones. Cocitation is another linkage that tracks the extent to which patents are jointly cited by applications in other areas. This can pinpoint the integration of two technologies to serve new applications.

- *Environment*: Military satellite sensors spawning civilian applications and transforming weather forecasting
- *Sports*: Graphite composites for aerospace being used in fishing rods, golf club shafts, and tennis rackets

Technological change propagates both in time and in space. An innovation produces effects felt in widening circles. Before the advent of the Internet and the dominance of multinational enterprises, the likelihood that an innovation would be adopted in a given time period declined with distance from the source in the absence of other forces. One model suggested an exponential decay (Sirinaovakul, Czajkiewicz, et al. 1988). However, in the information age, international diffusion is increasingly rapid, sometimes faster than diffusion within the country where the innovation began. For instance, Japanese firms brought home VCRs to market first, even though they were invented in the United States.

Adoption also depends in part on whether active efforts are made to propagate the innovation (e.g., by government agencies or producers) or to impede it (e.g., by restricting access to proprietary knowledge). It also depends on the infrastructure of the industry (manufacturing or service), firm size, capital requirements and availability, and how the innovation provides a competitive advantage.

Analyzing technological impacts leads to a more detailed knowledge of how technological changes can transform some technologies and eliminate others. Such changes, in turn, lead to substantial economic and societal impacts.

9.5.3 Analyzing Economic Impacts

As in all impact areas, the problem of drawing boundaries between what is considered (what is internalized) and what is ignored (externalized) is very important in analyzing economic impacts. Different boundaries may dramatically affect the outcome. What is an economic miracle with one set of boundaries may be an economic disaster with another. Either set of boundaries may be reasonable in terms of the forecast values and goals. Care in setting boundaries is critical.

Economic impacts on technology can be briefly described as an analysis of *who pays* (in the broad sense of providing resources) *and how much*? It is important to consider public sector contributions to the technological developments in the private sector. Such contributions may be in the form of subsidies or favorable economic regulations. Unfavorable economic regulations may create a disincentive to innovate. Economic impacts of technology can be briefly described as an analysis of *who gains, who loses, and how much?* Considerations should include, for example, jobs and income lost by a massive substitution of a new production technology for a mature one. Externalizing these job losses may lead to great economic gain for one party of interest, while internalizing them may lead to great economic loss.

Techniques used in economic analysis are treated elsewhere in this book (see Chapter 8). Approaches to economic forecasting and market analysis such as input-output analysis (Section 8.3.2) and cost-benefit analysis (Section 10.2) are among those that prove useful in economic impact analysis.

When boundaries are thoughtfully drawn, cost-benefit analysis can be especially helpful. However, the forecaster must keep in mind that the numbers used are usually ballpark estimates rather than precise data. An economic analysis should include valuing natural resource degradation, estimating the costs of pollution abatement, and considering the health problems arising from a new technology. While these are difficult to estimate, they should be internalized in the analysis. Exhibit 9.4 expands on these considerations.

9.5.4 Analyzing Environmental Impacts

Environmental impacts on a technological development may include those due to using environmentally friendly materials rather than ones that might be more technologically effective. Likewise, the environment in which the technology must

Exhibit 9.4 Measurement of Societal Benefits and Costs

Distribution: Data should be compiled to allow separate tabulations of benefits and costs for different stakeholders

Nonmonetized Effects: Data on noneconomic effects should be given a shadow monetary value (estimate) where feasible. If this is not feasible or acceptable, the effects should be noted.

Benefits Gained versus Costs Saved: In many cases, the categorization of effects is arbitrary. If an impact can reasonably be considered as either a cost or a benefit, this should be noted so that subsequent analyses can show both. Categorizing effects as benefits gained or costs saved in particular has a strong effect on the cost-benefit ratio. Increasing the numerator is quite different from decreasing the denominator even if the impact on the final cost-benefit ratio is the same. Computation of multiple economic criteria, along with the use of multiple discount rates, offers a better basis for decision making than cost-benefit analysis alone

Source: Modified from Sassone and Schaffer (1978) and Porter, Rossini, et al. (1980).

operate often can dictate its characteristics. For instance, a saltwater environment might dictate the use of a material that shortens the technology's life cycle.

Many new technologies have environmental impacts. Some are positive, such as the impact of catalytic converter technology on auto emissions. However, often the impacts are negative, such as the waste disposal problems posed by nuclear reactors. Pollution issues are not uncommon. Topical areas for environmental impact assessment include ecosystems, land use, water and air quality, noise, and radiation. Sometimes impacts from different areas are coupled, as when waste heat from power plants raises water temperature or decreases water quality, which in turn disrupts ecosystems. While these are complex issues, there is a substantial literature on environmental impact assessment that can be accessed to deal with specific problems and situations. The Global Development Research Center (2011) offers quick information about the environmental impact assessment process. Books that may be helpful include Wood (2002), Lawrence (2003), and Glasson, Therivel, et al. (2005).

Ecosystem impacts cross the topical areas listed above since organisms (including humans) depend on land, air, and water for their survival and prosperity and since their well-being can be negatively affected by noise and radiation. Assessing ecosystem impacts begins by obtaining baseline data on the systems impacted by the development. These ecosystems can be local, as with power plant impacts, or widespread, as with the possible impacts of cell phone radiation on the human brain and nervous system.

Once the baseline conditions have been established, making a map of the impacted populations (human, animal, and plant) is a good beginning. The map is a useful tool in estimating the effects of disturbances caused by the technology.

Sometimes higher-order impacts may be more significant than direct ones. Therefore, tracing the effects of effects may prove important. The process described uses both scanning (identifying affected populations) and tracing techniques.

Some technological developments impact land use. Such impacts may include water table changes, waste and pollutant accumulation, and destruction of current and future land use. The amount and character of the waste, pollutants, and hazardous materials generated by the technology must be estimated. Then the location and the long-term waste load can be assessed. Next, options for storage, collection, and disposal of the wastes are analyzed, and alternatives, such as recycling, are considered. Since toxic wastes can impact water and air quality, any coupling between them should be determined. In cases where land is rendered unusable in the short or long term (e.g., through strip mining or mountaintop removal technologies), the impacts on nearby ecosystems need to be considered as well as the long-term prospects for the land with or without remediation.

Water quality and quantity have become major issues as increased population and industrialization are making increasing demands on limited supplies. An analysis of potential water quality impacts should consider possible alterations to hydrologic characteristics, introduction of dangerous levels of physical properties, and changes to chemical and biological constituents. Quality concerns include:

- Hydrologic characteristics (e.g., flows, drainage patterns, and the condition of aquifers)
- Physical properties (e.g., temperature, turbulence, and solubility)
- Chemical constituents (e.g., dissolved oxygen, dissolved solids, nutrients, toxic chemicals, and pH)
- Biological constituents (e.g., algae, bacteria, weeds, and fish)

Water quantity impacts include overuse of the available supply impacting areas such as ecosystems and land use.

Air quality is of international importance since the atmosphere is not bounded by national borders. Over time, the number and stringency of air quality standards have increased, as have monitoring activities. There also have been improvements in quantitative modeling. The major issues in air quality impact assessment are to determine how much of the various pollutants is added and if these additions raise significant problems for ecosystems. It also is important to consider mitigating measures.

Many countries enforce air quality standards. For example, the U.S. Clean Air Act requires the Environmental Protection Agency (EPA) to set National Ambient Air Quality Standards for six common air pollutants (U.S. Environmental Protection Agency 2010b). These air pollutants (also known as *criteria pollutants*) are found throughout the United States. They are:

- *Particulate matter*, which includes a wide range of solid and liquid particles of various sizes: 2.5 to 10 micrometers from roadways and dusty industries; 2.5 micrometers or less in smoke and haze from forest fires, power plants, industries, and automobiles

- *Ground-level ozone* (O_3) created by a chemical reaction between oxides of nitrogen (NO_x) and volatile organic compounds (VOCs) in the presence of sunlight
- *Carbon monoxide,* mostly from motor vehicle exhaust
- *Sulfur dioxide,* mostly from fossil fuel combustion and power plants and other industrial facilities
- *Nitrogen oxides,* mostly from motor vehicles, power plants, and off-road equipment
- *Lead* from motor vehicles and industrial sources

These pollutants can harm human health, degrade the environment, and cause property damage. Of the six pollutants, particle pollution and ground-level ozone are the most widespread health concerns.

With the threat of climate change through global warming exacerbated by human activities, a serious worldwide effort is being made to curb greenhouse gases. Concern apparently was raised first by Arvid Hogbom in the early twentieth century based on his calculations of atmospheric carbon dioxide additions from burning coal (Flam 2009). After a long delay, the EPA has proposed regulations to deal with this source of air pollution (U.S. Environmental Protection Agency 2010a). The proposal addresses the emissions of the group of six greenhouse gases that may be covered by rules limiting their emissions:

1. Carbon dioxide (CO_2)
2. Methane (CH_4)
3. Nitrous oxide (N_2O)
4. Hydrofluorocarbons (HFCs)
5. Perfluorocarbons (PFCs)
6. Sulfur hexafluoride (SF_6)

Since the greenhouse gases have different warming potentials, their emissions are expressed in terms of a common metric to allow their impacts to be directly compared. International standard practice is to use the carbon dioxide equivalent. Emissions of gases other than CO_2 are translated into CO_2 equivalents by using their global warming potential. While the EPA is proposing carbon dioxide equivalent as the preferred metric, the proposal solicits comments on alternative measures.

Noise (unwanted sound) is most often associated with surface and air transportation, heavy industry, construction, and human activities (e.g., rock concerts). While noise is an annoyance, it also can cause physiological and structural damage. Noise assessment begins by establishing existing ambient noise levels. Noise levels during both construction and operation phases need to be considered. Noise sources should be described and the noise contours they produce estimated by time of day and duration (e.g., by using measurements, analogies, or computer models). Criteria for acceptable noise levels can be established by considering existing and proposed nearby land uses or local ordinances. No matter how they are established, community members should be involved.

Low levels of radiation are always present. They usually are referred to as *background radiation*. However, localized high levels of nuclear and/or electromagnetic radiation (e.g., Japan 2011) may cause serious health and ecosystem damage. After background radiation levels have been established, estimates must be made for changes caused by the technology. This will usually involve theoretical calculations or models.

9.5.5 Analyzing Social Impacts

Social impacts are the effects of technologies and technologically driven projects on the surrounding communities. They are sometimes difficult to separate from other impact areas such as economic, environmental, institutional, and health issues. Many social impact measures are covered in census reports in terms of the demography of a community. These include changes in the population, such as number, density, age distribution, income distribution, education, and employment characteristics. They also include changes in infrastructure such as housing and building stock, transportation facilities and use, educational institutions, and medical facilities. Changes in the tax base are important, as they directly impact public infrastructure. Also important are changes in cultural and recreational opportunities.

The breadth of social impacts can range from households when a new power plant is built to the world when technologies substantially extend the human lifespan. A typical set of social impacts are those related to the "boom and bust" phenomena caused by a major development localized in a community. Typically, growth rates at or above 15% per year cause major disruption in a community. It is important to realize that social impacts of a technology on the community usually produce impacts on the technology, usually expressed through institutional, political, and regulatory mechanisms.

To analyze the social impacts of a project, such as a large power plant, start with the impact identified for the development stage of the technology. Then determine the magnitude of the changes. For example, in the case of a new power plant, estimate the requirements in terms of money, people, and infrastructure for the construction and operation phases. These produce changes in the demography of the community, including population, income distribution, housing stock, medical facilities, roads, educational facilities, and businesses to support the additional population. Determine how meeting the needs these changes produce can be funded.

There is an extensive literature on social impact assessment that should prove useful (Barrow 2000; Becker and Vanclay 2003; Burdge 2004). A good resource is "Guidelines and Principles for Social Impact Assessment," which was prepared by the Interorganizational Committee on Guidelines and Principles for Social Impact Assessment (U.S. Department of Commerce 1994).

9.5.6 Analyzing Institutional Impacts

Institutions are groups of individuals who undertake collective activities that may be involved with developing, implementing, or operating the technology or may be impacted by those activities. Institutions may be formal organizations, such as government agencies and corporations, or they may be informal, such as a group of people freely associating for a common purpose (e.g., a neighborhood action group). Such institutions may produce impacts on new technologies, for example, by encouraging or restricting their development. All sorts of institutions may, in turn, be impacted by the technology.

Institutions may significantly change during the process of developing a new technology, especially if the technology is sufficiently disruptive to change such factors as organizational structure, employee qualifications, patterns of external interaction, and others. For example, over time, IBM shifted its production from electromechanical calculators and business machines to electronic computers in response to microprocessor technology developments. The impacts of these changes altered the makeup of its workforce as well as its R&D and production facilities.

Institutions may become stronger or weaker as a result of a new technology. For example, in the early 1950s diesel locomotive technology matured, and General Motors applied mass production techniques to build these locomotives. Steam locomotive manufacturers refused to adopt diesel technology or to replace single-unit and small-batch production technology with mass production. As a result of these decisions, the three major U.S. steam locomotive manufacturers, American, Baldwin, and Lima, went out of business.

Usually significant institutional changes are reflected in changes in the TDS that occur over time. Tracing the causal effects of such changes, perhaps by analogy to processes elsewhere or in different technological arenas, can provide a window into potential changes. Forecasters should always consider the possibility of disruptive institutional changes, as these tend to have the strongest ripple effects and to present the greatest opportunities and risks to the organization.

Interactions between public and private institutions can be extremely significant. Defense and aerospace contractors, drug manufacturers, and universities, among others, innovate with public funds or by using publicly funded research. Moreover, government economic assistance has been paramount in the survival of banks, insurers, and automobile manufacturers. Moreover, the government can be the major market for a technology development (e.g., the so-called military-industrial complex).

With the increased importance of internationalization, many impacts of technologies transcend national boundaries. Multinational firms manufacture and market in large numbers of countries. For instance, the impacts of General Motors' financial difficulties have produced major impacts on its Swedish subsidiary Saab. Multinational firms must deal with a variety of political systems whose laws, goals, organization, and operational procedures are quite different. Interactions between multinational companies and developing nations create risks and opportunities for both, including those in economic, environmental, and cultural realms. In cases such as these, attention to developing the TDS is important, as the stakeholders are quite diverse.

As with social impacts, institutions can have considerable effects on the technology. In many cases, institutions seek to avoid change caused by a new technology by inhibiting its development or use. These disincentives may be tangible (e.g., economic), but they also may be exercised through moral suasion. Like social impacts on technology, institutional ones are often exercised through political or regulatory mechanisms.

To analyze institutional impacts on a technological development, consider the institutional stakeholders identified in the TDS that can influence the project and the methods they can use to impact the project. Laws, regulations, and funding are typical ways in which institutions impact a development. Analyze the likelihood and nature of the impacts. Then use available data supplemented by expert opinion, when it can be found, to estimate the magnitude of the effects. Note that analyzing institutional impacts often requires that the analyst cross categories of knowledge such as economic, legal, and regulatory expertise in performing the analysis.

9.5.7 Analyzing Political Impacts

Political institutions may be key stakeholders in a new development. Politics translates values, taken as societal preferences, through public institutions into actions. Both political impacts on and of a technological development are based on implicit or explicit assessments of the potential effects it may produce. Political involvement and political interaction in the development of a technology should appear in the TDS.

Many of the most important current technologies arose from a perceived societal need promoted through the political process. For instances, nuclear energy, space technologies, the Internet, and advanced wound and trauma care all arose from actions by sectors of the U.S. government. Publicly funded research (e.g., the "war on cancer") has led to many developments in medical technologies and drugs as well, and the development of pollution control technologies, especially in California, has been stimulated by public concern for environmental quality.

One of the most important political sources of impacts on developments is ideology. Ideology can be described as a view that is based on faith. Thus, it cannot be refuted by empirical facts, which are simply dismissed or else reinterpreted to support ideological conclusions. Politicians and legislators are not bound by empirical facts or scientific evidence. What one person sees as an ideology another may perceive as factual, even scientific. Individual politicians may work in a world of belief that ranges from unregulated capitalism to socialism.

To analyze the political impacts on a technology, identify the political bodies that can make decisions about it from the TDS. Then determine the range of decisions that is possible, including "no decision." Finally, given the makeup and dominant point of view of these political bodies, estimate the likelihood and scope of the various possibilities. This estimation should be supported by expert opinion, as it has a strong subjective component.

Political impacts of technological developments can be approached in like manner by identifying the political bodies that may be impacted and the range of impacts that may occur. Again, estimations of impact likelihood and extent carry a strong subjective element that can be best addressed by expert opinion.

9.5.8 Analyzing Legal and Regulatory Impacts

Analyzing legal and regulatory impacts involves a sound understanding of the laws and regulations that impact or may be impacted by a technological development. Because of increased internationalization, this almost certainly will involve the laws and regulations of more than one country. The process of developing and utilizing a technology depends on laws and regulations that shape the options for organizing, funding, distributing, and using the technology—impacts on. Countries that have high regulatory and legal barriers to foreign firms are major cases in point. The TDS is a very useful map for identifying and dealing with these impacts.

Laws and regulations can drive technology development through subsidies and regulatory incentives to innovate. In the United States, subsidies of various sorts, including loans, grants, and tax credits, are widely used to promote alternative energy production. Regulations also have been used to hinder development, as with stem cell research and the development of stem cell therapies. It is important to realize that the regulatory environment can quickly change in response to political pressures. For instance, before the breakup of the AT&T monopoly, regulation of phone service by the U.S. government retarded improvements in the communications technologies available to the public. This logjam was broken by deregulation (itself a regulatory intervention), and improved technologies became readily available to users.

Law and regulation nearly always are reactive rather than proactive. Thus, the drug LSD had a period of unregulated use until the early 1960s, when drug enforcement interests successfully pushed for its regulation. It also is important for the forecaster to understand that the decision not to regulate can have major impacts as well. The U.S. government's decision not to regulate most new financial instruments developed after 1980 was the ultimate cause of the recent world financial crisis. Acharya and Richardson (2009) provide a detailed institutional analysis of the causes.

To analyze legal and regulatory impacts, use the TDS to identify the range of current laws and regulations that may impact or be impacted by a new development. This is largely a qualitative exercise of using expert opinion and informed judgment to identify how these laws and regulations may impact the development. A similar process may be used to determine what legal and regulatory changes may occur as a result of a new technology. An example is the question of sales taxes collection by Internet merchants. Another concerns the impact of widespread Internet use on privacy laws.

The Organization for Economic Cooperation and Development has developed a framework for regulatory impact analysis (Organization for Economic Co-Operation and Development and Directorate for Public Governance and Territorial Development 2010) that may prove useful in analyzing regulatory impacts.

9.5.9 Analyzing Behavioral, Cultural, and Values Impacts

Values may be described as conceptions of desirable states that guide judgments across specific objects and situations toward ultimate end states (Enk and Hornick 1983). This group of impacts involves those on or caused by the beliefs of individuals and the ways in which they interact with the world around them. Values, for instance, have slowed attempts at population control and have made abortion the driver of single-issue politics for a substantial number of people in the United States.

Human values are very diverse, as the news reminds us every day. Religion, which can be viewed as either personal or cultural, is a major source of values. There is little homogeneity in values held within a single country, much less worldwide. The term *culture wars* has become current to describe major differences in the values held by different groups of people in the same country.

Values change. Consider how your values differ from those of your parents and how your children's differ from yours. Values concerning such highly personal issues as sexual behavior, abortion, sexual preference, and desirable family size have been radically altered over the past decade, and they have changed differently for different people. Values can change in a number of ways, including (Rescher 1969):

- Acquisition or abandonment
- Increasing or decreasing importance/emphasis
- Increasing or decreasing standards for a value
- Widening or narrowing subscription to a value

Impacts of values on the development of technologies can be analyzed by determining the values of stakeholders described by the TDS. Currently held values may be studied directly by asking the people who hold them through surveys and open-ended interviews (Chapter 5) or indirectly by having them evaluate sets of scenarios (Chapter 7). The likelihood and magnitude of effects that these values may have on development of the technology can be estimated by techniques such as expert opinion and analogy to historical events. When studying values-related impacts, the forecaster can expect rather large uncertainties in their estimates.

To analyze the impacts of technologies on values, the same techniques may prove helpful. Historical studies of how attitudes about work and leisure were altered by technological developments may suggest patterns of change. In extreme cases, groups of individuals may entirely reject prevailing technologies and seek a return to values of an earlier and simpler society (e.g., the Smith and Luddite movements).

Cultural and behavioral technologies typically are generational. The technologies that one generation encounters and is comfortable with are different from those of the next. In the 1940s, for example, making a long-distance phone call was both complicated and costly. As a result, such calls were not commonly made or their length was severely restricted by the average person. Fast forward to the 2010s, when almost any phone on earth can be reached by keying a combination of numbers at little cost and young people walk busy streets texting friends. A generation is not necessarily a fixed period of time. Enormous changes may occur during periods of rapid technological change. Children born in the twenty-first century are rapidly becoming accustomed to information technologies that their parents and older peers encountered much later in life. Notions of privacy, contact with the physical world, ways of communicating, and multitasking are some of the changes observed. See Brad Stone's interesting article in the January 9, 2010, *New York Times* for a description of some of these impacts (Stone 2010). If developments in biotechnology accelerate, as has been the case with information technologies, an analogous set of impacts may appear.

Such impacts may prove difficult to analyze, as their likelihood depends on some notoriously difficult issues to predict, such as the extent and rate of adoption of new, disruptive technologies. Here also techniques such as expert opinion and analogy should be tried.

9.5.10 Analyzing Health-Related Impacts

Health-related impacts may be locally significant, as in a manufacturing process, or globally significant, as in the large-scale deployment of a technology. These impacts can be both positive and negative.

Positive impacts include the dramatic increase of life expectancy in developed countries during the twentieth century brought about by implementation of public health measures and by new drugs and vaccinations. Moreover, communications and transportation technologies have cut the time between accidents and emergency care. Information technologies provide hope for even more efficiently and effectively handling medical records. Such impacts may be studied using demographic projection and analogy.

When the impacts are negative, they may include failure of large-scale technological systems; discrete smaller-scale accidents; low-level delayed hazards such as cancers; and increases in infectious or degenerative diseases.

Areas impacted may range from individual workplaces to widespread communities. They may even be worldwide. Sources of health hazards may include high levels of nuclear and electromagnetic radiation and toxic chemicals, including carcinogens and mutagens. The World Health Organization site, World Health Organization (2010), may prove a useful source for issues relating to environmental and occupational health. Topics found there include sources and exposure levels, health risk groups and how they will be affected by exposure, definitions of the significance and acceptability of health impacts, and identification of mitigation measures.

9.6 IMPACT EVALUATION

Evaluation is the process of assigning value. For a forecaster, the value of a significant impact depends on the goals of the forecast. These goals may include those of supporters, users, or society at large as well as those of the developing organization. For a technology development, impacts may be evaluated on the basis of whether they help or hinder achievement of these goals and how they do so. Goals may vary, sometimes significantly, from stakeholder to stakeholder, depending on their perspective. Knowing how they vary is important in impact evaluation in order to gauge the stakeholders' reactions. The evaluation process should go further to determine how the development and use of technology will mesh with societal values, which themselves often are in a state of flux. Actions to enhance or mitigate impacts may be identified during and after impact analysis. However, choosing what actions to take and how to take them are elements of utilizing the output of the forecast and are discussed in Chapter 11, "Implementing the Technology."

In the private sector, possible new technology developments often are primarily evaluated on criteria related to the short-term return on investment. However, using only those criteria neglects strategic criteria that may affect the longer-term development and viability of the organization itself. For example, the Polaroid Corporation elected to play defense when confronted by the rapidly developing competitive technology of digital photography even though it had major economic and technological implications for the Polaroid process. The corporation neglected to act on reasonable forecasts of the growth of digital photography and its impacts on their single-line business. The impact of this decision on Polaroid's business proved to be disastrous.

So, how can impact evaluation be performed? Specific techniques are discussed in Chapters 10 and 11. In general, however, begin with the results of the impact identification process (Section 9.4), which include the:

- Values and goals of the organization developing the technology as they affect it
- Values and goals of organizations that impact on the technology, as well as those of organizations impacted by the technology
- Societal values and goals that may impinge on the technology

Next, *triage* the impacts into *significant, moderate,* and *minimal*. If there is any doubt about the significance of an impact, include it.

Then identify persons representing the important stakeholders in the development and utilization processes. Seek their perspectives on whether the significant and moderate impacts will be positive, negative, or neutral. Don't influence them by clarifying the values and goals of the forecasting team in advance. If stakeholders interact during the process, they may become more interested in engaging others, learn from one another, and enrich the evaluation process. It also is useful to gain stakeholders' perspectives about options to enhance or mitigate the impacts. These may be critical for successfully implementing the technology.

Consult Chapter 5 about ways to obtain expert and stakeholder opinions. Be certain to carefully review their input. An easier, but less reliable, approach to assessing stakeholder opinion is for members of the forecasting team to play the roles of stakeholders. However, valuable information may be lost by this simplification, and organizational biases are almost certain to skew the results.

The triage method mentioned above is a relatively quick and easy evaluation process. However, more structured processes can be developed on the basis of the qualitative and quantitative measurements (see Sections 5.4 and 11.4.1).

9.7 CONCLUSION

One might consider the innovation process to be threefold: figure out what can happen, make good things happen, and make the consequences as positive as possible. Forecasting and impact assessment should play major roles in this process. Impact assessment, even if it is a quick and dirty process, is important to learn the possible outcomes, especially those that are not obvious, indirect, or delayed. This knowledge is extremely important in managing technology development and use.

In forecasting, identification is the most important component of impact assessment. The forecaster must go beyond the obvious and expected impacts to consider those that are unanticipated and unintended. As has been suggested, impact analysis often can be done without overwhelming detail and precision. In many cases, simply sorting impacts according to significance and likelihood will prove sufficient.

Impact evaluation foreshadows management decisions in the development and utilization processes. It determines the relevance of impacts to the values and goals of the developer, stakeholders, and society. Thus, it opens the way to sound decision making.

This chapter ends with the mention of *sustainability*, an extraordinarily important area that has not been mentioned thus far. Practicing sustainability requires that the world's inhabitants adopt lifestyles that neither deplete the resources on which they depend nor destroy the people living on the planet. Governments and corporations as well individuals have slowly begun to adopt sustainable practices. Issues like using renewable energy and material sources and recycling nonrenewable materials are important. It should be mandatory to assess the sustainability of any new technology. While encouraging people to buy, buy, buy may be a rational short-term tactic, unnecessary consumption is the enemy of sustainability. If and how the world transitions to a sustainable society holds tremendous implications for technological development. Thus, these are critical issues in any forecasting or planning project.

REFERENCES

Archarya, V. V. M. Richardson (2009). "Causes of the Financial Crisis," *Critical Review*, **21**(2-3): 195–210.

Armstrong, J. E. and W. W. Harman. (1977). "Strategies for Conducting Technology Assessments." *Report to the Division of Exploratory Research and Systems Analysis, National Science Foundation*. Washington, DC, Department of Engineering-Economic Systems, Stanford University.

Barrow, C. J. (2000). *Social Impact Assessment: An Introduction*. London: Arnold.

Becker, H. A. and F. Vanclay, eds. (2003). *The International Handbook of Social Impact Assessment: Conceptual and Methodological Advances.* Cheltenham, UK, Edward Elgar.

Burdge, R. J. (2004). *The Concepts, Processes, and Methods of Social Impact Assessment*. Middleton, WI, Social Ecology Press.

Coates, J. F. (1976). "Technology Assessment: A Tool Kit." *Chemtech* **6**: 372–383.

Edelson, E. and M. Olsen. (1983). "Community Leaders' Perceptions of the Impacts of Energy Conservation Measures." In *Integrated Impact Assessment*, ed. F. A. Rossini and A. L. Porter. Boulder, CO, Westview Press: 187–201.

Enk, G. A. and W. F. Hornick. (1983). Human Values and Impact Assessment. In *Integrated Impact Assessment*, ed. F. A. Rossini and A. L. Porter. Boulder, CO, Westview Press: 57–71.

Flam, F. (2009). "The Cold, Hard Facts on a Heated Debate." *The Seattle Times,* 17 September 2009, p. 18.

Glasson, J., R. Therivel, et al. (2005). *Introduction to EIA: Principles and Procedures*. New York, Routledge.

Global Development Research Center (2011). "Resources on Impact Assessment." Retrieved 14 January 2011 from http://www.gdrc.org/uem/eia/impactassess.html.

International Association for Impact Assessment. (2010). "The Leading Global Network on Impact Assessment." Retrieved 1 September 2010 from http://www.iaia.org/.

Jones, M. V. (1971). "A Technology Assessment Methodology: Some Basic Propositions." *Report MTR6009 for the Office of Science and Technology*. Washington, DC, Mitre Corporation.

Lawrence, D. P. (2003). *Environmental Impact Assessment: Practical Solutions to Recurring Problems.* Hoboken, NJ, Wiley-Interscience.

Linstone, H. A. (1988). "Multiple Perspectives: Concept, Applications and User Guidelines." *Systems Practice* **2**(3): 307–331.

Lough, W. T. and K. P. White, Jr. (1988). "A Technology Assessment of Nuclear Power Plant Decommissioning." *Impact Assessment Bulletin* **6**(1): 71–88.

Luchsinger, V. P. and J. Van Blois, eds. (1988). *Spinoffs from Military Technology: Past and Future.* Management of Technology I. Geneva, Interscience Enterprises Ltd.

Organisation for Economic Co-Operation and Development and Directorate for Public Governance and Territorial Development. (2010). "Directorate for Public Governance and Territorial Development." Retrieved 1 September 2010 from http://www.oecd.org/document/12/0,3343,en_2649_34141_42247372_1_1_1_1,00.html.

Porter, A. L., F. A. Rossini, et al. (1980). *A Guidebook for Technology Assessment and Impact Analysis*. New York, North Holland.

Rescher, N. (1969). "What Is Value Change? A Framework for Research." In *Values and the Future*, ed. K. Baier and N. Rescher. New York, The Free Press: 69–109.

Sassone, P. G. and W. A. Schaffer (1978). *Cost-Benefit Analysis: A Handbook*. New York, Academic Press.

Sirinaovakul, B., A. Czajkiewicz, et al. (1988). "A Spatial Diffusion Model for Advanced Manufacturing Technology." In *Technology Management I*, ed. T. M. Khalil, B. A. Bayraktar and J. A. Edosomwan. Geneva, Interscience Enterprises: pp. 291–301.

Stone, B. (2010). "The Children of Cyberspace: Old Fogies by Their 20s." *The New York Times*, January 9. Retrieved 14 January 2011 from http://www.nytimes.com/2010/01/10/weekinreview/10stone.html.

U.S. Department of Commerce. (1994). National Oceanic and Atmospheric Administration, National Marine Fisheries Service. "Guidelines and Principles for Social Impact Assessment." Retrieved 1 September 2010 from http://www.nmfs.noaa.gov/sfa/social_impact_guide.htm.

U.S. Environmental Protection Agency. (2010a). "Fact Sheet—Proposed Rule: Prevention of Significant Deterioration and Title V Greenhouse Gas Tailoring Rule." Retrieved 1 September 2010 from http://www.epa.gov/NSR/fs20090930action.html.

U.S. Environmental Protection Agency. (2010b). "What Are the Six Common Air Pollutants?" Retrieved 1 September 2010 from http://www.epa.gov/air/urbanair/.

Wood, C. (2002). *Environmental Impact Assessment*. New York, Prentice-Hall.

World Health Organization. (2010). "Environmental Health." Retrieved 1 September 2010 from http://www.who.int/topics/environmental_health/en/.

10
COST-BENEFIT AND RISK ANALYSIS

Chapter Summary: Cost-benefit analysis provides a vital perspective to help decision makers make the best choices based upon the information and resources available to them. This chapter introduces concepts and tools to estimate the costs and benefits of new technologies. Since every technology implementation will have different chances of success or failure, the chapter extends cost-benefit considerations to incorporate uncertainty and to apply risk analysis.

The technology delivery system (TDS) provides a framework for looking at the future of an innovation, and preceding chapters have introduced concepts and tools for adding substance to that framework. Every TDS presents numerous opportunities for decisions. Technology managers must choose among alternative paths with the awareness that they are allocating scarce resources.

10.1 OPPORTUNITY COSTS AND CHOICES

Since resources are limited, it is not possible to tackle every technological challenge, even every desirable one. Nor is it possible to analyze and mitigate every potential impact. To make decisions wisely, managers need tools to allocate resources with discipline. For example, a new process that generates energy might seem attractive, but what opportunities must be foregone to implement it? Perhaps concentrating on marginal improvements to conserve energy would be more reliable and bring more rapid and significant financial returns. On the other hand, taking some risk now to adapt to a new energy source might allow the company to leapfrog over the competition and gain a dominant market position in a few years. What should the manager decide? The cost of choosing any opportunity is the value lost by not pursuing another. All decisions involve such *opportunity costs*.

Analytical tools will never completely answer such hard questions. There always will be uncertainty. However, cost-benefit and risk analyses can provide a disciplined way to think about such problems and to consistently make wiser decisions. Society also makes choices. Some of these choices are reflected in the costs facing firms and will be considered by their managers. Other costs and benefits are external to firms and will not be considered in making business decisions unless society takes steps to see that they are taken into account. From an economic point of view, it is ideal for all social costs and benefits of a project to be weighed. Chapter 8 pointed out that the social and institutional contexts are critical to the successful growth of innovations. This means that risks external to the firm also must be taken into account by managers. The techniques of cost-benefit and risk analysis described in this chapter, used in conjunction with the impact assessment procedures outlined in the previous chapter, will enhance the manager's ability to present a fair case for appropriate innovation and help it become reality.

10.2 COST-BENEFIT ANALYSIS

Cost-benefit analysis attempts to count all of the costs of projects and evaluate all of the benefits that result. In the simplest case, cost-benefit analysis directly compares outlays with the benefits that result. If you buy something for $100 today and you know you can sell it tomorrow for $120, it is easy to see that the cost was less than the benefit and the cost-benefit ratio of this little buy- and-sell project was less than 1. For a complex project involving a new technology, the measurement of costs and benefits will be a lot more complicated.

Cost-benefit analysis usually is associated with public projects. Projects like new highways should bring more benefits than the outlays of public money to complete them. The concept also can be applied to things other than construction. For example, new regulations may be subjected to cost-benefit analysis that includes both the public and private expenditures they imply as well as estimates of the financial good that they will do. Such studies can be controversial. Important variables often are not easily measured, and the resulting approximations can be easily affected by bias. For example, politicians and business interests may exaggerate the benefits of a dam while underestimating the cost of the environmental impacts that the project entails. Nevertheless, most societies recognize that evaluating costs and benefits is the right thing to do, even if the implementation is far from perfect.

10.2.1 Cost-Benefit Analysis within the Organization

Like societies, all organizations have limited resources and need to use them wisely. While for-profit firms generally call decisions about resources *capital budgeting* or just *financial management*, in reality they are cost-benefit analysis. Their goal is for additions to the company's value over time to exceed the assets that are used because of the decision. However, the framework and philosophy of cost-benefit analysis, particularly the emphasis on seeking out all costs and benefits, can be beneficial. There are a lot of examples of business failures that resulted from narrow accounting that failed to consider all the results of actions. This is especially true for the management of technology. Short-term accounting calculations are seldom adequate and often lead to underinvestment in innovation. Nor does existing financial information indicate competitive threats from new processes or new market opportunities.

There are reasons why firms fail to broadly assess the implications of their decisions. Investments in traditional activities seem well defined and certain, while devoting resources to change seems speculative. So, expenditures on marginal improvements in existing processes that quickly show increases in profits can be favored over superior but longer-term innovations. Investment in the development of even a clearly beneficial new technology may require years to produce a positive return. How does the decision maker know if the returns will be large enough to justify the investment? How long should a manager be willing to continue committing money before benefits begin? Will bankers or financial people within the organization approve the decision?

Chapter 8 described how markets and institutions sometimes can be slow to allow technology adoption. The problem is that money has time value; a dollar today is worth more than a dollar tomorrow. For example, if the interest rate is 5%, $100 in a certificate of deposit will grow to $105 in a year. It follows that $105 delivered one year from today could be said to be worth as much as $100 today. Similarly, $110.25 in two years is equivalent to $100 today, as is $115.76 in three years, because compounding interest at 5% annually would produce these amounts. Therefore, at 5% per year interest, the present value of $105 in one year is $100 today; $100 is also the present value of $110.25 in two years or $115.76 in three years. In general, the *net present value* (NPV) of returns in some future year t is given by

$$\text{NPV} = \sum_{1}^{n} \frac{(R_t - C_t)}{(1+i)^t} \tag{10.1}$$

where i is the *discount rate,* which works like the interest rate in the example above; R_t is the receipts of cash in year t; and C_t is the outlay of cash in year t. This simple formula is the basis of all capital budgeting decisions for which money must be spent today to secure expected returns tomorrow. By using this equation to look at the time pattern of expected costs (outlays) and receipts, the most attractive projects can be determined. In doing this, the discount rate might be the interest rate on borrowed money, but it also could be the cost of capital for the firm or a rate the firm sets to weed out less profitable projects. The NPV sometimes is called the *discounted present value*.

For example, assume that a firm could pursue any of three mutually exclusive technologies, A, B, and C, that have the pattern of net benefits $(R_t - C_t)$

TABLE 10.1 Hypothetical Net Benefits for Projects A, B, and C (in thousands of $)

Year	Project A	Project B	Project C
1	−600	−400	−700
2	600	220	190
3	600	250	220
4	−200	200	220
5	300	−300	210
6	−700	300	200

shown in Table 10.1. Suppose that Project A has a rather high initial outlay, say $1 million, with relatively low annual operating costs. Further, suppose that one reason for these low operating costs is that the process produces by-products that potentially are environmentally harmful but that are not currently regulated. In the first year, Project A will attract $500,000 of revenue with only $100,000 in costs for material, labor, taxes, and other operating expenses. When the $1 million initial outlay is included, net benefits are −$600,000 for the first year. The next two years will produce $750,000 in revenues and $150,000 in outlays to meet the product demand. In the fourth year, more restrictive environmental regulations are expected and major plant renovation will be needed. Outlays will exceed revenues by $200,000 that year, but the subsequent year again will bring net benefits of $300,000. The last year also will bring some revenue, but the plant's operating costs will be too high to allow it to continue production. Since demolition will require the disposal of toxic wastes, the cost of retiring the facility will be very high, resulting in a net outlay of $700,000.

Project B does not have the severe environmental problems of Project A, and the initial capital investment is not as large. Annual operating costs, however, are much higher. The first year results in net outlays of $400,000; net benefits in the next three years are $220,000, $250,000, and $200,000, respectively. In the fifth year, major modifications will be needed, making net benefits −$300,000. However, this work will extend the life of the technology for one more year and produce $300,000 of net benefits in the last year of operation.

Project C will require the highest initial outlay, but the facility is durable and has low environmental impacts. The process is costly, and the expected net returns are modest but consistent throughout the project's life, since no major modifications or renovations will be needed.

How does a manager select from among these three alternatives? If the time value of money were not important, the most profitable would be Project C. The sum of its net benefits is $340,000 compared to $270,000 for Project B and $0 for Project A. If the popular, *but dangerously simplistic*, notion of payback period were applied, Project A would appear best because it returns enough to completely cover the investment outlay in the second year. Project B requires almost three years to return its investment, and Project C takes more than four years. The best approach, however, is to examine the sum of the present values of each year's

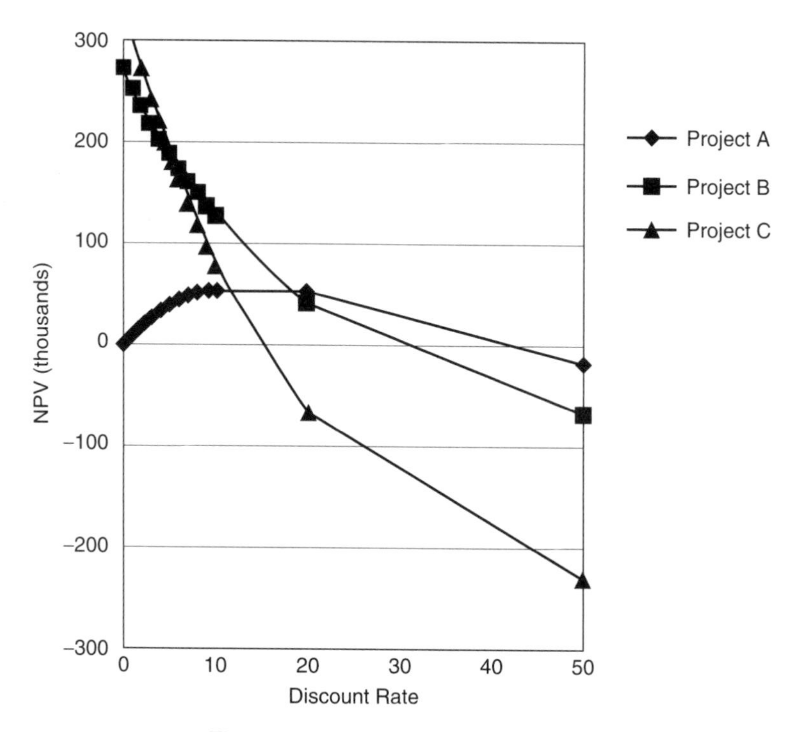

Figure 10.1. NPV of Three Projects

net benefits for each project—that is, to apply Equation 10.1 to each entry in Table 10.1 and calculate the total present value for each project.

Finding present values could be tedious, but spreadsheet software will do it quickly (see, for example, the NPV function in the formula tab of Excel). Table 10.2 shows the results of present value calculations for the net benefits of each project for discount rates ranging from 0% to 50% per year, and Figure 10.1 displays the same data. The results make it clear that *which of the three alternatives is best depends upon the discount rate*. From 0% to about 5% Project C is the best alternative, while Projects A and B become more profitable than C as the discount rate increases.

Obviously, choosing the discount rate for this analysis is critical. Conceptually, the rate should be determined by the firm's cost of capital, which in turn depends upon its financial structure, tax laws, credit conditions, credit ratings, and other variables. As a practical matter, however, the discount rate will be specified by the firm's finance officers if corporate funds are involved. For a small firm or an individual, the rate probably will be the interest rate on the loans needed for the project or the estimated cost of venture capital (often 50% per year or more). If the firm does not have to borrow, the rate will be determined by the best alternative rate available on the funds (that is, the opportunity cost).

Later discussions consider methods for estimating receipts and outlays and compare the net present value approach with other criteria. There also are

TABLE 10.2 NPV of Three Projects

Discount Rate (%)	Project A	Project B	Project C
0	0.0	270.0	340.0
1	10.3	251.6	306.4
2	19.2	234.3	274.7
3	26.9	217.9	244.9
4	33.6	202.5	216.9
5	39.3	187.9	190.5
6	44.0	174.1	165.6
7	48.0	161.0	142.2
8	51.2	148.6	120.1
9	53.8	136.8	99.2
10	55.7	125.7	79.5
—	—	—	—
20	53.6	40.5	–66.6
—	—	—	—
50	–17.0	–68.5	–228.4

descriptions of ways in which this approach can be adjusted for various levels of risk implied by the alternative projects. The purpose of this introduction is to stress that this type of financial analysis is at the core of balancing the benefits and costs of innovation. At times, the market may require innovation because of new competitive threats; at other times, government regulation may require it because of such things as environmental protection or workplace or product safety. Even these complications can be handled within this context.

10.2.2 Societal Stake and the Organizational Response

The preceding section alluded to the societal implications of innovation decisions. The TDS shows that technologies grow and develop within an environment that both impacts them and is impacted by them. Proper forecasting and management of technology takes into account this social interconnection. The assessment of social benefits and costs is at the heart of that accounting. Economic systems are set up to provide signals and incentives for individuals to do things that benefit the world as well as themselves. However, no system includes all the signals and incentives required, nor do decision makers always listen for them. Sometimes, this means that resources are misallocated. That is, sometimes too many resources are devoted to some activities and too little to others. This is a motivation and perhaps a justification for societies to impose restrictions on private behavior through legal sanctions, regulations, taxes, or other constraints. This occurs even when the society is primarily committed to the market mechanism because of market failures such as externalities in costs and benefits.

Since the individual or firm that develops a new technology is part of society, the benefits and costs from a decision to proceed are benefits and costs to society as well. In fact, under most circumstances, the correct private decision about innovation is also correct for the society. That is why societies decentralize such decisions rather than making them through a central authority. However, the private assessment of costs and benefits is not always sufficient because new technologies frequently lead to impacts beyond the firm—sometimes unexpected ones. Sometimes these broader impacts are positive and justify public subsidy of private developments; at other times, the broader social costs lead to regulation, control, or even prohibition of a new technology.

Within the broader context of impact assessment, social cost-benefit analysis can provide information to determine how society should react. The beginnings of cost-benefit analysis in the United States can be traced to water projects of the U.S. Army Corps of Engineers. Although the River and Harbor Act of 1902 and the Flood Control Act of 1936 mandated consideration of costs and benefits, it was the Planning, Programming, Budgeting System (PPBS) work of the 1960s that brought widespread public application of the concept (Thompson 1980). Many see such analysis as a politically neutral tool for analyzing expenditures or regulations. However, others view it as an essentially arbitrary application of quantitative analysis to questions that are largely qualitative. These disagreements, like the preferences for private versus government decisions, cannot be resolved here. The goal of this section is to provide the manager with a systematic way to look at the social aspects of decisions.

Mishan (1976) presents social cost-benefit analysis in a manner that is analogous to the discussion of private decisions that is introduced in the preceding section. For example, if there are three projects, A, B, and C, how can decision makers decide which should be pursued? The earlier discussion considered receipts and outlays for a firm over time and showed how the decision could be based on a comparison of discounted present values. However, private decisions often produce spillover benefits and costs. Although these are not considered in the firm's decision making, they must be included in social considerations of costs and benefits. The difference from the firm's considerations lies in the extent of the social benefits and opportunity costs that need to be considered—in essence, all impacts of the development on all parties. From a social point of view, the present value of social benefits should exceed that of social costs; that is, the cost-benefit ratio should be less than 1.

The starting point for social cost-benefit analysis is the concept of *Pareto optimality*. Simply stated, this says that if at least someone can be made better off without making someone else worse off, then a project is unambiguously good. Even when someone might be worse off, the criterion still applies if the gains of winners are sufficiently large to compensate losers enough to make them indifferent to the change. To ignore an opportunity to make Pareto improvements is to waste resources. However, three questions must be considered: Can benefits and costs be measured? If so, how will they be measured? And will those who lose really be compensated for their losses by those who gain?

Many simplistic approaches have been used in cost-benefit analysis to answer the first two questions. In fact, that is one reason why the concept is so

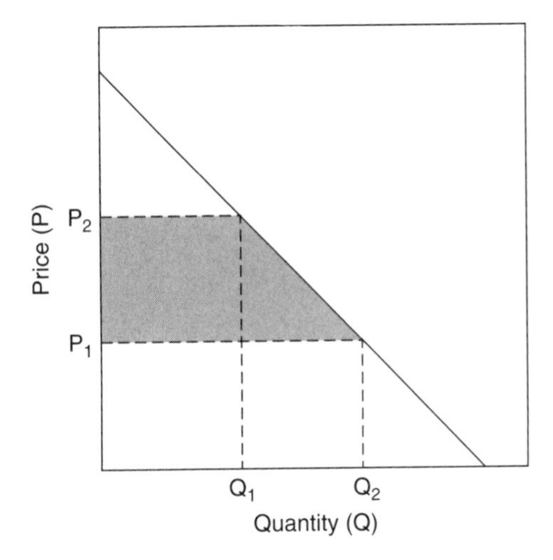

Figure 10.2. Market Demand and Consumer Surplus

controversial. Good cost-benefit analysis requires painstaking identification and measurement of changes in what economists call *consumer surpluses* and *rents* for labor and other resources.

Figure 10.2 shows the familiar market demand curve of elementary economics texts. The horizontal axis represents the quantity of a product purchased (Q); the vertical axis shows the price paid (P). The demand curve shows the highest price consumers are willing to pay for various quantities of the product or, alternatively, the maximum quantity they will buy at various prices. Suppose the market price of a product is P_1 and the quantity purchased is Q_1. If the price were lower, more could be sold. If the price were higher, less than Q_1 would be purchased. Note that, even at this price, there are consumers who would be willing to pay more. Since the price is P_1, these consumers are paying less than they would be willing to pay for all but the very last unit. The difference between what these consumers would be willing to pay and what they have to pay is called *consumer surplus*. The value of the consumer surplus in this case is shown by the triangular area above the line at P_1.

A new technology could increase the consumer surplus and thus produce a social benefit. For example, a new production process might lower the costs of the product and reduce the price to P_2. This would raise the quantity purchased to Q_2, since more consumers could afford and/or would be willing to pay for the product. The gain in the consumer surplus from introduction of the new technology is represented by the gray rectangle and triangle in Figure 10.2. These represent, respectively, the contribution of the lower price for those who were already buying and the consumers' surplus for new purchasers who did not purchase at the previous price. These are gains from the new technology that are independent of the profits gained by its innovators. So, a new process for delivering a product at lower costs can add value to people's lives. There is a lot of evidence of this

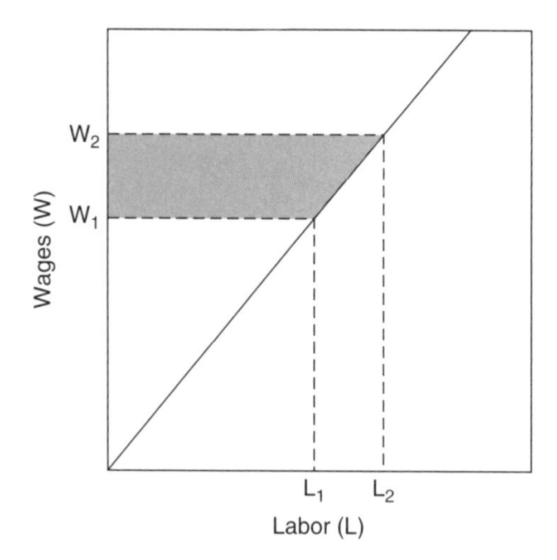

Figure 10.3. Market Supply and Producers' Surplus

in electronics and not so much in health care, although new technologies that have improved the quality and quantity of life are socially beneficial as well.

The benefits to labor from the new innovation also result from what they are willing to do and what the market dictates. Figure 10.3 shows a supply curve for labor. The higher the wage, the more labor will be offered to firms in this market. W_1 and L_1 represent the market wage and the amount of worker time that is used at that wage. At wages below W_1, there are still workers who would be willing to work. Thus, at W_1, workers are receiving more than the minimum required. The difference between what they are paid and what they would demand for their labor is called *rent* or *producers' surplus*. Just as consumers need not always pay what they are willing to pay, producers, like labor, can get more than the minimum they would accept. The amount of the producers' surplus is represented by the single gray area in the figure.

If technology increases the demand for factors like labor, then the wages in the market may move up to W_2. This will increase the producers' surplus that workers receive by the amount of the gray area shown in Figure 10.3. For example, new technologies to produce alcohol fuels from algae would increase the compensation of engineers with expertise to design the needed processes. A similar sort of analysis could be done for other resource owners, such as those who control land or natural resources. For example, the popularity of corn-based ethanol dramatically increased the prices of farmland in Iowa. If the changes in producers' surplus for all the individuals and businesses involved are added up, the sum gives an indication of the social benefits from the new technology on the production side.

This discussion has proceeded as though the effects of new technology always are social benefits and the surpluses produced are easily identified and quantified. Obviously, technological changes have social costs, and the costs and benefits often are hard to measure. For example, some workers may not be needed in the

Exhibit 10.1 Measurement of Social Benefits and Costs

- *Distribution*: Data should be compiled to allow separate tabulations of the benefits and costs for different stakeholders (e.g., the city versus the state or taxpayers versus indigent persons).
- *Nonmonetized Effects*: Data on noneconomic effects should be tabulated in suitable quantitative units if feasible (e.g., acres of land to be developed or number of housing units displaced); if this is not feasible or is unacceptable, qualitative measures should be used.
- *Benefits Gained versus Costs Saved*: In many cases, categorization of effects is quite arbitrary. If an impact reasonably can be considered in either category, that should be noted so that subsequent analyses can show things both ways.

These measures facilitate multiple analyses. For instance, regional cost-benefit comparisons can indicate if certain jurisdictions receive proportionately more of the benefits and pay less of the costs. This information can help decision makers derive compensation schemes.

Source: Based on Sassone and Schaffer (1978).

new production process and may find themselves unemployed. How does one calculate that social cost? If the workers can find work in another industry, the loss to them will be determined by the difference in wages. If they are unemployed, presumably their increased leisure time will have some value. Answering these questions will involve judgments about hypothetical opportunities and workers' values. Such speculations permit analysts, intentionally or unintentionally, to introduce their personal values, and these will affect the outcome. Exhibit 10.1 presents important measurement considerations.

Another issue that has not been addressed in the social context is the time value of money. The previous section showed that future costs and benefits had to be discounted to their present values for effective decision making by firms. For an individual, the opportunity costs of forgoing returns from investing capital make it clear that future returns differ from present ones. This argument does not apply in such a straightforward way for society as a whole.

Some argue that the social discount rate should be lower than the private rate because many returns on public projects are not evaluated in monetary terms. Further, the risk of failure on any one project is small compared to the size of government resources (Porter, Rossini, et al. 1980). If a lower discount rate is appropriate for public projects, shouldn't that same rate be used for the public evaluation of private projects? The answer is probably no, but there is a great deal of controversy about the proper social discount rate.

The *distribution* of costs and benefits is another problem inherent in cost-benefit analysis. For example, the analysis does not imply that winners *will* compensate losers, only that losers *could be* compensated if benefits exceed costs. Obviously, if there is no mechanism for compensation, then even large benefits over costs may not be sufficient justification for a project that makes poor people even poorer. The problematic nature of loser compensation is a central cause for skepticism about the application of cost-benefit analysis in the social context.

Another issue arises for social projects for which the costs and benefits are extended far into the future. Should the well-being of future citizens be discounted relative to the well-being of current ones? Since elections today cannot be decided by tomorrow's voters, the answer in a democracy may seem to be yes. However, constitutional democracies do not always operate solely on the principle of majority rule, and future citizens have rights that should be taken into account. Difficulties compound when technological decisions involve external effects (externalities) and public wins or losses. For example, it is not feasible to exclude some people from the good or isolate them from the bad. Technologies that may add to global warming, for instance, can harm everyone, even those who had no role in adopting the technology. Nor can even the most avid pacifists be excluded from the benefits of a strong military that keeps foreign enemies from endangering their families. Cost-benefit analysis for public decisions is complex. Fortunately, there is a large body of helpful literature that provides much more guidance, such as Gramlich (1990). Some imperfect techniques that have been used to estimate the values of externalities and public goods are presented in Exhibit 10.2.

Sustainable development has become an important goal for countries throughout the world. Decades ago, the Brundtland Commission was commissioned by the United Nations to consider the problems of environmental deterioration that seemed to accompany economic growth. This World Commission on Environment and Development (1987) recommended that private and public development decisions account for the full range of impacts over the life cycle of the development. For instance, an automaker contemplating whether to substitute plastic for metal body shells should consider:

- Environmental costs of obtaining the materials (e.g., petroleum production and refining to make plastics versus aluminum or steel production)
- Production impacts (e.g., relative job intensities, waste production, and toxic exposures to workers)
- Life cycles (e.g., expected lifetimes, recycling mechanisms, and the costs and hazards of final waste disposal)

Sustainable development decisions were intended to include the costs of all such impacts, from design through operation and disposal at the end of the life cycle. The goal was to preserve the ecosystem for future generations.

Sustainable development has not been without controversy. The current debates over carbon emissions limitation and global warming are illustrations of the scientific and policy disagreements that can arise. Moreover, considerable government intervention would be required given that many of these costs are routinely

Exhibit 10.2 Some Techniques to Estimate the Value of Externalities and Public Goods

These techniques are perhaps best presented via an example. Suppose that a private company wants to develop a wind farm offshore from an oceanfront community. The windmills will generate low-pollution energy and reduce dependence on petroleum imports. These impacts are positive public goods, but the loss of the natural ocean vista and potential effects on the ocean's wildlife could be considered negative external effects or public costs. How would one determine the value of the public costs of the wind farm?

1. Survey the willingness of those affected to pay to retain their ocean views. This usually requires disguising the purpose so that people won't exaggerate their own interests, but such disguises are not a simple matter.
2. Use gaming (e.g., parties concerned with the windmill's view obstruction could play through alternative scenarios and compensation schemes). Again, there is a danger that stakeholders will exaggerate their true valuations, and disguising the true purpose of the game may be quite difficult.
3. Use public referenda (e.g., allow voters to express their preferences among a set of alternative development and compensation schemes).

   All of the above schemes are open to criticism for their lack of realism. Using more independent data also may not be entirely accurate either, but it would probably be more objective. For example, one could
4. Draw an analogy to private goods (e.g., examine other residential property that had lost a view amenity and determine the impact on real estate values).

   Or
5. Look at legal settlements where views have been damaged.

   Or
6. Compare the market values in the same neighborhoods of houses with and without views.

There are problems with all these approaches. Willingness to pay is difficult to assess reliably, and it is always better to use observed behavior when possible. Furthermore, it is hard to separate willingness to pay from the distributional question of ability to pay. It is even less likely that one can properly evaluate and discount all future social costs and benefits, including external ones. For this example, how can the preferences for clean energy over untouched ocean ecology and view be assessed for those in the year 2035?

Source: Based on Porter, Rossini, et al. (1980).

externalized by technology producers. There has not been much evidence of such intervention. Poor nations, which many feel will be beneficiaries of avoidance of global warming, may view the carbon-limiting actions as efforts to hamper their industrialization.

In spite of the difficulties of getting precise answers to questions about social benefits and costs, the *process* of trying to find them is very useful. Using reasonable resources to weigh the public costs and benefits of an innovation can help avoid expensive problems later. Undesirable social impacts can reduce demand for a technology and invite costly regulation or even bans. There are no free lunches, and innovation will produce social costs as well as benefits. Technical managers will make better decisions if their view of the trade-offs between them is clearer. That is not to say that the decisions will be ideal. Problems that are likely to show up far in the future are unlikely to get much consideration. However, some effort to look at social benefits and costs should not be discouraged because the effort will not be perfect.

10.2.3 Cost-Benefit Analysis Methods

Earlier in this chapter, the concept of discounted present value was presented as a way of viewing costs and benefits over time in relation to today's decisions about innovations. The discipline of making such decisions is generally called *capital budgeting*. Although most people think of this term in connection with buildings and equipment, the twenty-first-century manager is just as likely to use it for decisions about the development of technologies.

This section provides more description of the application of the discounted present value model and related tools to analyze costs and benefits over time. Before the data and parameters of such analyses are discussed in more depth, there will be some discussion of two related concepts for deciding whether to proceed with an innovation: the internal rate of return and the payback period.

Internal rate of return (IRR) relies on the same information as the discounted present value discussed above. However, instead of specifying a discount rate, the IRR approach finds the rate that makes the discounted values of the outlays and the receipts just equal to zero. That is, Equation 10.1 is modified to

$$0 = \text{NPV} = \sum_{1}^{n} \frac{(R_t - C_t)}{(1 + i)^t} \qquad (10.2)$$

where *irr* is the rate that makes all of the discounted outlays and receipts sum to zero. Very profitable projects will have a high *irr*, while less profitable ones will have discounted sums equal to zero at a much lower internal rate of return. Unprofitable projects will not generate positive *irr*'s. Note that in Table 10.2, Project A has an *irr* of 0%. This is an unusual situation in which this project makes positive returns from the beginning and has costs (for adequate disposal perhaps) at the end. In Figure 10.1, Projects B and C must have *irr*'s of about 27% and 15%, respectively.

Computing *irr* is difficult, but spreadsheets will calculate the figure quickly and easily. Simply put the series of receipts less outlays for each period in the life of the innovation in an array and apply the IRR function as directed. This array can also be used with the NPV function and a specified discount rate to find the NPVs discussed in Section 10.2.1.

This IRR for a project can be compared to the cost of capital for the organization or to the "hurdle" rate set by the firm's financial officers for attractive projects. It also can be compared with that of other potential projects to determine which one is most attractive. This could be particularly useful where the life cycles of the projects differ.

Another rule of thumb for making project decisions is the *payback period*. Quite simply, the payback period is the number of years required for project returns to repay the initial investment:

$$\text{Payback period} = \frac{\text{Initial Investment}}{\Sigma \text{ Returns}} \tag{10.3}$$

So, if the project requires an initial outlay of $1 million and it produces net returns of $500,000 per year, then the payback period is two years.

While this approach is simple, it has real limitations for the management of innovations. It certainly can be used to justify the marginal benefits of continuous improvement initiatives, especially when their payback period is less than a year. Such projects often can be done within a budgetary cycle and may not even need capital budget approvals. However, disruptive technologies and even major innovations within technologies are unlikely to fare well under payback criteria. Even when benefits far outweigh costs, the payback period of significant technology change typically will be three to five years at best and can easily take far more time. NPV and IRR evaluate both the time value and size of the excess of benefits over costs in a much more effective way.

The capital budgeting framework, in particular NPV, generally is best for evaluating the benefits and costs of new technology. Therefore, the ensuing discussion focuses on some suggested ways to apply it. Private sector cost-benefit analysis is analogous to the development and analysis of pro forma statements in venture capital investing. However, this book uses the TDS as a context in which to consider much broader implications than those included in the typical discussion of venture capital. The discussion begins by considering the simplest model and then suggests ways to incorporate the lessons from earlier chapters about assessing the context, impacts, and responses. The good news is that the analysis of costs and benefits forces a realistic appraisal of the prospects for new technology. The bad news is that reality is complex and a lot of estimates will need to be made.

Creating long-range financial plans is hard. Although there are numerous books and websites that describe business and financial planning, some that even provide fill-in-the-blank approaches, there are really no answers that are both easy and useful. However, a short and useful guide to this topic is *Business Plans That Work* (Timmons, Zacharakis, et al. 2004). A more detailed treatment that can be used like a handbook is provided by *New Venture Creation: Entrepreneurship for the 21st Century* (Timmons and Spinelli 2008). Ultimately, however, the quality of the resulting plans directly depends on the quality of the assumptions on which the data are built.

The first part of the financial analysis is an estimate of the pattern of sales over the years. The starting point for this estimation is the discussion in Chapter 8 of economic and market analysis. Recall that the emphasis there was on identifying who the customers will be and determining what they would be willing to pay for the innovation. This is the basis for estimating the size of the market. Next, assumptions need to be made about the percentage of the market that will be captured by the innovation and how the sales will grow with time. For new technologies, growth likely will follow the S-shaped pattern discussed in Section 6.3. Thus, at some time, sales will reach their maximum and begin to decline. While one would hope that this will occur beyond the planning horizon for a new technology, the reality is that innovations rapidly become obsolete.

Three other important issues need to be addressed to get a realistic picture of the market performance of the new technology. First, many innovators underestimate the length of the sales cycle. It takes time for people and organizations to consider new products before making a decision to buy. The more disruptive the technology is, the longer it might take for potential buyers to embrace the concept and adjust the other aspects of their behavior to take advantage of the new potential it offers. A second, somewhat related, issue is discussed in *Crossing the Chasm: Marketing and Selling Disruptive Products to Mainstream Customers* (Moore 1999). While new technologies can be launched by sales to early adopters who are anxious to be trendsetters, the larger mainstream market requires considerable effort and resources to reach and convince. Neglect in planning for this can lead to failure. The third, and perhaps most important, issue is the effect of competition. If there is a market for an innovation, there probably are competitors. Moreover, their challenges will increase if the technology is successful. For some innovations, such as those on the Internet, the struggle for sales often ends in "winner take all" results. In other industries, such as medical devices and pharmaceuticals, patents are extremely important. Patent protection does not last forever, however, and obtaining a patent is seldom sufficient to protect a new technology. The progress of the early aircraft industry in the United States, for example, was seriously retarded because of the existence of multiple patents and inadequate cross-licensing. Henry Ford also experienced similar problems (Heller 2008, pp. 30–31). Assumptions about projected sales need to account for the role of competitors and how to manage their challenges.

To make realistic sales projections, one must know the technology, how long it will take to commercialize, the customers, and the industry. Every situation will have unique characteristics. However, two approaches can be helpful in making a good forecast and making it believable. First, look for analogies with earlier technologies whose growth patterns should be similar. The growth pattern for a new drug using nanotechnology may be similar to that for pioneering drugs from genetic engineering, for example. Second, a well-reasoned scenario of the life cycle of the technology (see Chapter 7) is important. It can be used to convince oneself and decision makers that the complexities of demand have been considered.

Once the demand projections have been completed, the outlays needed to implement the innovation must be estimated. In the early months, perhaps even years, outlays will exceed receipts. Estimates of the expenditures required to transform the innovation into a marketable product will be needed as well as estimates of the costs of producing and selling it. Good prototyping and extensive interactions with developers, potential customers, and operational personnel will be needed to anticipate issues such as preferences, manufacturability, and serviceability; these will require resources. Very few innovations avoid the need for continuous improvement and, occasionally, radical overhaul. Therefore, later periods in the product life cycle should include funding for ongoing R&D. Early assessments of the costs of service and support for the new technology also are needed. Since these often are the most profitable activities related to new technology, it is important to understand both their cost and revenue potentials. There are many situations in which underestimating the costs of satisfactory after-sales support led to the failure of new products.

Most capital budgeting textbooks provide sufficient information for you to complete the cost-benefit analysis for your organization. However, this book has emphasized that the broader considerations inherent in the TDS should be taken into account. Using the tools from earlier chapters, one can ascertain such things as the potential positive and negative impacts accruing to society from the development, production, and use of the new technology. The TDS also presents feedback loops that may either enhance or limit the prospects for the innovation. For example, a new technology for extracting oil from tar sands may lead to rapid population growth in remote areas. In these areas, lack of infrastructure to support industrial and commercial development and to provide public services could slow the flow of the new energy supply to the market and thus delay profitability. On the other hand, a new technology to make fuel from urban waste could create jobs in depressed areas, delight governments, and produce subsidies and tax credits. The environmental impacts caused by the production, use, or disposal of the new technology also must be considered. These often require contingency planning both for liability reasons and because of the social responsibilities of the organization and the people involved with it.

After receipts and outlays over time are estimated using traditional financial projections and the expanded considerations provided by the TDS, they can be arrayed in two columns of a spreadsheet to find the answers to Equations 10.1 and 10.2 using the NPV and IRR (formulas built into the spreadsheet). If the NPV is positive and higher than other alternative uses of resources at an appropriate discount rate and/or if the internal rate of return is sufficiently high, then implementation of the innovation probably should be pursued.

10.2.4 Economic Value Added

There are limitations to any concept that is used to evaluate innovations. Even though the NPV formula includes a discount rate determined by the opportunity cost of the capital being used, some finance professionals have argued that the real measure should be something called residual income from investment in new things. *Residual income* (RI) was first defined as the amount of net operating profit after taxes (NOPAT) less the required profit computed as the amount of investment times the required rate of return. Later, others took this concept further and introduced *economic value added* (EVA). In this approach, one should adjust the outlays that are really investments in the future rather than costs of current activities. The most obvious example is R&D expenditures that expand the potential of the organization. Rather than treating them as outlays for current operations, the calculation removes them for each period. However, such a case could also be made for advertising expenses, amortization of goodwill for strategic acquisitions, and perhaps even investment in the professional training of employees. Obviously, such adjustments need to be apparent and the case for them needs to be strong. For more details on the calculation of EVA, see Damodaran (2010).

The reason for using EVA is that traditional accounting measures can penalize innovations, especially those that have high but delayed returns. Executives often reject new technology projects because they are focused on stock market reactions and perhaps on the criteria for their own compensation, which may emphasize annual or even quarterly financial results. This can lead to temporary rewards for the executives and perhaps shareholders at the expense of the firm's future. This is particularly true for technology-based industries, where continual technology development is essential.

10.2.5 Earned Value Management

Another concept that can be important in the analysis of costs and benefits of innovation is *earned value measurement* (EVM). While NPV and EVA are meant to determine if a project should be undertaken, EVM is meant to monitor the project and the development of the technology as it proceeds. The concept has been especially applied in the defense industry, where very large, complex projects must be carefully managed to minimize cost overruns and to ensure adequate returns to contractors. However, the concepts are applicable to project management and review for any technology development where discipline and accountability are important. Obviously, keeping track of these measures will have costs, and simple projects that are well understood or quickly completed may not need them. A very good discussion of EVM was prepared by Booz, Allen and Hamilton under a contract with the U.S. Department of Energy and is available at Booz Allen Hamilton (2010).

Basically, EVM requires that a project plan include *measures of value* produced from such things as achieving important milestones. Budgets are laid out over time, and an effort is made to tie expenditures to these measures of value. To put it very simply, EVM examines expenditures and results to see if half of the value of the project has been achieved when half of the money is gone. For example, the demonstration of a working prototype of a new technology might be determined to be worth half of the project cost. Thus, when half of the money has been spent, managers would be accountable for showing that the prototype has the necessary performance attributes.

10.2.6 The Balanced Scorecard

Evaluating technology development projects with consideration of wider issues than purely financial ones seems very appropriate in light of the discussions of the TDS. Broader measures of performance like the *balanced scorecard* began in the context of the performance of executives in organizations (Kaplan 1996). However, the underlying idea of the multiple dimensions of success can apply to choices in pursuit of innovation. A summary of the application of the balanced scorecard can be found at Balanced Scorecard Institute (2010).

The basic idea is that the evaluation of activities needs to include internal processes, organizational learning and growth, and customer perspectives, as well as traditional financial performance measures. These factors seem very applicable in evaluating the benefits and costs of innovation. In addition to NPV, IRR, and even EVA, it can be important to know how the new technology will affect, and hopefully improve, internal processes or even if it is compatible with them. In addition, it is sometimes important to pursue technology developments that may not be financially favorable at present to learn about new applications of technology that may be needed to be competitive in the future.

10.3 ACCOUNTING FOR RISK AND UNCERTAINTY

Thus far, discussions have assumed that the costs and benefits of alternative projects are known with certainty. However, managers never have perfect foresight. At best, they have a somewhat subjective, probabilistic assessment of the likelihood of various outcomes. Obviously, introducing risk can change the choice among alternative projects. For instance, if the expected present values of two projects are equal, the one that is more likely to succeed is preferable. Depending upon the risk preferences of the decision maker, projects with high risk may be rejected in favor of others for which profits will almost surely be lower but losses are unlikely. The following distinction between risk and uncertainty is important:

- *Risk*: An action can result in more than one outcome with different probabilities of occurrence (e.g., every weather forecaster in town agrees that there is a 50% risk of rain tomorrow).
- *Uncertainty*: An action can result in more than one outcome, depending on external conditions of unknown probability or for which there is very little confidence in the probability estimate (e.g., one weather forecaster predicts that there is a 50% chance of rain a week from tomorrow).

10.3.1 Accounting for Risk within Organizations

Efforts to implement new technologies involve both risk and uncertainty. If the innovation is a minor change that will lead to lower production costs for an established product, then both the risk and the uncertainty of the outcome may be low. On the other hand, adopting a completely new technology or developing a new product for which market acceptance is unknown involves a much lower likelihood of success, and even that low probability is subject to a lot of uncertainty.

A practical way to deal with risk and uncertainty is to use higher discount rates. The theory behind these higher rates is beyond the scope of this discussion but is an important topic of entrepreneurial finance (see, for instance, Smith and Smith 2000, pp. 234–252). Actually, discussions of determining the discount rate adjusted for risk can be found in most books about corporate finance, capital budgeting, and engineering economics. The suggestion here is that the types of technology changes discussed in this book are much more closely aligned with investments in new entrepreneurial ventures than with traditional corporate investments in new equipment and facilities.

The basic idea of adjusting the required rate of return is to alter the net present value for the fact that the project is risky and the expected cash flows over time are uncertain. The required rate increases with the riskiness of the venture. For example, a 10-year U.S. Treasury bond had a yield of 3.7% per year in March 2010. At the same time, a corporate bond with an A rating (indicating a relatively safe investment) was yielding over 5.7% because it was viewed as more risky than the government bond. The corporate bonds of a company with a high level of debt or uncertain prospects are even riskier. In fact, these securities, sometimes called *junk bonds*, may have been paying as much as 7% per year or more. Rates for venture capital loans to emerging firms were no doubt considerably higher. The additional interest can be viewed as a reward for the added risk of these instruments over those of the U.S. government. A similar risk premium often is used to adjust the financial assessment for a project. As Table 10.2 shows, increasing the discount rate lowers the present value of a project. Hence, if the discount rate of one project is increased significantly to account for its risk, then other projects may become more attractive.

Although this approach to introducing risk into the analysis is simple, it has some serious drawbacks. For instance, adding a risk adjustment to the discount rate reduces the NPV and thus the relative attractiveness of a project. However, the effects of this adjustment are compounded with time. For instance, the present value of $100 in one year at a 10% discount rate is $90.91, compared to $87.72 for a risk-adjusted discount rate of, say, 14%. However, after 20 years, the present value for the higher discount rate will be less than one-half that for the lower rate. Such a compounding of risk adjustment is severe and is unlikely to be appropriate for many projects. Another problem with simple discount rate adjustment is that it implicitly assumes the independence of returns in each time period. In reality, however, early success might greatly increase the certainty of net benefits in subsequent years. Conversely, early losses could completely destroy the project's future.

One solution to these problems is to adopt a scenario strategy. This approach will complicate Table 10.1, since instead of a single column of net benefits for each project, there will be several. For instance, Project A might have one column that portrays terrible losses, another column that depicts terrifically high benefits, and others in between. Perhaps the stream of benefits depicted in Table 10.1 is the most likely (i.e., has the highest probability), and lower probabilities can be assigned to the other outcomes. Projects B and C could have similar or quite different outcome distributions. Sometimes there may be nearly equal probabilities of great success or terrible failure and almost no chance of a middle ground. The array of possibilities depends upon the nature of the project and the judgment of the manager who makes the decisions.

There is another alternative that can explicitly change the risk from year to year. Just as actuaries estimate premiums based on the probability of events and their costs, analysts of new technologies could explicitly add the expected values of possible events to the expected receipts and outlays for each period. For instance, suppose your company is considering constructing a new facility to produce electric power by advanced solar cell technology. Experts provide the following estimates:

- Annual chances of an accident are 1 in 1 million.
- Annual chances of sabotage are 3 in 1 million.
- The costs of either an accident or sabotage are valued at $1 billion.

These estimates translate into an expected cost of each of:

$$\begin{aligned}\text{Expected value} &= \text{Likelihood} \times \text{Magnitude}\\ &= [(1 \times 10^{-6}) + (3 \times 10^{-6})] \times \$1\ \text{Billion} = \$4000\end{aligned}$$

This annual expected value cost can be combined with other costs to determine NPV with Equation 10.1. The risks of events in this approach need not be the same every year, so one can overcome the disadvantage of compounding encountered when the discount rate is adjusted for risk. This is still a scenario approach, but it may be simpler than the multiple outcomes approach suggested above.

The choice of the best project using either scenario approach is no longer quite so clear. Under conditions of certainty, selection of a discount rate identifies the highest present value and thus indicates a clear direction for the decision maker. With uncertainty, however, somewhat arbitrary values must be applied, and even then the results can be ambiguous. Although there may be cases in which the outcomes for one project clearly dominate those of others, it is more likely that the best choice will depend on how committed the manager is to risk aversion. Usually, projects with the highest potential rewards also carry high risks. Different individuals may make equally rational choices based on differing preferences for risk and return. As the discussion broadens to include the social dimensions of new technology, the quest for an unambiguously good choice becomes even harder. Note that, even within the firm, there is less than complete objectivity when decisions are made under uncertainty. The additional uncertainty introduced by the implications of a good TDS approach adds to the complexity. However, it also helps to avoid ignoring impacts and reactions in the broader context that can either hinder or destroy the prospects of the innovation.

Dealing with risk in project decisions is often treated by modifying the NPV analysis with the methods described above. Another aspect of risk is the potential for decision makers to change course during the process of developing and implementing. Smith and Smith (2000, pp. 89–99) described the use of *available options* in the strategic planning of entrepreneurship and innovation. Referring to the earlier work by Trigeorgis (1996), they suggest the following examples of real options:

- Option to defer—timing on new product introduction
- Option to expand or contract—scaling the business or the technology application
- Option to abandon—stopping a project when failure looks more likely
- Option to stage investment—spreading out the need for cash for development
- Option to switch inputs or outputs—changing raw materials or products
- Option to grow—extending the brand name to other products.

All of these options have value because they reduce the risk involved in committing to the implementation of a new technology. They can be evaluated to some extent by using the decisions trees described in Section 6.7.1. Furthermore, these options are analogous to financial options such as puts and calls, which can be traded in securities markets. Smith and Smith (2000) provide a primer on options in an appendix. Like financial options, the ability to postpone decisions until more information is available reduces risk and, therefore, adds value to the project. Stephen Gray and Cambell Harvey provide a good comparison of financial options and real options in projects in a series of PowerPoint slides (Gray and Harvey 2010). Huchzermeier and Loch (2001) examine real options use in the management of R&D projects. They consider uncertainties arising from project payoffs, project budgets, product performance, market requirements, and project schedules, and they offer advice on the appropriate timing of making commitments rather than preserving flexibility in projects. Further information and recommendations for specific books on real options, capital budgeting, and project evaluation are available at Real-Options.org (2010).

As with financial options, the value of real options increases with the riskiness of the investment. However, there also are differences. For instance, real options do not have a market in which to trade. Thus, option-valuing techniques can overstate their value. Moreover, real options are often interdependent, so a choice of one can affect the value of others. Nevertheless, the concept of real options can be useful in comparing the advantages and disadvantages of alternative projects.

10.3.2 Accounting for Risk—the Social Dimension

The preceding section described handling the financial implications of a firm's risks in introducing a new technology. New technologies also imply risks for society. Social risks must be addressed in making decisions about new technology because of their moral implications as well as because of the negative effects that bad impacts can have on the firm. Societies usually perceive risk in a far more subjective manner than expert risk assessors. Subjective or not, these perceptions shape people's responses to the introduction of new technologies. Therefore, it is important for technology forecasters and decision makers to understand how these perceptions are formed.

People respond to risks in complex and often very different ways. For instance, people who willingly take serious risks in hang gliding or mountain climbing may be unwilling to eat processed foods because of perceived health risks. Other people prefer to drive rather than fly, even though the risk per traveler mile is much lower for airline travel. In a classic paper, Chauncey Starr (1969) compared various risks and reached the following conclusions:

- The public is willing to accept voluntary risks roughly 1000 times greater than involuntary ones.
- The risk of death from disease is a "psychological yardstick" for the acceptability of risk.
- The acceptability of a risk is roughly proportional to the cube of the real or imagined benefits of the activity.

Later, Starr and Whipple (1984) compared financial risks such as those described in an earlier section with the health and safety risks that surround technology. They noted that information on risks is often based on "fuzzy estimates." Exhibit 10.3 identifies several perceptual factors that affect individual risk judgments.

Another problem facing the technology decision maker dealing with societal risks is that society rejects the idea of balancing monetary variables against basic values. Given that the major focus of risk is human mortality, analysts naturally seek a dollar value for human life to complete their cost-benefit calculations. Society finds that unacceptable, notwithstanding the fact that it implicitly makes such decisions all the time. A whole literature has emerged on the value of life (see Exhibit 10.4).

Starr and Whipple (1984) also noted that beyond the fuzziness of scientific study results lies the political dimension of decisions about technology. The issue may not be the magnitude of the benefits, risks, and costs, but rather who reaps the benefits and who bears the risks and costs. Furthermore, there is the problem of the "myth of abundance" described by Freeman and Portnoy (2006, p. 3):

> *Even when it knows better, the public likes to be told that its government is working to eliminate all environmentally transmitted risks. Sensing this, politicians shy away from analytical approaches based on the premise that resources are finite and priorities have to be set.*

Exhibit 10.3 Selected Perceptual Factors Affecting Risk Evaluation

- *Probability Squeeze*: People tend to overestimate risks from low-probability events (e.g., death from a nuclear power plant accident) and to underestimate those from high-probability ones (e.g., developing heart disease from smoking).
- *Sense of Control*: The willingness to tolerate risk skyrockets when exposure is voluntary or controllable (as per Starr's observations).
- *Dread*: Several factors cluster as the opposite of sense of control (e.g., catastrophic, uncontrollable, inequitable, and high risk to future generations).
- *The Unknown*: This composite factor reflects unobservable effects, unfamiliar risks, and delayed effects. When a technology combines dread with the unknown, perceived risk is greatest (Slovic, Fischoff, et al. 1986).
- *Omission Over Commission*: Government agencies, for example, are biased against innovation because the public encourages them not to take chances with personal health (e.g., the U.S. Food and Drug Administration tends to avoid the introduction of a potentially harmful new drug at the expense of opportunities for that drug to reduce health hazards). The legal system biases companies in the same direction.
- *Economics Be Damned*: Safety and health measures taken by government often bear little relationship to cost-benefit trade-offs because Americans do not like to confront lives-for-dollars choices.

Exhibit 10.4 The Value of Life

What is a human life worth? Methods to make such judgments and the resulting estimates vary widely. Kahn (1986) gathered a number of estimates based on wage/risk trade-offs by workers in the United States and the United Kingdom. The key to the estimates can be illustrated by this example. If a worker wants an additional $800 per year to compensate for an increased chance of dying in that year of 1 in 10,000, then the value of life is calculated as $8 million Overall, 1 person in 10,000 will die and each of the 10,000 will be paid $800 to bear this risk, for a total payment of $8 million. Kahn balanced eight labor market studies, considering their leanings toward overestimating or underestimating the value of life, to arrive at a best estimate of $8 million in 1984 dollars for an American life. Three questionnaire surveys also supported an estimate of about this magnitude, and this value was far greater than that typically used in policy analyses. Since the Consumer Price Index has doubled since 1984, one might expect that this figure would be $16 million or more in 2010.

One of the most critical issues about new technology is relating estimated risk to the risk perception and the acceptability of that risk. This issue has not yet been satisfactorily resolved. The perception of risk by impacted communities typically shows less variation than the scientific and technical estimates of risk. In public decisions, the perception of risk drives decisions. Thus, the acceptability of *perceived risk* is the most important go/no-go criterion in a decision about implementing a new technology. One graphic news story about a tragedy can mean the end of an acceptable risk ... even if the news story is false.

There is an abundance of distrust about decisions relating to new technology. In public communications, businesses usually minimize the risks and emphasize the benefits to counter the "no risk is acceptable" attitude. There is understandable reluctance to say that any risk is acceptable since that brings charges of insensitivity or of a cover-up. Discussions of risks frequently become polarized, and decision makers often dismiss lay views as ignorant or irrational (Shrader-Frichette 1985). However, the public may perceive that potential benefits are small and that both benefits and risks will be unfairly distributed. Since each side typically relies on experts to support its view, it is hardly surprising that the public is skeptical of all experts. Therefore, it sometimes will not support a new technology that, according to experts, produces benefits outweighing its risks. At the same time, the public often embraces technologies and products that experts advise against (e.g., tanning beds, cigarettes, and illegal drugs). The only hope of avoiding this impasse seems to be to involve users and stakeholders in the forecast, decision, and implementation processes.

Recognizing the issues surrounding risk perception is vital, but it does not give the technology manager an overall assessment of how people will react. That assessment is vital to determine an organization's decision to develop a new technology and, if it does proceed, how to accommodate risk concerns. The most important approach has been emphasized repeatedly in this book: *involve potential users and stakeholders in the forecasting and development processes*. Ask them often about risk and listen to their answers. Two approaches can provide hard evidence on perceived risk:

1. *Revealed Preferences*: Seek analogous cases and statistical data on human responses to similar implementations of technology (e.g., property value effects for similar plants in similar communities or the level of accidental death accepted with a similar technology). People's behavior reflects their judgment of an acceptable balance between the benefits and risks of current technologies.
2. *Expressed Preferences*: Survey those directly affected to determine their perceptions. Public hearings, voting, and other means may be used to elicit public attitudes.

The first approach draws on past behavior but requires extrapolation to the current situation. Moreover, the value assumptions that underlie revealed preferences have been questioned (Shrader-Frichette 1985). For example, if people accept dangerous jobs at relatively low wages because they do not have alternative jobs with higher wages, has their behavior revealed value preferences or their level of desperation? Or does ignorance about the health impacts of a technology interfere with revealing true preferences for a safer alternative? Furthermore, there are questions about whether current behavior properly considers long-term consequences, especially if the consequences will be borne by future generations. While the second approach of expressed preferences concretely relates to the new technology, it relies on expressed opinions and extrapolation to real behavior. There is a lot of evidence that people may not do what they say they will do. For example, many people may advocate fuel-efficient and less polluting cars but still buy large SUVs.

The U.S, Environmental Protection Agency (EPA) website provides a risk portal that outlines issues in risk assessment and management (U.S. Environmental Protection Agency 2010). In EPA terminology, *risk assessment* provides information about impacts, while *risk management* deals with actions to avoid or mitigate bad impacts and to enhance good ones. While their primary concern is the environment, the factors the EPA proposes in assessing risks are similar to those of the TDS:

- Scientific factors provide the basis for the risk assessment and include information drawn from toxicology, chemistry, epidemiology, ecology, and statistics.
- Economic factors inform about the costs of risks, mitigation/remediation options, and distributional effects.
- Laws and legal factors define the basis for the EPA's risk assessments, management decisions, and, in some instances, the schedule, level, or methods for risk reduction.
- Social factors deal with issues that may affect the susceptibility of individuals or groups to risks (e.g., income level, ethnic background, community values, land use, zoning, health care availability, lifestyle, and psychological condition).
- Technological factors consider feasibility, impacts, and the range of risk management options.
- Political factors address interactions among branches of the federal government, as well as interactions with state and local governments, special interest groups, and perhaps, foreign governments.
- Public value factors reflect the broad attitudes of society about environmental risks and risk management.

Risks to human health are critical concerns in making technology decisions, and health risk assessment provides a good template for handling social risk in general. Paustenbach and Keenan (1988) described four parts to a health risk assessment:

- *Hazard identification* seeks to determine whether a product, by-product, or process is carcinogenic (causes cancer), a developmental toxicant (causes birth defects), or a reproductive toxicant (reduces the possibility of pregnancy) or has other adverse health effects.

- *Dose-response assessment* examines the amount of the substance necessary to produce adverse effects. This often is expressed in terms of the *risk-specific dose* (the dose necessary to produce a risk of 1 in 1 million, for instance).
- *Exposure assessment* deals with who is affected and how. This requires knowing where the impacts are likely, their timing, and the characteristics of the people, plants, animals, water, air, and soils that are impacted. Since substance producers ultimately are liable for impacts, even if they hire someone else to dispose of wastes, exposure assessment clearly is a management responsibility.
- *Risk characterization* uses information from the previous steps to describe what can happen. This is the most important and difficult part of the risk assessment.

In the United States, the Centers for Disease Control (CDC) play an important role in developing information for risk assessments. However, CDC protocol specifically forbids it from quantifying risks (U.S. Environmental Protection Agency 1985). The EPA and most firms would prefer some quantification to assist in making judgments about the management of risk. As noted, it is hard to make quantitative judgments about health without seeming insensitive, and society resists placing a value on human health.

Cost-benefit analysis must be integrated with the TDS model. As complex as that might be, risk and uncertainty are inherent in many dimensions of the TDS, especially the impacts of the new technology. Freeman and Portnoy (2006) concluded that the analysis of ventures with environmental risks should explicitly consider all risk dimensions (e.g., their voluntary or involuntary nature) and the distribution of the costs and benefits among elements of society. Mishan (1976) suggested that risk could be introduced into social analysis by using subjective probabilities and presenting arrays of outcomes (e.g., the most likely outcomes together with their lower and upper bounds). These are steps in the right direction. Addressing social issues while planning for technology development, no matter how complex and uncertain that may be, is much more effective than reacting to impacts after implementation.

10.4 CONCLUDING THE FOCUSING PHASE

This book began by providing the background of management decision making about technology. In a dynamic and complex world, managers must plan for effective innovation. Planning requires that managers also have access to forecasting information. The TDS was presented as a way of framing the multiple dimensions to be forecast so that good decisions can be made. That framework emphasizes that the context of good planning extends well beyond the organization and its industry to include the world that will be impacted by the technology. Understanding this context is an ongoing process, and modifications will be made to the forecasts and plans as better information becomes available. In the end, all of this is meaningless if it does not lead to better decisions. Technology managers will never know as much as they would like, but their effectiveness depends upon their willingness to decide.

Chapters 4 through 6 continued an earlier theme of more expansive thinking to describe the various ways of gathering and using relevant information. These chapters described tools to help the technology manager explore the range of information that might be relevant and then analyze its implications. Chapter 7 shifted the emphasis to focusing the information and analysis for use in the important task of making decisions. Scenarios were shown to be a powerful way of bringing a lot of information and analyses together to project implications for the technology and for the organizations that implement it. Chapter 8 dealt with the vital questions of the economic viability of innovations and, more specifically, with identifying the markets to adopt and sustain them. Chapter 9 returned to the broader context of the TDS to suggest ways to analyze the impacts of technologies and the role of those impacts in making decisions. Finally, this chapter returned to ways to bring all the information together again to determine the costs and benefits of decisions in a world of risks and uncertainty.

The choices surrounding technology and its implementation under conditions of risk and uncertainty are not simple, nor are they certain. They involve judgment and values as well as knowledge and techniques. Proper consideration of whether a technology should be developed must include an impact assessment that evaluates potential risks to the environment and to the public. This is the only path to reliable financial decisions. The manager who makes decisions about technology needs to take everything into account. Therefore, the proper procedure may be to first perform the internal investment analysis. If it is positive, then a social cost-benefit analysis that considers society's risks and uncertainties and its likely responses should be performed. Anticipating all the impacts of a technology probably is not possible, and trying to assess society's reactions will slow the decision-making process. But in the long run, it is the best way to allocate the resources of the organization and those of the global society within which it functions.

REFERENCES

Balanced Scorecard Institute. (2010). "What Is the Balanced Scorecard?" Retrieved 1 September 2010 from http://www.balancedscorecard.org/bscresources/aboutthebalancedscorecard/tabid/55/default.aspx.

Booz Allen Hamilton. (2010). "Earned Value Management Tutorial, Module 5: EVMS Concepts and Methods." Retrieved from http://www.science.doe.gov/opa/pdf/FinalModule5.ppt.

Damodaran, A. (2010). "Economic Value Added." Retrieved 1 September 2010 from http://pages.stern.nyu.edu/~adamodar/New_Home_Page/lectures/eva.html.

Freeman, A. M. I. and P. Portnoy. (2006). "Economics Clarifies Choices about Managing Risk." *The RFF Reader in Environment and Resource Management*, ed. W. E. Oates. Washington, DC, Resources for the Future **95:** 15–20.

Gramlich, E. M. (1990). *A Guide to Benefit-Cost Analysis*. Upper Saddle River, NJ; Prentice-Hall.

Gray, S. and C. R. Harvey. (2010). "Real Options in Project Evaluation." Retrieved 1 September 2010 from http://faculty.fuqua.duke.edu/~charvey/Teaching/BA456_2006/Real_options_in.ppt.

Heller, M. (2008). *The Gridlock Economy: How Too Much Ownership Wrecks Markets, Stops Innovation and Costs Lives*. New York, Basic Books.

Huchzermeier, A. and C. Loch. (2001). "Project Management Under Risk: Using the Real Options Approach to Evaluate Flexibility in R&D." *Management Science* **47**(1): 85–101.

Kaplan, R. (1996). *The Balanced Scorecard: Translating Strategy into Action*. New York, McGraw-Hill.

Kahn, S. (1986). "Economic Estimates of the Value of Life," *IEEE Technology and Society Magazine* **5**(2): 24–31.

Mishan, E. J. (1976). *Cost-Benefit Analysis*. New York, Praeger.

Moore, G. (1999). Crossing the Chasm: Marketing and Selling Disruptive Products to Mainstream Customers. New York, HarperCollins.

Paustenbach, D. and R. Keenan. (1988). "Health Risk Assessment in the 1990's." *Hazmat World* **1**(3): 48–56.

Porter, A. L., F. A. Rossini, et al. (1980). *A Guidebook for Technology Assessment and Impact Analysis*. New York, North Holland.

Real-Options.org. (2010). "Real Options in Theory and Practice." Retrieved 1 September 2010 from http://www.real-options.org/.

Sassone, P. G. and W. A. Schaffer. (1978). *Cost-Benefit Analysis: A Handbook.* New York, Academic Press.

Shrader-Frichette, K. (1985). *Risk Analysis and Scientific Method*. Dordrecht, The Netherlands, D. Reidel.

Slovic, P., B. Fischoff, et al. (1986). "The Psychometric Study of Risk Perception." In *Risk Evaluation and Management,* ed. V. Covello, J. Menkes, and J. Mumpower. New York, Plenum Press: 3–24.

Smith, R. and J. Smith. (2000). *Entrepreneurial Finance*. Hoboken, NJ, John Wiley & Sons.

Starr, C. (1969). "Social Benefit versus Technological Risk." *Science* **65**(3899): 1232–1238.

Starr, C. and C. Whipple. (1984). "A Perspective on Health and Safety Risk Analysis." *Management Science* **30**(4): 452–463.

Thompson, M. S. (1980). *Benefit-Cost Analysis for Program Evaluation*. Beverly Hills, CA, Sage.

Timmons, J. and S. Spinelli. (2008). *New Venture Creation: Entrepreneurship for the 21st Century*. New York, McGraw-Hill.

Timmons, J., A. Zacharakis, et al. (2004). *Business Plans That Work*. New York, McGraw-Hill.

Trigeorgis, L. (1996). *Real Options*. Cambridge, MA, MIT Press.

U.S. Environmental Protection Agency. (1985). *The Use of Risk Assessment in Regional Operations*. Washington, DC, U.S. Government Printing Office.

U.S. Environmental Protection Agency. (2010). "Risk Assessment." Retrieved 1 September 2010 from http://epa.gov/riskassessment/basicinformation.htm#arisk.

World Commission on Environment and Development. (1987). *Our Common Future*. Oxford, Oxford University Press.

11 IMPLEMENTING THE TECHNOLOGY

Chapter Summary: This chapter considers the major issues surrounding strategic planning for forecast implementation and for selecting and managing the vision that is chosen. The chapter discusses the need for continued forecasting, forecasting implementation issues, and strategic planning approaches. Selection of alternative designs or implementations of technology is considered. Technology roadmapping, an important technique for strategic coordination, is reviewed. The chapter concludes with a discussion of the challenges of forecasting an often deeply uncertain future.

A forecast not used in implementing a technology is an important management tool wasted. Employing the forecast to inform decision making during implementation is the culmination of the process of exploring, analyzing, and focusing. The forecast must be linked to the larger context within which it was done to develop the credibility that is essential for its timely and decisive use. Moreover, if it is to be effective, the forecast must be compellingly communicated to decision makers as well as to others inside and perhaps outside the organization. This chapter considers the major issues surrounding strategic planning for forecast implementation and for selecting and managing the vision that is chosen.

11.1 FORECASTING CONTINUES

Forecasting activities do not end with the start of the implementation phase. If the technology continues to interest the organization, then monitoring (Section 4.2) should continue, even if only at a low level to pick up new developments that may affect the technology (Rossini, Porter, et al. 1976). It is important to follow up "game-changing" information, do additional forecasting if necessary, and factor the results into the implementation process. Significant new developments may require an expanded forecasting effort. This might be viewed as a feedback loop to some phase of the completed forecast. The message is clear: Never cease learning about the technology as implementation proceeds.

With new information in hand, the forecaster can ask: "How does this alter the forecast and the implementation plan?" Typically, significant new information may support the development of the technology, inhibit it, or present serious alternatives to it. What is learned by analyzing the new information must be used to modify forecast conclusions. The practical outcome is that the implementation plan can be altered to account for new circumstances. Without an ongoing forecasting effort, this cannot be accomplished.

11.2 IMPLEMENTATION ISSUES

Beginning to implement the development of a new technology is a significant step for most organizations. Issues surrounding the implementation will vary, depending on the nature of the organization (i.e., whether it is a corporation, a government agency, or a nonprofit), the scope of the implementation (large or small, disruptive or incremental), the time frame, and the relationships with partner and/or competitive organizations. Typically, however, several major issue areas emerge. These include strategic planning, selecting alternative implementations of the technology, identifying participants in the process, scheduling implementation activities, managing the process, and effectively communicating with constituencies.

Many of these activities have been described in detail in earlier chapters. For instance, scheduling activities are covered in Section 3.5. Others, such as organizational and communications structures (Section 3.3), involve issues that are essentially shared by forecasting and implementation teams. The following sections begin to address remaining important issues.

11.3 STRATEGIC PLANNING FOR TECHNOLOGY IMPLEMENTATION

The technology delivery system (TDS) provides a framework for the long-term views of strategy in general or for specific strategies, like that for intellectual property. Strategies must be implemented, and planning shows how resources can be applied over time to achieve them. Planning can range from brief entrepreneurial business plans intended to attract investors to much more elaborate plans for large, complex organizations. In all cases, it is not so much the agreement of subsequent actions with those planned that makes the plan valuable. Instead, it is the process of creating a plan that has a plausible chance for success and the understanding that it produces that justify the effort.

Exhibit 11.1 Planning Framework

1. *Forecast the Technology*: This is the starting point of technology planning and has been the subject of this book thus far.
2. *Analyze and Forecast the Environment*: Anticipate key factors in the organization's environment.
3. *Analyze and Forecast the Market/Use* (Chapter 8): It is imperative to include direct contact with potential customers.
4. *Analyze the Organization*: Understanding the strengths and weaknesses of your organization is critical.
5. *Develop the Mission*: Establish overall objectives, specific target objectives for the planning period, and measurement criteria. Include as many participants as possible to achieve a sense of ownership.
6. *Design Organizational Actions*: Develop a consensus strategy limited to a few key actions.

Source: Based on Maddox, Anthony, et al. (1987).

Existing firms seldom do complete business plans for innovations, but they too must have processes to plan for implementing new technologies. The framework for such planning is suggested in Exhibit 11.1, which is based on Maddox, Anthony et al. (1987). While these steps aren't magic, they are a systematic way to manage technologies for organizational success. Since very few technologies are sufficiently compelling by themselves to change the future, managers must have processes to develop them and provide the collateral assets that will make their implementation successful. Processes like the one described in the exhibit help keep the organization focused on those technologies that create real opportunities.

Developing strategy and planning is an ongoing process. Ideally, strategies and plans always are on the table for possible revision. The TDS is the starting point for the strategic planning process. The first four activities of the planning framework depend on a TDS that is carefully developed and constantly revised. The final two activities can be accomplished during the implementation phase using the forecast and the TDS as the principal bases. As mentioned repeatedly, it is important that the views of key stakeholders outside the organization are solicited and carefully considered in implementation planning. Thus, communication is critical (see Section 3.3.2).

11.4 SELECTING FROM AMONG ALTERNATIVE IMPLEMENTATIONS OF THE TECHNOLOGY

The forecast typically offers more than one viable option for implementing the technology. Selecting among them occurs after the mission is developed but before organizational actions are designed. From the mission statement and the forecast, specific questions arise such as:

1. What are the implementations from which the selection is to be made?
2. Who will be involved in the selection, and what roles will they play?
3. What criteria will be used for the selection, and how will they be weighed?
4. How will the criteria to be measured?

Answering these questions helps structure the selection process.

The answer to the first question is provided by the focusing phase of the forecasting process. It implies that the alternatives are technically feasible, economically viable, and involve no "show stoppers" resulting from larger societal issues. The answers to the second and third questions are determined by the decision-making process of the implementing organization. However, it is always a good idea to get the views of stakeholders outside the organization even if they aren't involved in the selection process. Answering the final question depends on judgments about the criteria that are compelling to the implementers and the decision makers.

Identifying participants in the selection process is critical. Here much of the information provided in Section 5.1 will help. There is an obvious trade-off between a broadly based team that includes all significant stakeholders and a tightly cohesive group that can move quickly. The former may produce an unfocused process, while the latter may develop one that is too narrowly focused. It is safest for the selection group to be broader than the implementation group so that important issues outside immediate implementation are considered. The selection group probably should include individuals who:

- Can justify, defend, and criticize the potential alternatives. These may include the original forecasters, especially those conversant with the societal context.
- May lead the development and production processes for the alternatives.
- Have analyzed or can analyze the economic implications of the alternatives.
- Are general and technical managers within the implementing organization.
- Are potential users of the technology in its alternative forms.

The selection group must determine the criteria to be used in making the selection and how these criteria will be measured and weighted. Appropriate criteria might include technical excellence, potential for timely implementation, economics, ease of production, range of utilization, and the acceptability of the alternative to users and the broader society. Initially, a broad range of criteria should be considered and then carefully narrowed to finalize a widely accepted set.

Weighing the importance of the criteria is the next step. In some cases, it is very helpful to clarify and compare the values and criteria of the various stakeholders. Policy capture (Hammond and Adelman 1976) is one way to do this, especially if values and criteria sharply diverge for different stakeholders (Exhibit 11.2). Clarification, however achieved, can help establish and weigh evaluation criteria.

Exhibit 11.2 Policy Capture

The county landfill is almost full. Siting a new landfill, incineration, and a recycling proposal each generate heated debate. Suppose the two dominant considerations are costs (C) and environmental protection (E). A number of specific scenarios are devised that implicitly cover the full range of possible levels for C and for E. Stakeholders are asked to participate in a policy capture exercise in which they give preference scores for each scenario (on a 1 to 100 scale). This process yields the data in Table 11.1 on each stakeholder's preference regarding each scenario.

TABLE 11.1 Stakeholder Preferences for Policy Scenarios

	Levels			
Scenario	C	E	Stakeholder A Preference	Stakeholder B Preference
1	30	51	99	10
2	94	72	5	40
3	78	87	40	90
4	60	75	35	60
5	12	23	70	20

This could be extended over additional scenarios, stakeholders, or considerations. A multiple regression program is then used to calculate the weightings that the stakeholders have implicitly given C and E (by statistically associating the C and E values with the preference values over the set of scenarios). Chapter 5 discusses how to perform regression calculations. This information can capture the extent to which each party values C and E, for instance:

$$\text{Stakeholder A's preference function} = 77.5 - 1.5C + 0.9E$$
$$\text{Stakeholder B's preference function} = -23.7 - 0.2C + 1.3E$$

In other words, Stakeholder A prefers low-cost alternatives (the negative coefficient, −1.5, indicates that A downgrades alternatives with high cost). Stakeholder A secondarily factors in high environmental protection; Stakeholder B emphasizes environmental protection and only slightly considers low cost.

Two-dimensional plotting of the scenarios against the C and E axes can further clarify the choices. For instance, some of the options may dominate others; Scenario 3 offers better E and lower C than Scenario 2. This could simplify the choices by showing that the only reason for favoring a dominated choice would be personal interests, especially NIMBY—not in my backyard!

There are pitfalls in applying policy capture, including failure to include all pertinent factors, sensitivity to the presentation, time demanded of participants, strategic misrepresentation, representative sampling concerns, and nonlinearities (Crews and Johnson 1975; Mitchell, Dodge, et al. 1975).

It may also help stakeholders with different views better understand each other and possibly even work out acceptable trade-offs. Candidate criteria below a certain level of significance and support can be eliminated. Whatever criteria and their weights are settled upon, the sum of the weights must be 1.

Having identified and weighed the importance of the evaluation criteria, the selection group can now address how they will be measured. Each criterion presents individual measurement issues and perhaps differing units of measurement as well (e.g., dollars, hours required to produce). Some uniform way to measure how well each alternative satisfies the criteria must be established. Ideally, a numerical scale can be used even if values are only estimated or based on judgments. Various measurement schemes are discussed in the following section.

11.4.1 Measurement

Table 5.11 in Section 5.4 presented the four levels of measurement and their operational uses.

Interval and ratio measures clearly are the most attractive for measuring the achievement of alternative criteria. However, they cannot always be used, especially when measurements are based on subjective judgments. Ordinal measures often are the best available. These can be handled in several ways.

Rating and *ranking* are two important approaches to creating ordinal scales. Ratings compare a measurement against a standard or a set of standards, while rankings give relative indications among a set of alternatives (e.g., 5 = Very Strong, 4 = Strong, 3 = Neutral, 2 = Weak, 1 = Very Weak).

Scales with an odd number of values (as above) allow raters to opt for a neutral or middle position, while scales with an even number force them to express a judgment. Scales can range in precision from binomial (0 or 1; yes or no) to as fine a gradation as desired (e.g., 1 to 1000). For subjective ratings, a simple scale is apt to be most helpful. The difference between 637 and 638 on a scale from 1 to 1000 is impossible to interpret in most cases.

An interval rating scale based on subjective, component judgments is an interesting alternative. Freeman, Frey, et al. (1982) used the Futures Forgone (FF) index to compare alternatives in 106 potential U.S. natural vegetation communities (PNCs) for each of 10 activity categories (e.g., wood harvest, tree life forms). The FF index was calculated for each PNC as

$$FF = \frac{(\text{Base year total} - \text{Projeccted year total})}{\text{Projeccted year total}} \tag{11.1}$$

Projected year totals could be derived either from quantitative trend projections or from subjective expert estimates.

In a *ranking,* individuals may be asked to judge one alternative as being higher or lower than another on a particular criterion with or without an option for equality. Refinements can take many forms. Sharif and Sundararajan (1984), for instance, use the more precise *analytic hierarchy process* scaling (see

TABLE 11.2 Weighted Decision Matrix

Weight	Criterion	Alternative A	Alternative B	Alternative C
0.3	D ($)	0.2	0.5	0.3
0.6	F (functionality)	0.6	0.2	0.2
0.1	R (reliability)	0.1	0.4	0.5

Section 11.4.3) to compare technological alternatives. Sometimes it is important to measure stochastic (probabilistic) information separately. For example, the Futures Group (1975) devised a matrix scheme that arrays impact likelihood versus desirability.

Regardless of whether rating or ranking is employed or what scaling scheme is used, the issue is to determine how well each alternative meets a criterion. For some criteria, only a qualitative determination can be made based on judgment and using an ordinal scale. However, no matter how determinations are made, the weight for each criterion needs to be applied to the results for each alternative.

Eventually, it is necessary to compare alternatives. One approach is to use a linear additive weighting model. Table 11.1 supposes that there are three alternative technologies (A, B, and C). Three selection criteria (D, dollars; F, functionality; and R, reliability) are to be applied. The consensus of the selection group about the relative weights for each criterion and the measure of how well each technology fulfills each are shown in the cells of the table. The relative weights assigned to criteria should sum to 1.0. A linear additive calculation of the table entries yields a total for A of

$$A = 0.3(0.2) + 0.6(0.6) = 0.1(0.1) = 0.43$$

Similarly,

$$B = 0.31 \text{ and } C = 0.26$$

The linear additive weighting model facilitates sensitivity analysis. For instance, a stakeholder could use the data in Table 11.1 to see that changing the weights for reliability to 0.3 and functionality to 0.4 results in Alternative B being favored.

Simpler ways to combine criteria and alternatives are possible (e.g., equally weighting all criteria; binary scoring, in which an alternative does or does not meet minimal requirements for each criterion), but these seem to use the available information less fully and produce no great computational savings.

Nominal measures can sometimes be useful, and in some cases they can be converted to sets of binomial variables. For *ordinal measures*, numerical manipulations may yield relative results without giving absolute information. Appropriate statistical manipulations for ordinal variables include computation of rank order correlations and nonparametric inference tests. However, it is advisable to use no more sophistication than is needed to achieve an accurate ranking. Too much sophistication can produce results that are not intelligible or believable to decision makers.

When it is necessary to rank alternatives using only human judgment, pairwise comparison is a useful approach. Each alternative is compared with one other, then with a second alternative, and so forth. This simplifies the judgments required, but the number of comparisons involved can be very demanding. For instance, it will require $(n - 1)!$ judgments for a set of n alternatives for each criterion considered. A matrix of pairwise comparisons can be constructed and consolidated to an ordering of the factors (Sharif and Sundararajan 1984). The technique of Interpretive Structured Modeling, discussed in the next section, is a useful facilitator for pairwise comparisons.

11.4.2 Interpretive Structural Modeling

Interpretive structural modeling (ISM) is an approach that can be used to simplify generating pairwise comparisons and conveying the results when more than two alternatives exist (see Watson 1978). The approach is effective for mapping contextual relationships between all pairs of elements in a set. ISM can be useful in applications as varied as constructing decision trees (Section 6.7.1) and scheduling implementation (Section 3.5). The process requires the forecaster to define a set of elements and a relationship between them that is to be investigated. The outcome can be displayed as a diagram that shows the direction of the relationship between elements (a digraph).

In this case, the elements are the alternative implementations of the technology and the relationship is "preferable to." ISM participants compare each pair of alternatives in turn and try to reach a consensus as to which is preferable. The discussion that accompanies these comparisons is by far the most important part of the process. From that discussion, an $n \times n$ matrix can be built on which the n alternatives are arrayed along both row and column headings. The matrix is helpful in structuring the discussion. The values in the matrix cells are:

- 1 if the row alternative is preferable to the column alternative
- 0 if the column alternative is preferable to the row alternative

The diagonal cells, of course, are blank. The comparison process is considerably streamlined by assuming transitivity (i.e., if A is preferable to B and B is preferable to C, then A is preferable to C), which eliminates the need for many comparisons. In some cases, discussion of specific alternatives may raise questions about transitivity. This presents an opportunity to go deeper into the understanding of the alternatives. An outcome matrix for a four-alternative process is shown in Exhibit 11.3.

Once the matrix is complete, each alternative can be rank ordered based on the number of 1's in its row. Alternatively, a diagram in arrows directed toward the least preferred alternatives can be drawn (Exhibit 11.3). In the exhibit, the order of preference of the alternatives from most to least is 1, 4, 2, 3.

Exhibit 11.3 ISM Example for Four Alternatives

A graphical representation of the outcome matrix (digraph) is presented in Figure 11.1, where the direction of the arrow indicates, for instance, that alternative 1 is preferable to all others and 3 is the least preferable alternative. Transitivity would allow the arrow from 1 to 4 to be deleted.

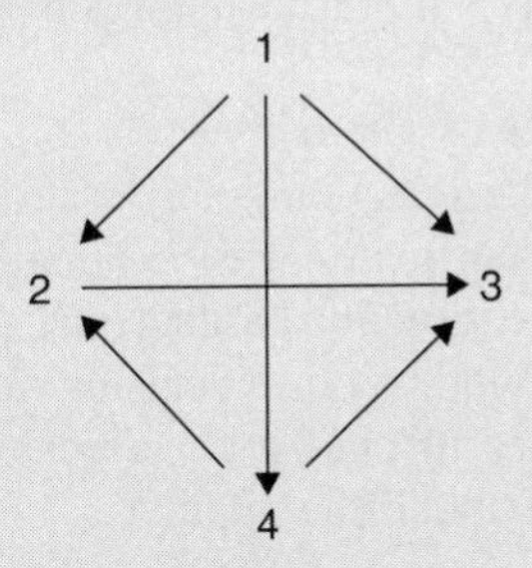

Figure 11.1. Digraph Representation of Outcomes

TABLE 11.3 Outcome Matrix

Alternative	1	2	3	4
1	X	1	1	1
2	0	X	1	0
3	0	0	X	0
4	0	1	1	X

11.4.3 Analytic Hierarchy Process

Another technique that may prove useful in comparing alternatives is the analytic hierarchy process (AHP). AHP was created to structure complex judgments (Saaty 1980; Saaty and Kearns 1985). The process involves four stages:

1. Systematizing the judgments (about the alternatives) into a hierarchy or tree
2. Performing elemental pairwise comparisons
3. Synthesizing those pairwise judgments to arrive at overall judgments
4. Checking that the judgments combined are reasonably consistent with each other

A basic (and enjoyable) tutorial on AHP by can be found at Wikipedia (2010).

11.4.4 Wrap-Up

The previous sections have dealt with what is called *multiple objective decision making*—that is, with multiple decision makers applying multiple criteria to select from among multiple alternatives. A wide range of techniques and support software is available for such activities (see Makowski 2010).

Simple, clear approaches have proven to be just as effective as complex ones, and they are more easily communicated to decision makers (Ascher 1978). Thus, it is perhaps best to use as simple an approach as possible to complete the selection process.

Typically, analyzing a set of alternative possible implementations of a technology requires laying out the selection goals, criteria, and subcriteria, clarifying them and the alternatives from material developed by the forecast, and making a documentable choice by a broadly based team. To ensure wide support for the result and simplify communication to decision makers and stakeholders, only techniques that are necessary to clarify the process and make the participants comfortable with the results should be applied.

11.5 TECHNOLOGY ROADMAPPING

Technology roadmapping is a needs-driven planning approach that was developed to help identify, select, and develop alternatives to satisfy a set of product needs (Garcia and Bray 1997). This document includes a brief example of a roadmap. There is a more graphic example at Phaal (2010b). The approach can be applied within an organization or in collaboration with organizations that have a strong interest in a specific technology area. There are three phases to the process (Garcia and Bray 1997):

Phase I. Preliminary activity

1. Satisfy essential conditions.
2. Provide leadership/sponsorship.
3. Define the scope and boundaries for the roadmap.

Phase II. Development of the Technology Roadmap

1. Identify the "product" that will be the focus of the roadmap.
2. Identify the critical system requirements and their targets.
3. Specify the major technology areas.
4. Specify the technology drivers and their targets.
5. Identify technology alternatives and their time lines.
6. Recommend the technology alternatives that should be pursued.
7. Create the technology roadmap report.

Phase III. Follow-up activity

1. Critique and validate the roadmap.
2. Develop an implementation plan.
3. Review and update.

These phases reveal that technology roadmapping involves a broad range of activities, many of which overlap features of the exploring, analyzing, and focusing processes proposed in this book. Similarities between the two include a bounded general framework, an overall goal or direction, a focus on implementing alternatives to achieve the goal, and an implementation plan. Differences include the seemingly lower emphasis on societal issues and impact assessment of roadmapping, its stricter a priori definition of an acceptable outcome, and the potential for a very long time horizon. Some of the well-known instances of roadmapping were performed by consortia of organizations (e.g., the Semiconductor Industry Association) and often entailed a very broad cooperative decision-making process. An Internet search for roadmapping software will produce a substantial collection of packages. A list of public domain roadmaps and a list of publications on the process of roadmapping may be downloaded from Phaal (2010a).

11.6 SUMMARY AND CONCLUDING OBSERVATIONS

A technology forecast usually will identify alternative technologies that offer promising futures for the organization. These alternatives must be carefully considered before the final implementation is selected. This requires careful development and application of selection criteria and measures. Because of the complex nature of the selection task and the diverse character of team members, scheduling becomes an important issue (Section 3.5).

Implementing a technology generally involves multiple units of the organization developing it as well as stakeholders with significant interest in the development. Thus, those engaged in the process must have a wide range of capabilities and interests (e.g., forecasting, societal and impact analysis, R&D, production, marketing, and finance). Because of the wide range of capabilities and interests, the management structure and communications patterns chosen for the team are critical and must be given conscious thought (see Section 3.3).

Successfully developing and executing an implementation plan requires selling it to participants no less than to management, financial supporters, potential users, and external constituencies. Communication with external stakeholders is an important means of obtaining input and building support for the final plan. Since not everyone engaged in the process will be technically sophisticated in every area that typically is addressed, communications should be as transparent and jargon free as possible. Attention to the forms in which decision makers and stakeholders prefer communication to be cast will pay handsome dividends.

REFERENCES

Ascher, W. (1978). *Forecasting: An Appraisal for Policy Makers and Planners*. Baltimore, Johns Hopkins University Press.

Crews, J. E. and G. P. Johnson. (1975). "A Methodology for Trade-Off Analysis in Water Resources Planning." *ISTA Journal* **1**: 31–35.

Freeman, D. M., R. S. Frey, et al. (1982). "Assessing Resource Management Policies: A Social Well-Being Framework with a National Level Application." *Environmental Impact Assessment Review* **3**: 59–73.

Futures Group, The (1975). *Technology Assessment of Geothermal Energy Resources Development.* Glastonbury, CT, The Futures Group.

Garcia, M. L. and O. H. Bray. (1997). "Fundamentals of Technology Roadmapping." Albuquerque, NM, Strategic Business Development Department, Sandia National Laboratories.

Hammond, K. R. and L. Adelman. (1976). "Science, Values, and Human Judgment." *Science* **194**: 389–396.

Maddox, N., W. P. Anthony, et al. (1987). "Creative Strategic Planning Using Imagery." *Long Range Planning* **1987**(5): 118–124.

Makowski, M. (2010) "Multi-Objective Decision Support Including Sensitivity Analysis."Retrieved 1 September 2010 from http://www.iiasa.ac.at/~marek/ftppub/MM/eolss_mcma.pdf.

Martino, J. P. (1983). *Technological Forecasting for Decision Making*. New York, McGraw-Hill.

Mitchell, A., B. H. Dodge, et al. (1975). *Handbook of Forecasting Techniques*. Stanford Research Institute Report to the U. S. Army Corps of Engineers Institute for Water Resources. Springfield, VA, National Technical Information Service.

Phaal, R. (2010a). "Roadmapping Resources." Retrieved 1 September 2010 from http://www.ifm.eng.cam.ac.uk/ctm/trm/resources.html.

Phaal, R. (2010b) "Technology Roadmapping." Retrieved 1 September 2010 from http://www.unido.org/fileadmin/import/16963_TechnologyRoadmapping.pdf.

Rossini, F. A., A. L. Porter, et al. (1976). "Multiple Technology Assessments." *Journal of the International Society for Technology Assessment* **2**: 21–28.

Saaty, T. L. (1980). *The Analytic Hierarchy Process: Planning, Priority Setting, Resource Allocation.* New York, McGraw-Hill.

Saaty, T. L. and K. P. Kearns. (1985). *Analytical Planning: The Organization of Systems*. Oxford, Pergamon Press.

Sharif, M. N. and V. Sundararajan (1984). "Assessment of Technological Appropriateness: The Case of Indonesian Rural Development." *Technological Forecasting and Social Change* **25**: 225–237.

Watson, R. H. (1978). "Interpretive Structural Modeling: A Useful Tool for Technology Assessment." *Technological Forecasting and Social Change* **11**: 165–185.

Wikipedia. (2010). "Analytic Hierarchy Process." Retrieved 1 September 2010 from http://en.wikipedia.org/wiki/Analytic_Hierarchy_Process.

12 MANAGING THE PRESENT FROM THE FUTURE

Chapter Summary: This chapter gives a distilled summary of what has gone before in this book as well as providing a basis for moving beyond it. The chapter discusses the overall approach to technology forecasting. Reflections on selecting methods and techniques are provided. The need for multiple alternative perspectives is considered. Forecasters can learn from previous forecasts and assessments. The need for strategic vision is considered.

The title of this final chapter (Smits, Rossini, et al. 1987) says it all. This book has presented a process and tools to do just that—manage the present from the future. Forecasts provide glimpses of the future and then become vehicles for making management decisions in the present. The narrowed range of probable futures presented by a forecast augmented by continued updates gives a sound basis for moving forward to implementation.

12.1 THE OVERALL APPROACH

The process outlined in this book consists of three forecasting phases—exploring, analyzing, and focusing—followed by implementation that uses the forecasting outcomes to move the technology forward to utilization. To review:

- *Exploring* casts the broadest possible net consistent with forecast boundaries and resources to sweep in all possibly relevant information and to make sure that all potentially critical areas are included. The process of developing the technology delivery system (TDS) begins in this phase.

- *Analyzing* identifies areas worthy of further study and delves more deeply into them. The TDS is developed further, the scope of attention is narrowed, and detail is substantially increased.
- *Focusing* selects the few most promising alternatives and lays them out, along with possible impacts, in the greatest detail possible within the forecast bounds. In this phase, the TDS should provide both broad information about the technology and deep, highly focused information about its most promising aspects.

Implementing deals with the postforecast phase, in which the technology is either pursued or dropped. Forecasting does not cease when implementation begins. It continues at a reduced level to monitor developments that might prevent, enhance, or alter the implementation of the technology. Implementing continues throughout the life of the technology. Knowing when to move on is a highly desirable outcome of the process.

The flow of the process is straightforward: survey the field, narrow and deepen it as the process proceeds, select the most desirable courses forward, and follow them. Don't hesitate to change as circumstances change. End the process in a timely way.

12.2 SELECTING METHODS AND TECHNIQUES

This book offers methods and techniques to support the three-phase forecasting approach. Some of them will be used in most cases; others will be only occasionally useful.

12.2.1 Using the TDS and the Major Families of Techniques

Constructing and maintaining the TDS or some other map of the critical elements of the technology and its societal context are essential for any forecast. This should be done at a level of detail necessary to meet the needs of the forecast. The TDS is not static. Forecasters and implementers should track its changes over time until the technology is of no further interest.

Monitoring is a very basic technique that takes on many forms. It should continue at varying levels throughout the forecast phases, during implementation, and beyond. How it is used depends on the nature of the forecast.

Expert opinion gathering, in its simplest form, monitors what experts are saying in publicly available sources. It begins in the exploring phase. While gathering expert and stakeholder opinions by structured approaches, such as surveys and panels, may be helpful in the exploring phase, it likely will be even more useful in analyzing, focusing, and implementing.

Trend analysis, in its various forms, is an important technique. Although generally accessible projections are useful in exploring, specific trend analyses will be even more useful in later phases. Since available data often are scarce, it frequently is necessary to use an empirically based growth model. It is important to have a sound rationale for this selection. Here are a few guides for some of the models discussed in Chapter 6. Straight-line growth is the default choice. S-shaped curves usually represent the market penetration of a new technology extremely well. Exponential curves are seldom appropriate except in the case of semiconductor capacity (i.e., Moore's Law). There is no theory underlying these selections.

Modeling is perhaps the most difficult family of approaches to use. Very simple models that help organize thought and analysis typically are the most useful models. Sophisticated models demand soundly based, accurate data and often are time-consuming and expensive to build and run. Unfortunately, there is a tendency for computational, analytical, and mathematical sophistication to give spurious validity to results based on input of doubtful quality. When data are lacking, they must be replaced by assumptions. But judgments are subjective and imprecise, and model output can be no more valid than its input. Using complex models for the sake of using them merely revisits territory long ago discredited.

Scenarios are one of the most useful and versatile techniques for forecasting and communicating forecast results. They make complex analyses accessible to the nontechnical user by synthesizing large amounts of related information into easily grasped word and/or graphic narratives. Thus, they often are the best technique for synthesis and communication. A small number of scenarios can provide an excellent basis for comparing alternatives and choosing from among them.

12.2.2 The 80–20 Rule

The 80–20 Rule says that 80% of results are produced by spending 20% of the resources that would be needed to produce complete results. The principle, if not the percentages, is most useful for forecasters and developers of new technologies. The message is clear.

In the technology development process, do the basic forecasting *first* to get the results you can quickly and cheaply produce. *Then* determine the most important additional results that could be obtained within available time and resource limits. *Use selectivity and focus to identify what is truly important to know.* Ask yourself "What additional important work can be done reasonably well within the study constraints?" Do it and quit for now. Remember that "perfect "can be the enemy of "useful" in forecasting. No technology development ends until the technology is abandoned. There may be other days to refine analysis and planning.

12.3 ALTERNATIVE PERSPECTIVES

This section discusses temporal perspectives on the decisions and actions undertaken in the present to facilitate technology development. To make intelligent decisions, managers must interpret situations in light of the information that they consider important. Their temporal perspective—past, present, or future—colors the information they select. While all three perspectives are important, inappropriate reliance on any of them can lead to trouble.

Looking Back from the Present: The most common temporal perspective is to interpret the future in light of past experience. This tends to lock the decision maker into the framework of past organizational goals, strategies, and experiences. Its view of the future emphasizes continuity as an extrapolation of that past. "The future will be just like the past except more so." But sometimes it isn't, and when it isn't, the impacts can be devastating.

Focusing on the Present: The "now" orientation stresses immediate solutions to pressing problems. However, rewards based on current and near-term profit and loss discourage a long-term perspective. American management, including R&D, has come under fire for focusing on short-term payoffs within annual budgeting cycles. Obsession with today's problems and opportunities can blind one to tomorrow's problems and opportunities.

A Futures Perspective: Obviously, this book advocates a futures perspective. However, no competent decision maker can ignore the past or the present or place implicit confidence on forecasts of a specific future. Nevertheless, forecasts offer an important roadmap for what alternative futures might hold. While the journey may not follow the proposed route, only the foolhardy embark on a journey without a map.

Forecasting methods are empirical substitutes for a working theory of sociotechnical change that simply does not exist in anything approaching a predictive form. The techniques presented in this book can be used to uncover a range of likely futures. Thus, they allow the forecaster to use tentative information about the future to inform present decisions that, in turn, will affect the future. Be careful, though, not to become dependent on only a small number of techniques. Applying several of them will enrich results and their credibility. This section revisits a few techniques from a futures perspective.

Monitoring might be considered to be a present-oriented, almost universally applied technique. However, it gains a future orientation when the forecaster filters and structures the information gathered to indicate likely future developments. The best-selling book *Megatrends* (Naisbitt 1982) and its successor, *Megatrends 2000* (Naisbitt and Aburdene 1989), illustrate the use of monitoring in a very simple form in the present to predict possible futures.

Trend extrapolation uses data about the past to anticipate future developments. If taken too literally, trends can tie the forecaster too tightly to the past, so that breakthroughs or barriers disruptive of trends may be missed.

Other techniques, such as expert opinion methods and scenarios, can directly engage the future. For instance, asking experts to predict the situation in the year that technology X will be commercially available presumes that they have a tacit model of the future that they can tap. Likewise, scenarios can be used to creatively explore future possibilities.

No matter how information about possible futures is generated, it must be related to the present. What actions can be taken or planned now to address potential futures? In other words, what can be done to "manage the present from the future?"

Technology forecasting purports to predict possible future changes in technology, even though each technology is different, nontechnical environments dramatically vary, and there is no set of standard variables to describe it. There are only sound approaches, such as the three phases suggested here, and even they utilize empirical techniques whose worth under all possible conditions is uncertain. There is no substitute for human judgment to limit uncertainty.

12.4 LEARNING FROM PAST FORECASTS AND ASSESSMENTS

Judgment can be strengthened by examining the past to improve present performance in anticipating the future. Thus, it is instructive to examine the results of past forecasts to see if they were successful in limiting uncertainty. Also see Section 2.1.4 for common forecasting errors.

Reviewing old forecasts shows that many were too ambiguous to tell if they proved to be accurate. However, George Wise (1976) gathered a large number of forecasts that he could judge to be right or wrong and sorted these into 18 technological areas (e.g., computers, factory automation, new materials). He aggregated predictions for each area, reporting the average percentage right by area. These ranged from 18% right for 22 housing technology forecasts to 78% for 18 new materials forecasts. The median by area was 45% right; the interquartile range was 38% to 51%. That is, one-quarter of the areas fared worse than 38% and one-quarter did better than 51%. This rough estimate helps put forecasting expectations into perspective. They are far short of certainty but much better than chance. But remember, technology forecasts are intended to paint a landscape of possible futures, not a portrait of a certain future.

Michael Scriven (1967) draws a useful distinction between *formative* and *summative* evaluation. Formative evaluation considers a study while it is underway and can be very helpful in guiding the forecast. Formative procedures often are quite informal—for instance, providing early forecast drafts to experts and decision makers to tell if critical elements are missed or driving forces are misunderstood while there is time to make corrections. Their participation in focusing the study also can enhance their acceptance of it later on.

Summative evaluation reviews a study after it is done. This means judging a study in terms of its validity and utility. The interrogation model developed by Martino (1983) provides a good set of questions to ask about a forecast (see Exhibit 11.4 in Section 11.6).

Forecast objectives also are important. For instance, *The Limits to Growth* (Meadows, Meadows, et al. 1972) was roundly criticized in regard to the validity of its attempt to model world dynamics. Yet, the stated objective of the study was to get world leaders to think about potential worldwide environmental disasters. It succeeded spectacularly in meeting this objective, whether or not the model itself had the least validity.

Linstone notes that there are explicit forecast perspective(s): technical (e.g., engineering), organizational (or societal), and personal. Each emphasizes different goals, modes of inquiry, ethical bases, planning horizons, and other characteristics. It is critically important for technology forecasts to blend these perspectives to avoid ill-founded optimism, pessimism, or just plain myopia. In an illustration based on Linstone (1988), a manager faces a decision about whether to enter a new business area. She has a detailed cost-benefit analysis from *technical* staff, indicating that the area is ripe for development. To this she adds discussions with the *organization's* department heads to determine the extent of support for or opposition to such an expansion. Then she talks to a friend whose company is involved in a different aspect of the target market and draws on his *personal* intuition, experience, and advice. She then integrates these different, possibly conflicting, perspectives to arrive at a decision.

Drawing on previous observations and those of the authors, Exhibit 12.1 provides 12 recommendations for technology forecasting.

Exhibit 12.1 A Dozen Recommendations for Technology Forecasting

1. *Get the technology right*: Understand the technical domain sufficiently to address the essential functions at the right level of aggregation.
2. *Pick the right parameters*: Technological parameters must pertain to the decision to be made; data must be available.
3. *Get the context right*: Identify the institutions involved, socioeconomic influences, and critical present and future decision points.
4. *Beware of core assumption drag*: Technical myopia or ideological fixation can cause the decision maker to miss qualitative changes from past patterns.
5. *Beware of the Zeitgeist*: Challenge the conventional wisdom; try out alternative perspectives; do not allow the forecast to merely mirror the prevailing mood.
6. *Keep the time horizon short*: Frequent, lower-cost forecasts are preferable to more substantial, less frequent ones.
7. *Do it simply*: Invest study resources in reducing the most critical uncertainties; rarely will these respond to elaborate modeling.
8. *Use multiple approaches*: Seek convergence from diverse approaches with complementary strengths.
9. *Perform sensitivity analyses*: Examine how the forecast would change if initial conditions, important variable levels, functional relationships, or milestones change.
10. *Provide uncertainty estimates where possible*: Give ranges of parameter projections over time and prediction (confidence) intervals.
11. *Take the middle path*: Balance between far-out forecasts with low probabilities and ultraconservative ones (often offered by committees).
12. *Ensure broad involvement*: Involve stakeholders and decision makers in every phase of the process.

12.5 VISIONS

Bringing the future home to the present requires images of the future. That is what forecasts are intended to provide. The future is very much an open system with fuzzy, impenetrable boundaries in both space and time. Images of the future must accommodate uncertainty and be adaptable, yet provide focal points to guide present actions. They also must include both contexts and goals. Such visions can be used by decision makers to guide present decisions and actions.

A vision of the future is a believable and easily communicable story about the future of a technology that is both concrete and open-ended—a real "grabber" that invites buy-in. It is concrete in the sense that it can carry one from the present reality to the future by providing a credible narrative. It is open-ended in that it is a bold outline without subheadings and closed definitions that clearly points to plausible futures.

Creating visions combines observation, analysis, intuition, and imagination. Formal and informal observations of sociotechnical development patterns provide the raw material to drive a vision. But they need to be interpreted and structured according to the conceptual framework of organizational beliefs, assumptions, and goals. The tools presented in this book can be applied directly to observations of the present and past. Intuition thrives on rich input and builds on creativity (Chapter 4), the results of analysis, and, perhaps, more speculative futurist writings such as science fiction. Imagination fleshes out the patterns that intuition generates with rich images.

Interaction with stakeholders and decision makers is absolutely vital to generate credible visions that will be adopted. Their worldviews, assumptions, needs, and goals must be incorporated, and they need to understand how the vision was created. Decision makers must be convinced of the efficacy of the methods and believe in the capability of the developers. Thus, good communication is essential.

The magic in using visions of the future as bases for a present action derives from perspective. Decision makers should see present actions through the eyes of someone in the future captured in the vision. As advocated in sports psychology, they should first imagine the key actions needed for success and then perform them. The act of seeking to grasp the future moves individuals and institutions into a proactive posture in dealing with the world. No longer passive in the face of sociotechnical change, they can become actors in an ongoing drama whose script they are writing. This book was written to empower this perspective on change.

12.6 A FINAL WORD

This book's message of exploring, analyzing, focusing, and implementing a future is one that departs from seat-of-the-pants organizational management. Forecasting involves limiting uncertainty about the future, not making predictions about specific events on which you can bet. It provides the bases of well-informed, carefully analyzed actions that can move a technology or an organization into the future with a favorable outcome. Thinking about the future with intelligence and skill is the best way to enter it. The development of this book, from its first edition, has been an open-ended process. It is up to you to progress from the starting point it provides!

REFERENCES

Linstone, H. A. (1988). "Multiple Perspectives: Concept, Applications and User Guidelines." *Systems Practice* **2**(3): 307–331.

Martino, J. P. (1983). *Technological Forecasting for Decision Making*. New York, McGraw-Hill.

Meadows, D. H., D. L. Meadows, et al. (1972). *The Limits to Grow*th. New York, Universe Books.

Naisbitt, J. (1982). *Megatrends: Ten New Directions Shaping Our Lives*. New York, Grand Central Publishing.

Naisbitt, J. and P. Aburdene (1989). *Megatrends 2000*. New York, Morrow.

Scriven, M. (1967). "The Methodology of Evaluation." In *Perspectives in Curriculum Evaluation*, ed. R. Taylor, R. Gagne, and M. Scriven. Chicago, Rand McNally: 39–83.

Smits, S. J., F. A. Rossini, et al. (1987). "Managing the Present from the Future: The Challenge of Preparing for Technology in Rehabilitation." *Journal of Rehabilitation Administration* **11**(4): 121–130.

Wise, G. (1976). "The Accuracy of Technological Forecasts: 1890–1940." *Futures Research Quarterly* **8**: 411–419.

APPENDIX A

CASE STUDY ON FORECASTING DYE-SENSITIZED SOLAR CELLS

Chapter Summary: This chapter demonstrates the technology forecasting process with a new emerging technology. The chapter discusses dye-sensitized solar cells as a potential next-generation solar technology. Each of the phases and many of the methods discussed in this book are examined in light of solar cells. A technology forecast of dye-sensitized solar cells is provided.

This case study illustrates the application of several technology forecasting methods. While it takes note of the book's chapter flow, not all of that material is addressed. The case study applies the forecasting approach for new and emerging science and technologies (NESTs) outlined in Table A.1 but not described elsewhere in the book.

The topic of the study is dye-sensitized solar cells (DSSCs), which is an advanced form of solar cell (photovoltaic). DSSCs are a subset of nano-enhanced solar cells that leverage molecular-scale (nano) material properties to enhance photovoltaic performance. This presentation emphasizes approaches and methods rather than presenting a complete forecast. It draws heavily upon four papers (Guo, Huang, et al. 2009, 2010; Guo, Porter, et al. 2009; Guo, Xu, et al. in press).

A.1 FRAMING THE CASE STUDY

Chapters 1 through 3 set the stage for the study. Chapter 1 emphasized that technical innovation is intimately entwined with social systems. Table A.1 reinforces this notion, highlighting the interplay among multiple technical and social factors throughout the analyses.

TABLE A.1 Framework to Forecast NEST Innovation Pathways

I. Understand the NEST and its technology delivery system (TDS)	Step A: Characterize the technology Step B: Model the TDS
II. Profile R&D and link to potential applications	Step C: Profile R&D Step D: Profile innovation actors and activities Step E: Determine potential applications Step F: Engage experts
III. Project and assess prospective innovation pathways	Step G: Lay out alternative innovation pathways Step H: Explore innovation components Step I: Perform technology assessment
IV. Report	Step J: Synthesize and report

Chapter 2 aligned forecasting with a broader planning orientation. It noted that the selection of forecasting methods should address multiple factors affecting the nature and effectiveness of technological innovation. Multiple methods are almost always in order in this process.

Chapter 3 related technology management and planning orientations to managing the forecasting project. The technology forecast was emphasized as an aid to managerial or policy decision making, which implies that forecasters must focus on key issues and address them in a timely manner. The case study presented here, however, was carried out to learn about an emerging technology and to devise effective means to forecast its prospects. Thus, the urgency and precise foci of a forecast performed to inform corporate decisions are lacking. Rather, the study speaks to scholarly audiences. Nonetheless, it offers good examples of database mining and methodological applications to help forecast and assess a technology.

Chapter 4 dealt with understanding the context by depicting the TDS and stakeholder analyses, which corresponds to Steps B and D in Table A.1. The chapter also considered creatively exploring technological and application possibilities (Step E) and monitoring (Step C).

A.1.1 Characterizing the Technology

Step A in Table A.1 (Characterize the technology) is covered in Section 4.1.2. When forecasters are not subject matter experts and their audience may have an uneven understanding, technology characterization is a vital first step. For the DSSC case, this required *iterations* spanning several months. These involved monitoring to understand nano-enhanced solar cells and to identify local experts. Intermediate analyses explored nano-enhanced thin-film solar cells. Further work suggested that DSSCs were a promising type of solar cell to consider and revealed basic references for the technology. That understanding, in turn, led to refined database searching and then to the analyses highlighted here.

A.1.2 Dye-Sensitized Solar Cells

Renewable energy is increasingly viewed as critically important. Solar cells, or photovoltaics, convert the sun's energy into electricity. Worldwide, the solar energy market is increasing and has achieved a yearly growth of 26–46% during the last decade. Studies suggest three generations (Gs) for solar cells (Green 2003; Conibeer 2007). The present mainstay is the "first1-G" silicon solar cell, which accounted for 90% of the market in 2004. However, silicon cells are costly to manufacture and have efficiencies limited to about 14% in most production modules and up to 25% in the laboratory. Their cost per unit of power is at least several times higher than that of fossil fuel combustion (Institute of Nanotechnology 2006).

Many advanced technology solar cells are being investigated. "Second- G" units emphasize thin-film solar cells. One group of these cells uses nanotechnology attributes to boost performance. Among the technologies employed by "third-G" units are DSSCs, which were introduced by O'Regan and Grätzel (1991).

In DSSCs, light is absorbed by the dye and excites an electron to a higher energy level. This excited electron is injected into nanoparticles (usually titanium dioxide, TiO_2) and travels to one of the DSSC electrodes by hopping from one particle to another, thereby generating a current. The now positively charged dye undergoes an electrochemical reaction that shuttles "the hole" to the counter-electrode, where it is reduced back, and the cycle repeats (Aydil 2007). DSSCs are especially efficient in converting sunlight into electricity. Research actively continues on a variety of nanomaterials. Although DSSC commercialization is in its infancy, many anticipate promising opportunities.

A.2 METHODS

A series of analyses extending some 18 months suggest that the following methods warrant attention here:

- Expert opinion (interactive; not formalized)
- Multipath mapping (related to roadmapping)
- TDS depiction (with stakeholder analyses)
- Tech mining (empirical monitoring)
- Science overlay mapping
- Trend analyses
- Cross-charting
- Social network analyses

A.2.1 Engaging Experts and Multipath Mapping

Engaging experts (Step F, Table A.1) is very important throughout a technology forecast when in-depth knowledge is not prevalent in the forecast team. While

Chapter 5 treated expert opinion largely in terms of systematic approaches to compiling the views of multiple experts, this academic study informally engaged a few local experts. To gain an understanding of nano-enhanced solar cells, the forecasters were greatly aided by Prof. Jud Ready, PhD student Chen Xu, and other Georgia Tech colleagues. These experts identified seminal literature and relevant articles and reviewed draft reports on the technology and its prospective development. Such technical expertise also strongly contributed to Step G.

A group workshop (Chapter 4) was held late in the study. Participants included several faculty members and graduate students engaged in solar cell R&D and several involved with technology and policy analyses. The focus was on exploring possible innovation pathways by which advanced solar cell technologies might contribute to commercially successful products. Concerns included the following: which pathways show greatest promise; what barriers need to be overcome; who the stakeholders are; and what unintended impacts might be foreseen. The model for this *multipath mapping* was developed by Robinson and Propp (2008)—gathering a diverse group of experts (business, social, and technological) to explore pathways forward.

The workshop began with a brief review of the developmental maturity of various solar cell technologies with which DSSCs would compete, R&D highlights, the TDS, and the cross-charting technique. But the focus was on identifying viable innovation pathways and then on identifying possible stumbling blocks to innovation. Finally, the group initiated a (too) brief technology assessment by brainstorming potential unintended, indirect, and delayed impacts and possible means for mitigating them.

Figure A.1 shows a key slide used to prompt discussion. Although it is rough (e.g., distribution along the temporal dimension is only suggestive) and incomplete (e.g., only a few prominent alternatives are shown at each level), it was effective. The vertical axis suggests a progression by which special materials at the base levels enable structural/functional gains that, in turn, fuel emerging products. Those products might, or might not, target multiple or individual applications. Other possible layers, such as alternative (competing or complementing) technologies, market sectors, and target users, are not shown. Using PowerPoint projection, whiteboard, and flip-chart sheets, the group enthusiastically generated possible innovation paths, with no clear winner.

A.2.2 Developing the TDS

The TDS, which is prominently featured in the text, provided the core model for considering the DSSC innovation process. Step B (Table A.1) puts a priority on understanding the technology in its social context early in the analytical process (Section 4.1.2). Steps D, G, and H build upon this understanding. A paper by Guo, Xu, et al. (in press) focuses on building the TDS model for DSSCs. That process drew heavily on the forecasters' monitoring work. Profiling the most active players in R&D publication, patenting, and business activities contributed to the stakeholder analyses (Section 4.1.3) and to populating the TDS model.

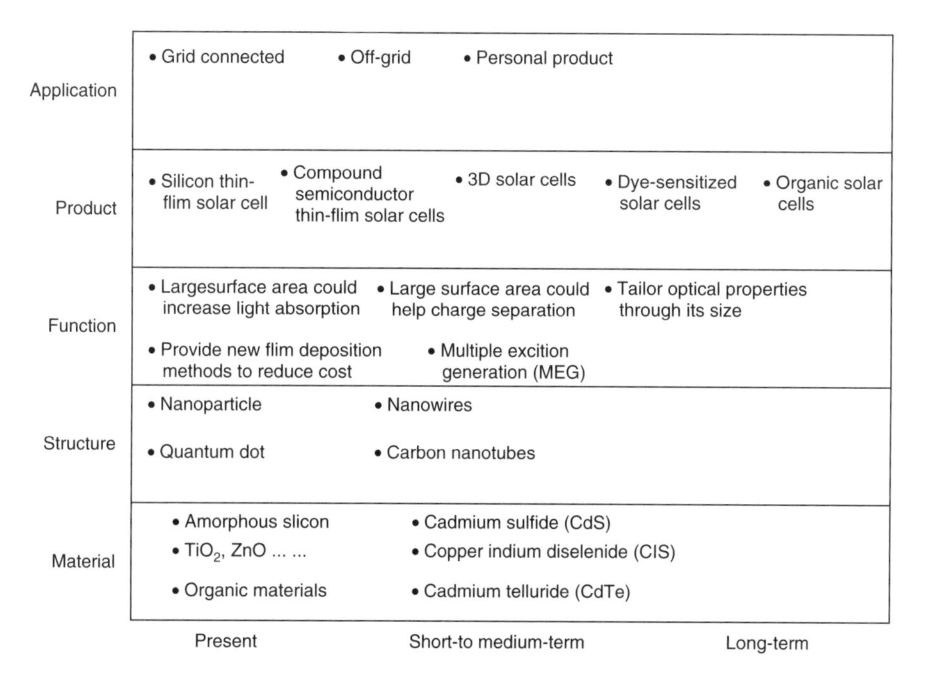

Figure A.1. Nano-enhanced Solar Cell Attributes by Time Array

Cross-charting, to be described shortly, helped associate "upstream" (research) efforts with potential "downstream" (applications) utility.

In this case, the target DSSC has been actively researched for some 13 years, but the cells are just entering the market. That market includes a very diverse set of ongoing activities ranging from relatively mature and new solar cells through other renewable sources to nuclear and fossil fuel–based energy sources. The market is complicated by the messy mixing of energy sources and delivery devices. For example, solar cells are both an energy source and an energy delivery technology, whereas fuel cells only deliver energy that has been derived by some other means.

Figure A.2 shows a compressed TDS that identifies types of key players and some innovation drivers and blockers of note. Consider the two parts of the TDS, the:

1. Technological enterprise that seeks to "deliver" a novel product or process
2. Contextual forces and factors acting upon that enterprise's efforts

The figure depicts the *technological enterprise* along the horizontal dimension. Starting at the left, the TDS seeks to ascertain R&D emphases that may be precursors of potential enhanced DSSCs, alternative technologies, and improved functionality. Such *technology push* is recognized as one driver of technological innovation. Ongoing monitoring is in order to watch for breakthroughs in contributing, complementary, or competing technologies.

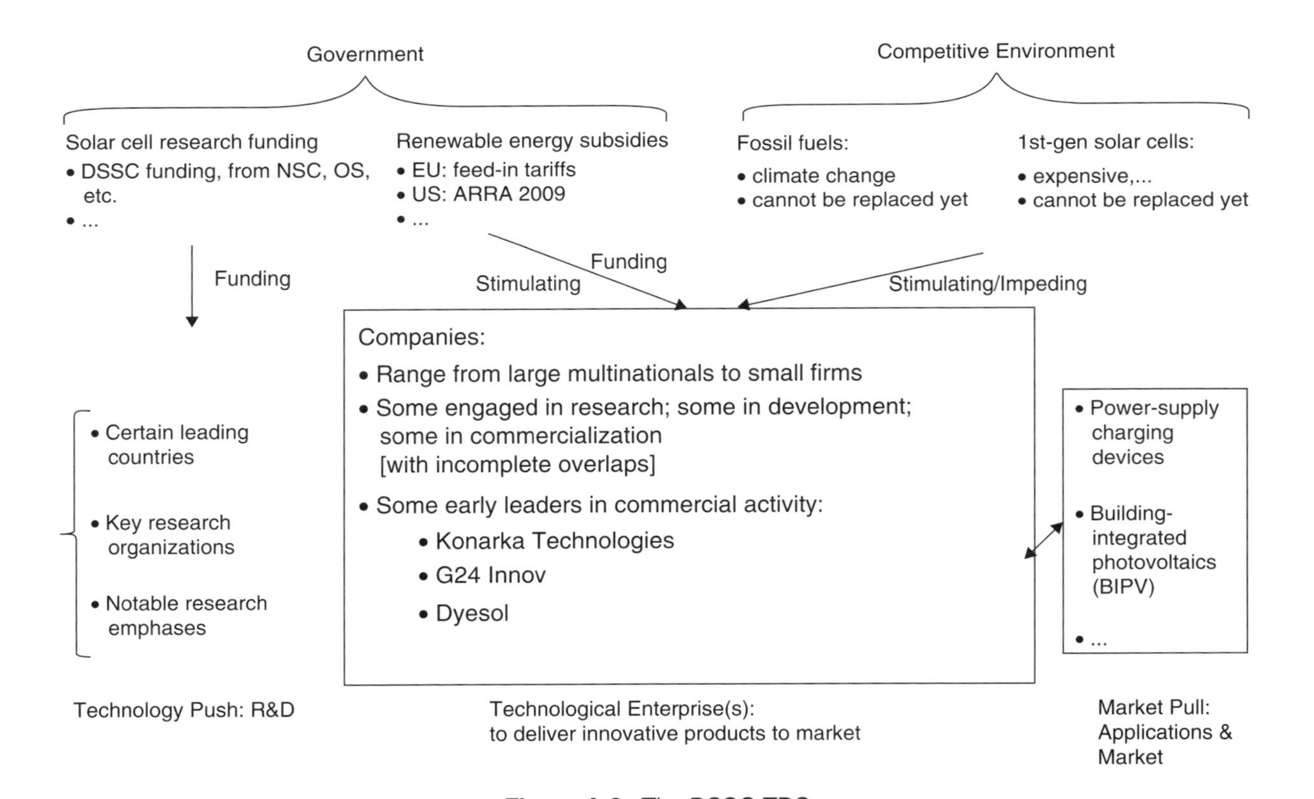

Figure A.2. The DSSC TDS

At the right side appears information on "Applications & Markets" that keys on what could be done with DSSCs. It spotlights the distinction between power supply devices, building integrated photovoltaic applications, and other applications. This area represents *market pull.* Currently, there is little demand for DSSCs, but there appears to be a growing customer base sympathetic to renewable energy. That suggests a potential edge for solar cells versus nonrenewable energy sources but tough competition for DSSCs within the renewable energy family. Importantly, DSSCs seem blessed with a combination of technology push and market pull (although that pull is not just for DSSCs).

The core enterprise resides between push and pull. What organization(s) would translate the advancing DSSC technical capabilities into viable products successful in the marketplace? Tech mining activities identified a diverse set of actors. An examination of research publications pointed to a mix of large multinationals and smaller companies. However, company policies vary from encouraging to prohibiting research publication, so any set of actors identified from publications will be incomplete. Likewise, DSSC patenting shows a mix of large multinationals and specialty firms. A third information source, Factiva (which covers indicators of business activity), also shows a mix of large and small firms. Interestingly, the active players in one venue are not necessarily the same as those in the others. For instance, Hayashibara Biochem Labs and Sumitomo Osaka Cement Company are active research publishers with no patenting activity and minimal business attention. Samsung is most active in patenting but has no publishing or Factiva notice. Dyesol Ltd. and G24 Innovations are two of the most prominent organizations in the business of data collection about business activity but show no publishing or patent activity. Sharp, Konarka, and Sony show notable activity across the board. The astute technical analyst should develop knowledgeable personal contacts to help interpret these patterns and to monitor which companies eventually step forward as key players in DSSC commercialization.

The second objective of sketching out a TDS is to identify influential *contextual forces.* Figure A.2 emphasizes two classes of these forces—government and competitors. Governmental is highly supportive, funding R&D and providing subsidies for adopting renewable energy sources. However, that support is not especially partial to DSSCs. Rather, it seeks to advance alternative renewable energy technologies of many types. Not spotlighted, but always a concern, is whether DSSCs would pose risks that would prompt onerous regulation (e.g., might TiO_2 materials exposed to years of electrical charge fluxes be hazardous?).

The competitive climate cannot be neatly summarized in the simple sketch. However, increasing energy demands, climate change concerns, and other social trends (e.g., concerns about importing oil) suggest long-term support for renewable energy sources. Traditional solar cells hold an overwhelming market share at present. Furthermore, these solar cells derive from silicon-based technology that includes the powerful infrastructure on which the semiconductor industry is built. Alternatives such as DSSCs face a great struggle to penetrate the main market segments even as silicon solar cells continue to dominate established markets.

While there are other notable contextual factors to consider in DSSC commercialization, this study is intended to be suggestive rather than exhaustive in describing the TDS. What has been presented would need to be augmented by considering the likely actions of financial and regulatory agencies as well as potential environmental concerns and public acceptance.

A.2.3 Tech Mining (Chapter 5) and Science Overlay Mapping

Chapter 4 discussed monitoring techniques, including tech mining, while Chapter 5 considered an array of information resources and useful indicators that could be developed from data secured from them. In this case study, these concepts are applied to the portions of R&D profiling covered by Steps C and D (Table A.1). None of the analysis techniques described in Chapter 6 were applied.

In *Tech Mining,* Porter and Cunningham (2005) suggest that a technology forecast should start by setting forth the key questions to be answered. They offer a list of 39 typical "Management of Technology" questions, but these are somewhat specific for present purposes. Rather, let's consider a few ways that R&D profiling can address "who," "what," "where," and "when" questions. Usually, it takes considerable knowledge of the technology and its TDS to answer questions about "how" and "why."

Chapter 5 denotes six important types of information resources. Databases that focus on technical or contextual content are subsets of these categories. Key databases used in the DSSC study included:

- Science Citation Index (SCI) of the Web of Science—abstracting basic research from over 12,000 journals
- EI Compendex—emphasizing engineering and applied research
- Derwent World Patent Index
- Factiva—abstracts content from newspapers, magazines, and newswires (Dow Jones 2010)

Such databases compile, format, and augment information and usually provide the results in the form of field-structured abstract records. They offer an accessible and extremely rich resource for technology intelligence, forecasting, and assessment. However, they are not sufficient in themselves; they should be augmented with additional Internet searches to tap diverse and recent advances and expert input.

These databases were searched using a two-tier approach. First, the R&D data sets were compiled (Porter, Youtie, et al. 2008); then a further search was performed for records pertaining to thin-film solar cells or, more particularly, DSSCs. The search algorithm is "Dye sensiti" or "dye-sensiti" or "dssc." This captures both American and English spellings (*sensitized*; *sensitised*); variations on those roots; and the common acronym, DSSC. The search retrieved 2168 abstract records (of which 1918 were journal articles) of research publications treating

DSSCs, indexed by SCI. Basic data cleaning was performed on those records to consolidate variations of country, institution, and author names. This process also elicited key terms from a combination of three fields—author keywords, Keywords Plus (based on titles cited by these publications), and title phrases (extracted via Natural Language Processing). Text mining software tailored to exploit field-structured records, mainly applied to patent and R&D databases, was used (Search Technology 2010a). This facilitates a range of analyses but is not needed to access the SCI content as such.

The seminal DSSC article (O'Regan and Grätzel 1991) was followed by two Gratzel-coauthored pieces in 1994, but no others were captured by the search through 1996. In 1997, 22 publications emerged, and the number increased steadily through 2009 (470 for approximately 80% of the estimated eventual year total).

Table A.2 illustrates the application of a basic tech mining tool, using a profiling macro (in VantagePoint, Search Technology 2010a) to display chosen data fields for a set of entities of special interest. This begins to get at a combination of "who, what, where, and when." In this example, the top researcher affiliations are listed, with three additional fields broken out:

- *Countries*: The top country is the organization's locale; the other countries are where coauthors resided. For instance, note that the last-listed organization, located in Sri Lanka, collaborates heavily with colleagues in Japan.
- *Subject Categories*: Web of Science (SCI) categorizes journals into these groups; they provide a useful set of research areas. Here, they give a sense of the research emphases and concentrations. Note that the Chinese Academy of Sciences (CAS) and the two Swiss organizations are quite diversified. In contrast, the U.S. National Renewable Energy Lab concentrates heavily in physical chemistry. For more detail, key terms (keywords and title phrases) could be shown.
- *Percent since 2007*: 60% or more of the publications by CAS and Korea University date from 2007; in sharp contrast, 17% or less of those by the U.S. National Renewable Energy Lab and the Sri Lankan Institute of Fundamental Studies are that recent. This suggests that Asian R&D is placing increasingly higher priority on DSSCs than the leading U.S. research organization studying them.

Other fields could be broken out, such as authors, number of times cited, and so on. For instance, one might break out data fields for the top countries, selected topics (e.g., subject categories, particular topics), leading authors, or time periods (to discern changing emphases).

Comparative analyses can be particularly informative. Imagine that you are exploring DSSC research on behalf of the U.S. National Renewable Energy Lab. Benchmarking their research activity against that of another leading research organization could be useful. Various graphical presentations could be used

TABLE A.2 Breakout for the Top 10 DSSC Research Organizations

Affiliation	Countries	Subject Categories (top five)	% Since 2007
Chinese Academy of Science (156)	China (156) Japan (8) Switzerland (7) India (3) USA (3)	Materials science, multidisciplinary (43) Chemistry, physical (41) Chemistry, multidisciplinary (35), Physics, applied (18) Nanoscience and nanotechnology (17)	60% of 156
Swiss Federal Institute of Technology (95)	Switzerland (95) UK (14) China (11) Italy (9) South Korea (6) USA (6)	Chemistry, physical (39) Chemistry, multidisciplinary (30) Materials science, multidisciplinary (27) Chemistry, inorganic and nuclear (17) Nanoscience and nanotechnology (13)	43% of 95
École Polytechnique Fédérale de Lausanne (77)	Switzerland (77) UK (13) Germany (7) USA (7) Netherlands (6) Japan (6)	Chemistry, multidisciplinary (32) Chemistry, physical (28) Materials science, multidisciplinary (26) Physics, applied (18) Nanoscience and nanotechnology (13)	48% of 77
National Institute for Advanced Industrial Science and Technology (69)	Japan (69) China (2)	Chemistry, physical (29) Materials science, multidisciplinary (25) Chemistry, multidisciplinary (15) Energy and fuels (10) Chemistry, inorganic and nuclear (6)	41% of 69
University London Imperial College Science, Technology and Medicine (62)	UK (62) Switzerland (15) Spain (13) Brazil (4) Netherlands (4) Germany (4)	Chemistry, physical (25) Chemistry, multidisciplinary (22) Materials science, multidisciplinary (15) Chemistry, inorganic and nuclear (7) Nanoscience and nanotechnology (6) Physics, applied (6) Physics, condensed matter (6)	37% of 62

Korea Univ (54)	South Korea (54)	Chemistry, multidisciplinary (19)	67% of 54
	Switzerland (9)	Materials science, multidisciplinary (12)	
	India (7)	Electrochemistry (11)	
	China (4)	Energy and fuels (8)	
	Spain (2)	Chemistry, organic (7)	
		Chemistry, physical (7)	
Osaka Univ (51)	Japan (51)	Chemistry, physical (28)	29% of 51
	Sri Lanka (4)	Materials science, multidisciplinary (16)	
	China (2)	Chemistry, multidisciplinary (12)	
	South Korea (2)	Nanoscience and nanotechnology (7)	
		Energy and fuels (5)	
Kyoto Univ (48)	Japan (48)	Materials science, multidisciplinary (19)	48% of 48
	Thailand (6)	Chemistry, physical (13)	
	China (4)	Chemistry, multidisciplinary (11)	
	Finland (2)	Nanoscience and nanotechnology (8)	
		Electrochemistry (8)	
National Renewable Energy Laboratory (47)	USA (47)	Chemistry, physical (31)	17% of 47
	South Korea (4)	Chemistry, multidisciplinary (9)	
	Switzerland (2)	Materials science, multidisciplinary (8)	
	Israel (2)	Nanoscience and nanotechnology (4)	
		Chemistry, inorganic and nuclear (3)	
Institute of Fundamental Studies (43)	Sri Lanka (43)	Materials science, multidisciplinary (18)	16% of 43
	Japan (19)	Chemistry, physical (11)	
	USA (3)	Energy and fuels (10)	
		Physics, applied (7)	
		Physics, condensed matter (6)	

to spotlight relative strengths and weaknesses (e.g., dual-bar charts). Guo and colleagues' four DSSC papers offer several national program comparisons, for instance by plotting activity trends together (Guo, Huang, et al. 2009, 2010; Guo, Porter, et al. 2009; Guo, Xu, et al. in press).

The top five countries appearing in 2168 SCI publications were:

1. Japan 418
2. China 413
3. United States 281
4. South Korea 245
5. Switzerland 186

This is somewhat surprising. General nanotechnology research in SCI shows the United States and China strongly setting the pace. Solar cells find strong U.S. and Indian research activity. Switzerland's DSSC stature is remarkable, especially in that 154 of the DSSC papers involve M. Gratzel as an author (globally the leading author).

Tech mining also promotes the use of *innovation indicators*, which, you will recall from Chapter 5, are derived measures that blend information to highlight issues of managerial importance. Figure A.3 combines quantity (number of publications) with a measure of influence (number of citations of these publications by others) through part-year 2008. Such composite indicators offer additional insight. In this case, one might note that Switzerland, with fewer publications than the other four leading DSSC countries, has accrued the most citations. Probing further in the extended data set (through 2009), over 2500 of those citations are to one paper (O'Regan and Grätzel 1991). The data also show that, to date, the growing set of Chinese and South Korean publications are less frequently cited than those of other leading research producers.

The geographic location of research activity can highlight hotbeds of activity and help the forecaster understand requirements and relationships. Stephen Carley, of the Georgia Tech nanotechnology group, has developed macros that expedite the process of locating researchers based on the latitude and longitude of their institutional affiliation. Resulting geo-maps are generated through Google Earth or Google Maps, depending on the coverage and the desired representation (Search Technology 2010b).

Another form of mapping locates research activity among the disciplines. Figure A.4 shows the locus of the DSSC publications indexed for 1991–2009. This approach utilizes the approximately 175 Web of Science subject categories associated with SCI journals (the nodes shown as faint background in the figure). Factor analyses of the citation activities among subject categories yields 14 "macro-disciplines" shown by the labels in Figure A.4. For more information on such science overlay mapping, or to make your own, visit Georgia Institute of Technology (2010).

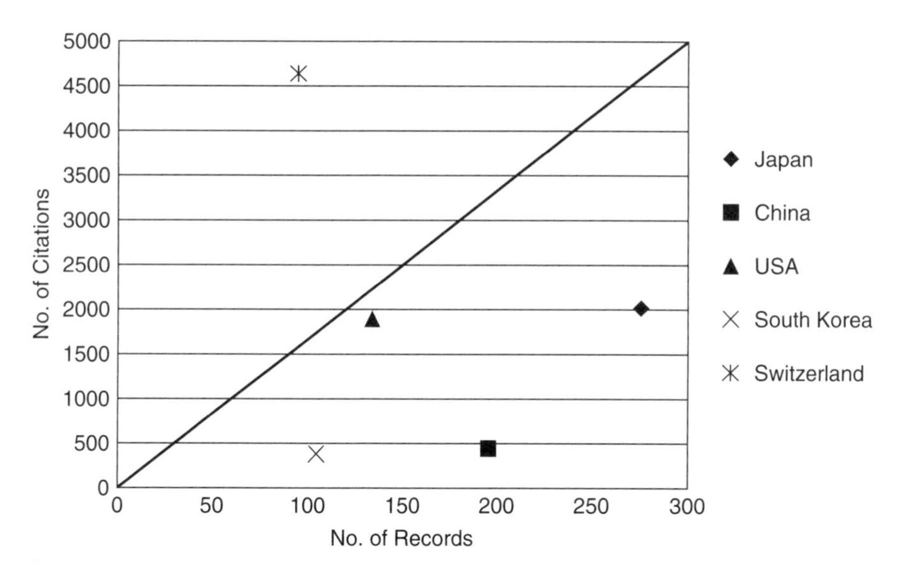

Figure A.3. Quantity and Influence of DSSC Publications (SCI) for the Top Five Countries

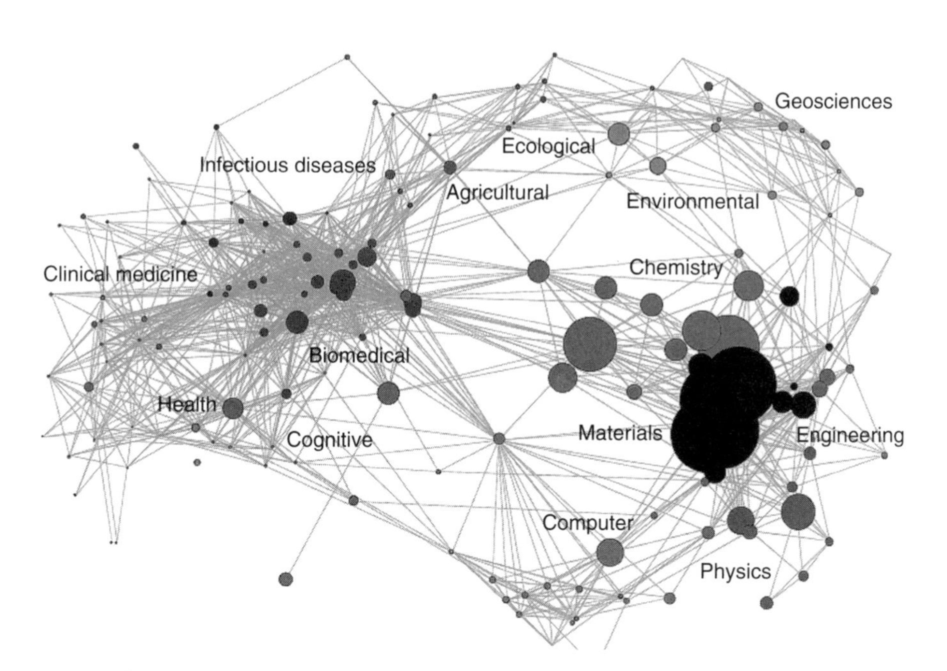

Figure A.4. DSSC Research Overlaid on the Base Map of Science

A.2.4 Trend Analyses

Tech mining provides many candidate elements that one could plot against time, including:

1. Overall research publication activity based on the prime database for these analyses (SCI) (see Figure A.5).
2. Comparison of the SCI trend with that seen in, for instance, a more applied database (e.g., EI Compendex), patents (Derwent World Patent Index used in this case study), and/or the business-oriented database (Factiva used here). Overlaying these in one chart helps assess whether activity is concentrated in research, heavily vested in development (patenting), or extending into commercialization, as indicated by business media attention.
3. Subtopic trends—these may suggest useful breakout. As examples:
   - Various trends presented in previous papers, including comparisons of research activity by countries. Guo, Huang, et al. (2010) note that China is continuing to increase its share of nano-enhanced thin-film solar cell R&D.
   - The categories of organizations doing research, such as academic, government, and nongovernment organizations. This information could be used to track trends in R&D activity by each category. Increasing corporate attention, for instance, may indicate approaching commercialization (Guo, Huang, et al. 2009).
   - The trends in materials or techniques being studied. For instance, trends show that TiO_2 receives the most attention among nano-enhanced

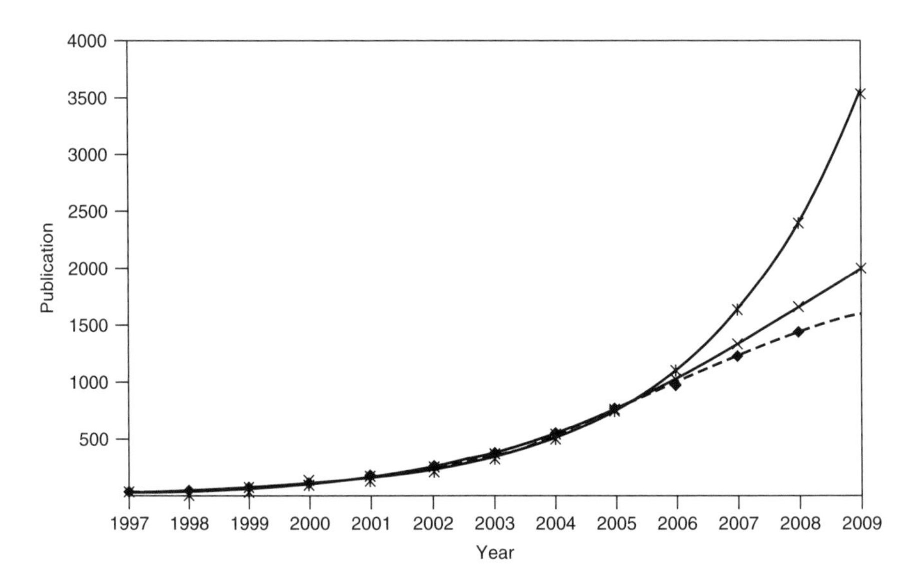

Figure A.5. DSSC Research Publication Trends

thin-film materials, but attention to zinc oxide (ZnO) has been increasing more rapidly in recent years (Guo, Porter, et al. 2009).

Figure A.5 shows research publication activity on dye-sensitized materials. Since these materials are rarely mentioned in the SCI nanotechnology data set prior to 1997, the figure begins with that year. The cumulative data from 1997 on (as per a 2008 search), are indicated by the diamonds in the figure. Using the trend macro associated with this book, the plot shows a Fisher-Pry curve (dashed line) fit to the data using a limit of 2000 because of the short time horizon. The curve fits these data extremely well ($R^2 = 0.997$). This is perhaps surprising since Fisher-Pry is primarily a market penetration model.

The search for DSSC records was updated (April 2010). This yielded higher values for 2006–2007 and a more solid estimated value for 2009. The Web of Science continues to add records for a given year at a diminishing rate for the next several years. The earlier value projected using this Fisher-Pry model was 1613, while the more solid estimate was 2282—way off! Therefore, two additional extrapolations were modeled:

- Fisher-Pry with the limit raised to 3000—the middle curve (its projected 2009 value of 1981 is still low)
- Exponential growth—the top curve with X's (its projected 2009 estimate of 3548 is far too high)

These results require a cautionary note. Despite the fact that the initial Fisher-Pry model provided an excellent representation of the information available in 2008, that did not guarantee an accurate trend projection.

A.2.5 Cross-charting and Social Network Analyses

Huang, Guo, and Porter developed a technique they called *cross-charting* to associate technical advances with potential uses. The four Guo papers cited in the References section present several variations of this technique (Guo, Huang, et al. 2009, 2010; Guo, Porter, et al. 2009; Guo, Xu, et al. in press). Figure A.6 is one such variation charting from nanomaterial development to DSCC application.

The cross-charting approach involves extracting key development features proceeding from research through commercial adoption and then showing how these features could interconnect. So, for example, if Quantum Dot contributions in Figure A.6 pointed more strongly to integrated photovoltaics for the construction industry than to power supply generation devices in general, that could help distinguish their potential innovation pathways. Then one could zoom in to study which organizations evidence interest in integrated photovoltaics for the construction industry and their association with the Quantum Dot research performers.

Recall that one potential market indicated by the TDS (Figure A.2) was integrated photovoltaics for the construction industry. That application is a focus to

Nanomaterial	Functions		Dye-Sensitized Solar Cells	Product Advantages	Potential Market
	Enhanced Light Absorption		Nanoparticle Cells		
Nanoparticles	Enhanced Charge Separation			Choice of Color	
Quantum Dots	Reduced Recombination				
	Tailored Optical Properties		Nanowire Cells	Flexibility	Integrated Photovoltaics for the Construction Industry
Nanowires	Multiple Excitation Generation				
Carbon Nanotubes	Reduced Electron Transport Time			Rigidity	
	Performance as Dye		Quantum Dot Cells		

Figure A.6. Cross-charting: From Nanomaterial to Application for DSSCs

which selected "Product Advantages" point in Figure A.6. DSSCs are especially attractive for these applications because their color can be easily varied by choosing different dyes and cells can be built on flexible substrates. This has already been demonstrated (Matson 2007). These DSSCs could suit both grid-connected and off-grid systems.

Those characteristics, in turn, are linked back to particular DSSC types. One type, Quantum Dot–enhanced DSSCs, is of particular relevance. But it should be noted that other nanotechnology capabilities also could contribute to such features. The Quantum Dot emphasis is here carried upstream to particular functions deriving from various nanomaterials.

Alternatively, cross-charting could focus on a particular technical subtopic and then work downstream to consider functional benefits and potential applications. If this analysis were done for a commercial organization, the goal might be to identify potential licensees in order to pursue commercial opportunities. Or, one might start in midstream. For instance, if particular optical properties were of paramount interest (a "Function" in Figure A.6), one could investigate who was pursuing pertinent research and also who was engaging in commercial operations for which this could be beneficial.

Cross-charting pursues leads provided by the rich information resources retrieved from the various databases to offer glimpses up- and/or downstream. The technique combines text mining with analyst judgment. Key terms extracted from research records may suggest functional values that technology experts can help filter to discern what could be important. The challenge is to link

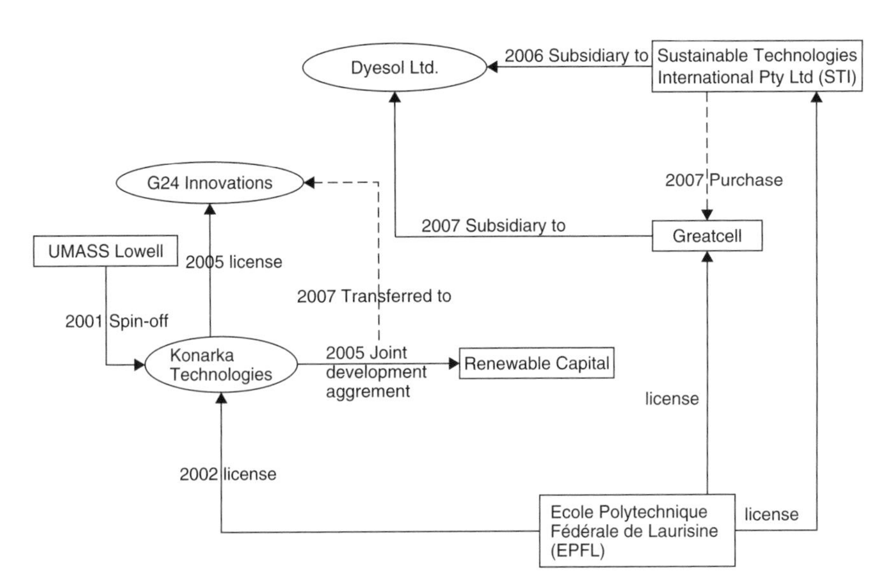

Figure A.7. DSSC Corporate Connections

these weak signals to content from other data types. For instance, the Factiva business-oriented records discussing DSSCs offer leads on application interests and companies pursuing them.

Another technique can help sift through the information on such organizations involved with DSSCs. As noted in Section A.2.2, different organizations are highlighted by the searches of DSSC publishing, patenting, and citations in the business press. Using software such as VantagePoint, one can generate visual "maps" showing various forms of collaboration. For instance, Guo, Xu, et al. (submitted) depicts the top research companies with the universities with which they coauthor. Such information is useful intelligence as one explores partnering. Further network analyses could ferret out potentially important links. Guo, Xu, et al. (submitted) generated a network map showing coauthoring links of the active research companies. Those point out some interesting industry–university associations. To illustrate, consider the three companies most prominently noted from the Factiva search as pursuing commercial production of DSSCs: G24 Innovations, Konarka Technologies, Inc., and Dyesol Ltd. Figure A.7 shows relationships among these and other organizations. G24 Innovations is producing DSSCs commercially, whereas Konarka is licensing related intellectual property. The Australian company, Dyesol Ltd., has notable ties to École Polytechnique Fédérale de Lausanne, a Swiss university that is especially prominent in DSSC publishing (second only to the Chinese Academy of Sciences).

A.3 THE REST OF THE STORY

Section A.2 described a number of methods applied in one case analysis. While quite rich and diverse, these methods do not cover all of the key elements of a comprehensive technology forecast. This section draws upon the chapters in this book to indicate what else might be done to round out the DSSC technology forecast.

Chapter 6 introduced various modeling and simulation approaches. One could imagine applying system dynamics to help elucidate energy technology market dynamics in conjunction with pricing and environmental considerations. Decision analyses and options analyses might be used to explore the value of various policy actions to increase the attractiveness of renewable energy to various consumers and other stakeholders.

Chapter 7 begins with material related to the focusing phase of forecasting. Guo et al. (submitted) performed a limited market needs assessment. Drawing on other sources (c.f. Macias and Teske 2008), they note various ways in which one could partition the solar cell market. One partition that differs from those noted in the TDS (Figure A.2) is between electric grid–connected solar cells (over 90% of the current sales) and off-grid systems. The latter could be designed to provide power for communities in the developing world that lack access to electric grids or to power a wide range of consumer products.

One might also want to examine energy technology roadmaps to garner ideas on solar cell prospects, competition, and forces to watch (Chapter 11). For instance, a national solar technology roadmap (Matson 2007) suggests that DSSC technology has special potential for lightweight, portable power-supply charging devices for consumer electronics and military applications (e.g., mobile telephones and military garments).

A.3.1 Market Forecasts

The interplay of politics and economics in technologies is formidable, and the prospects for expanded solar cell commercialization are especially sensitive to it. The economic toolset provided in Chapters 8 and 10 would be critical in a complete DSSC analyses. For many stakeholders (e.g., small companies), this would mean searching for existing open source analyses done by others or market reports sold by various research organizations. A quick Google scan for market research reports that treat solar cells turns up a number of such reports, including:

- Global Market: *Current and Next Generation Solar Cell and Related Material Market Outlooks*, Fuji-Keizai USA, Inc., 2007, 58 pages, $998
- *Solar PV (Photovoltaic) Cell Market: Potential Opportunities*, Koncept Analytics, 2007, 21 pages, € 722
- *Solar Cell (Photovoltaic) Equipment Industry Report,* ResearchInChina, 2009, 156 pages, $2200
- *Marketbuzz® 2010: Annual Global PV Industry Report*, $3995SolarBuzz, 2010, 301 pages.

A technology forecaster might acquire one or more such market studies. Of course, any user should always evaluate the critical assumptions and estimates upon which these reports are based.

A.3.2 Scenarios

Scenarios (Chapter 7) could add a critical dimension to the assessment of DSSC prospects and to the presentation of study results. One might employ a workshop to elucidate multiple innovation pathways like the one described earlier (recall Figure A.1). Results of the workshop could be synthesized to generate a few alternative developmental scenarios. Such scenarios could be helpful in considering:

- Components needed for successful commercial innovation (e.g., electric grid adaptations, energy prices)
- Policies that could boost the attractiveness of DSSCs in particular markets (e.g., renewable energy usage requirements) or that could impede DSSC market share growth (e.g., a major commitment to nuclear energy)
- Sensitivity to factors that appear to exert especially strong influence on the prospects for solar cell and DSSC development and use
- Communicating results and alternatives to decision makers

A.3.3 Technology Assessment

Technology assessment has two different meanings. It can refer to (1) evaluation of competing technological alternatives or (2) impact assessment (Chapter 9). Guo, Xu, et al. (in press) compiled technology evaluation sources to examine how DSSCs compare with competing energy technologies in terms of cost, efficiency, applicability, and sustainability (Bossert, Tool, et al. 2000; McConnell 2002; Grätzel 2003). A few of the observations are given in the following paragraphs.

DSSCs have unique advantages. The cost, the most important one, is about \$2/watt, which is 50% less than that of silicon-based solar cells (\$3/watt). Production facilities are much cheaper than those of silicon-based solar cells. The major materials in DSSCs (ZnO and TiO_2) are much more biocompatible than silicon. DSSCs also offer light weight, flexibility, transparency, and color options that are very attractive. Moreover, DSSCs can be used directly to produce high-energy chemicals from sunlight. Such "photosynthetic" devices solve the problem of finding sufficient energy storage. R&D also suggests that solid-state electrolyte and long-life-cycle sensitizers hold strong appeal in terms of delivering more reliable performance.

DSSCs compare less favorably with silicon-based and other solar cells in other respects. For instance, DSSCs are presently less efficient (usually only 5–8% efficiency in the lab). Recent research on increasing efficiency focuses on using quantum dots to convert higher-energy (i.e., higher-frequency) light into multiple electrons; solid-state electrolytes to improve the temperature response; and changing the doping of TiO_2 to better match that of the electrolyte being used. Potential efficiency appears to be 20% (Grätzel 2003). Solid-state electrolytes also could solve the problem of long-term cell stability. With the increase in efficiency and stability, DSSCs could be an excellent replacement option for existing technologies in "low-density" applications, such as mobile phone chargers. The technology comparison above could be greatly elaborated. Comparison is most useful for technology management when devices are compared with respect to their suitability for particular applications.

Impact assessment to identify potential unintended, indirect, or delayed effects of introducing DSCCs (e.g., potential environmental or health implications of TiO_2 particles; DSSC production, distribution, and eventual disposal) is very important (Chapter 9). However, this remains to be done.

A.3.4 Further Analyses and Communicating Results

This case study was not intended to inform organizational decision processes. Before commercialization of DSSCs is contemplated, thorough economic, market, cost-benefit, and risk analyses (Chapters 8 and 10) would be demanded by decision makers. The concerns described in Chapter 11 also would need to be addressed.

Communicating the analyses and the forecast result is critical to ensuring that they are useful in decision making. It is vital to provide results in ways that are familiar and preferred by target groups. Multiple and interactive modes always should be considered. In this instance, results were disseminated to multiple audiences with different emphases. In addition to the four Guo et al. references noted, selective results have been directed to a data mining audience (Porter, Huang, et al. in press).

Communications in this case study heavily relied on visualizations. These are a matter of taste and audience expectations. Four of the figures presented involve conceptual representations—words in some relationship to each other (Figures A.1, A.2, A.6 and A.7). In some instances, a Web-based presentation of such results might be a valuable option. That might provide effective simulation (e.g., showing a distribution evolving over time). The messages here are that communication is important and that the forecaster should consider the potential of multiple media.

REFERENCES

Aydil, E. S. (2007). "Nanomaterials for Solar Cells." *Nanotechnology Law and Business* **4**(3): 275–292.

Bossert, R. H., C. J. J. Tool, et al. (2000). *Thin-film Solar Cells: Technology Evaluation and Perspectives*. Petten, The Netherlands: Netherlands Energy Research Foundation ECN.

Conibeer, G. (2007), "Third Generation Photovoltaics." *Materials Today* **10**(11): 42–50.

Dow Jones. (2010). "Search, Additions, Indexing—Dow Jones Factiva." Retrieved 5 September 2010 from http://factiva.com/sources.asp.

Georgia Institute of Technology (2010). "Measuring and Mapping Interdisciplinary Research." Retrieved 5 September 2010 from http://www.idr.gatech.edu/.

Grätzel, M. (2003). "Dye-sensitized Solar Cells." *Journal of Photochemistry and Photobiology C: Photochemistry Reviews* **4**: 145–154.

Green, M.A. (2001), "Third Generation Photovoltaics: Ultra-high Conversion Efficiency at Low Cost." *Progress in Photovoltaics: Research and Applications* **9**(2): 123–135.

Guo, Y., L. Huang, et al. (2009). "Profiling Research Patterns for a New and Emerging Science and Technology: Dye-sensitized Solar Cells." Presented at the Atlanta Conference on Science and Innovation Policy, Atlanta, GA.

Guo, Y., L. Huang, et al. (2010). "Research Profiling: Nano-enhanced, Thin-film Solar Cells." *R&D Management* **40**(2): 195–208.

Guo, Y., A. L. Porter, et al. (2009). "Nanotechnology-enhanced Thin-film Solar Cells: Analysis of Global Research Activities with Future Prospects." Presented at the meeting of the International Association for Management of Technology (IAMOT), Orlando, FL.

Guo, Y., C. Xu, et al. (in press). "Composing a Technology Delivery System for an Emerging Energy Technology: The Case of Dye-sensitized Solar Cells." *Technovation*. Institute of Nanotechnology (2006). "Road Maps for Nanotechnology in Energy." Nanoroadmap (NRM) Project Working Paper. Retrieved 5 September 2010 from www.nanoroadmap.it/roadmaps/NRM_Energy.

Macias, E., and S. Teske. (2008). "Solar Generation V," European Photovoltaic Industry Association (EPIA) and Greenpeace, Retrieved 19 January 2011 from http://www.greenpeace.org/international/Global/international/planet-2/report/2008/9/solar-generation-v-2008.pdf.

Matson, R. (2007). "National Solar Technology Roadmap: Sensitized Solar Cells." U.S. Department of Energy, Retrieved 5 September 2010 from http://www1.eere.energy.gov/solar/pdfs/41739.pdf.

McConnell, R. D. (2002). "Assessment of the Dye-sensitized Solar Cell." *Renewable and Sustainable Energy Reviews* **6**: 273–296.

O'Regan, B. and M. Grätzel. (1991). "A Low-cost, High-efficiency Solar-cell Based on Dye-sensitized Colloidal TiO_2 Films." *Nature* **353**: 747–740.

Porter, A. L. and S. W. Cunningham. (2005). *Tech Mining: Exploiting New Technologies for Competitive Advantage*. Hoboken, NJ, John Wiley & Sons.

Porter, A. L., L. Huang, et al. (in press). "Forecasting Innovation Pathways for Newly Emerging Science and Technologies." *Technological Forecasting and Social Change*.

Porter, A.L., Youtie, J., Shapira, P., and Schoeneck, D.J. (2008). "Refining Search Terms for Nanotechnology." *Journal of Nanoparticle Research* **10**(5): 715–728.

Robinson, D. K. R. and T. Propp. (2008). "Multi-path Mapping for Alignment Strategies in Emerging Science and Technologies." *Technological Forecasting and Social Change* **75**: 517–539.

Search Technology. (2010a). "VantagePoint: Turn Information into Knowledge." Retrieved 5 September 2010 from http://www.thevantagepoint.com/.

Search Technology. (2010b). "Welcome to Search Technology," Retrieved 5 September 2010 from http://sites.google.com/a/searchtech.com/vpinstitute/.

INDEX

D

E

F

G

J

K

L

M

Q

T